1-Dimensional Metal Oxide Nanostructures

T0179156

Advances in Materials Science and Engineering
Series Editor: Sam Zhang

This **Advances in Materials Science and Engineering** series is designed to illustrate the many new and exciting challenges in the expansive and interdisciplinary field of materials science and engineering. The scope of this series is broad, with reference works and textbooks offering insight into all aspects of materials development. The titles in this series deliver authoritative information to professionals, researchers, and students across engineering and scientific disciplines. Each volume offers a comprehensive approach covering fundamentals, technologies, and applications and includes real-world examples where appropriate.

Biological and Biomedical Coatings Handbook
Processing and Characterization
Edited by Sam Zhang
Aerospace Materials Handbook
Edited by Sam Zhang and Dongliang Zhao
Thin Films and Coatings
Toughening and Toughness Characterization
Edited by Sam Zhang
Semiconductor Nanocrystals and Metal Nanoparticles
Physical Properties and Device Applications
Edited by Tupei Chen and Yang Liu
Advances in Magnetic Materials
Processing, Properties, and Performance
Edited by Sam Zhang and Dongliang Zhao
Micro- and Macromechanical Properties of Materials
Yichun Zhou, Li Yang, and Yongli Huang
Nanobiomaterials
Development and Applications
Edited by Dong Kee Yi and Georgia C. Papaefthymiou
Biological and Biomedical Coatings Handbook
Applications
Edited by Sam Zhang
Hierarchical Micro/Nanostructured Materials
Fabrication, Properties, and Applications
Weiping Cai, Guotao Duan, and Yue Li
Nanostructured and Advanced Materials for Fuel Cells
San Ping Jiang and Pei Kang Shen
Hydroxyapatite Coatings for Biomedical Applications
Edited by Sam Zhang
Biological and Biomedical Coatings Handbook, Two-Volume Set
Edited by Sam Zhang

For more information about this series, please visit: https://www.crcpress.com/Advances-in-Materials-Science-and-Engineering/book-series/CRCADVMATSCIENG

1-Dimensional Metal Oxide Nanostructures

Growth, Properties, and Devices

Edited by
Zainovia Lockman

CRC Press
Taylor & Francis Group
Boca Raton London New York

CRC Press is an imprint of the
Taylor & Francis Group, an **informa** business

CRC Press
Taylor & Francis Group
6000 Broken Sound Parkway NW, Suite 300
Boca Raton, FL 33487-2742

First issued in paperback 2020

ISBN-13: 978-1-138-57752-7 (hbk)
ISBN-13: 978-0-367-65688-1 (pbk)

Library of Congress Cataloging-in-Publication Data

Names: Lockman, Zainovia, editor.
Title: 1-dimensional metal oxide nanostructures : growth, properties, and devices / edited by Zainovia Lockman.
Other titles: One-dimensional metal oxide nanostructures : growth, properties, and devices
Description: Boca Raton, FL : CRC Press/Taylor & Francis Group, 2018. | Series: Advances in materials science and engineering | Includes bibliographical references.
Identifiers: LCCN 2018034537 | ISBN 9781138577527 (hardback) | ISBN 9781351266727 (ebook)
Subjects: LCSH: Nanostructured materials. | Metallic oxides.
Classification: LCC TA418.9.N35 A13 2018 | DDC 620.1/15--dc23
LC record available at https://lccn.loc.gov/2018034537

Visit the Taylor & Francis Web site at
http://www.taylorandfrancis.com

and the CRC Press Web site at
http://www.crcpress.com

1-Dimensional Metal Oxide Nanostructures

Growth, Properties, and Devices

Edited by
Zainovia Lockman

CRC Press
Taylor & Francis Group
Boca Raton London New York

CRC Press is an imprint of the
Taylor & Francis Group, an **informa** business

CRC Press
Taylor & Francis Group
6000 Broken Sound Parkway NW, Suite 300
Boca Raton, FL 33487-2742

First issued in paperback 2020

ISBN-13: 978-1-138-57752-7 (hbk)
ISBN-13: 978-0-367-65688-1 (pbk)

Library of Congress Cataloging-in-Publication Data

Names: Lockman, Zainovia, editor.
Title: 1-dimensional metal oxide nanostructures : growth, properties, and devices / edited by Zainovia Lockman.
Other titles: One-dimensional metal oxide nanostructures : growth, properties, and devices
Description: Boca Raton, FL : CRC Press/Taylor & Francis Group, 2018. | Series: Advances in materials science and engineering | Includes bibliographical references.
Identifiers: LCCN 2018034537 | ISBN 9781138577527 (hardback) | ISBN 9781351266727 (ebook)
Subjects: LCSH: Nanostructured materials. | Metallic oxides.
Classification: LCC TA418.9.N35 A13 2018 | DDC 620.1/15--dc23
LC record available at https://lccn.loc.gov/2018034537

Contents

Preface

Transition metal oxides have unusual yet useful electronic properties with applications in many potential and emerging fields, specifically in environment purification and energy devices. When fabricated in 1-dimensional (1-D) nanostructures, metal oxides like titanium dioxide, zinc oxide, copper oxide, zirconium dioxide and iron oxide have shown interesting size-dependent and structure-related properties. 1-D nanomaterials are a set of materials of less than 100 nm size in two out of three dimensions. They are comprised of elongated structures of lengths in micron or even millimeter scale but diameters of less than 100 nm. Typical examples are nanowires, nanotubes and nanorods. There have been many efforts to synthesize 1-D metal oxide, but a protocol which is cost-effective, efficient and robust is desired. Oxidation of metal falls into this category, and this book focuses on direct oxidation (thermal or anodic) of metal surfaces that can generate 1-D morphology.

Thermal oxidation occurs when metals are exposed to high temperatures in an environment with an excess of oxygen. Oxidation can also happen in aqueous solutions by reaction with oxygen molecules that are produced by water decomposition. For this, growth can be evoked by developing an electric field across the oxide layer, inducing migration of anions or cations (anodic oxidation), whereas for thermal oxidation growth occurs by a thermally activated diffusion process. Both thermal and anodic oxidation can result in the formation of surface oxide when the processes are properly controlled. Often thermal oxidation results in a layer of oxide with morphology comprised of nanowires whereas anodic oxidation creates anodic film with a nanotubular structure.

Chapter 1 of this book begins by defining, classifying and describing 1-D nanostructures in general. Then a discussion of surface oxidation for oxide nanowires formation shows the possibilities of forming nanowires by a direct oxidation process. More examples and in-depth analysis of oxidation for nanowires formation are presented in Chapters 2 and 3, focusing mostly on copper oxide, zinc oxide, iron oxide and tungsten oxide nanowires. These chapters include the possible mechanisms of nanowires formation.

Chapter 4, on the other hand, delves into an alternative process for nanowires or nanorods formation: chemical based. It also introduces readers to other typical methods for 1-D oxide nanostructures formation. Chapter 4 also aims to give an overview of applications of zinc oxide nanostructures in various technological fields. More on applications of nanowires and nanotubes can be found in Chapters 5 and 6, respectively. These chapters describe how heavy metal, especially chromium(VI) compounds, can be removed from water on either iron oxide nanowires (fabricated by thermal oxidation of iron) or zirconium dioxide nanotubes (fabricated by anodic oxidation of zirconium). Heavy metals are toxic contaminants often found in industrial wastewater. Some of them are carcinogenic and mutagenic, and hence the ability of 1-D metal oxides to remove them is noteworthy.

1-D metal oxides have also been applied in medical diagnosis, detection of pollutants and sensors. Considering that sensors are becoming more relevant, Chapters 7 and 8 are devoted to 1-D oxides for sensing devices. Examples give readers an overview on the possibility and importance of 1-D metal oxides as sensor materials.

As mentioned, Chapter 6 of this book is on anodic oxidation of zirconium for zirconium dioxide nanotubes formation. Apart from zirconium, metals like titanium can be anodized to produce surface oxides comprised of aligned and uniform nanotubes as well. This process is simple yet effective for fabricating self-ordered nanotubes. Titanium dioxide nanotubes have been seen as a prime candidate for energy devices: solar cell, hydrogen cell, batteries and fuel cells. Therefore Chapter 9 of this book is devoted to the topic of anodization for titanium dioxide nanotubes formation and on the use of these nanotubes for energy devices.

Chapter 10 includes a special topic on metal oxide-carbon nanotubes hybrids. The aim of this chapter is to show that metal oxides do not have to stand alone, as they can be "coupled" with carbon nanotubes to form a hybrid material with tailored properties. Apart from a transition metal oxide, it appears to be necessary to include in the fabrication of silicon oxide nanowires a local anodic oxidation process, which then can be etched for silicon nanowires formation. This is described in Chapter 11 of this book.

This book would not have been possible if not for Professor Sam Zhang Shanyong, who helped in initiating and introducing me to the efficient teams at CRC Press and Taylor & Francis. For that I thank Prof Sam and editor Allison Shatkin at CRC Press for editing and then producing this book. I have learned so much throughout the whole process. Research on oxide 1-D nanomaterials is evolving and expanding very rapidly, making it impossible to cover every aspect in one book. Nevertheless, this book serves as a general overview to those who would like to venture into oxidation processes for 1-D oxides formation and get a glimpse of some of the typical applications of the oxides. I have been compiling and reviewing chapters received from the contributors over many months. I am deeply indebted to all of them for their effort in preparing high-quality manuscripts and hope that they enjoyed writing them as much as I enjoyed reading them. I am extremely grateful for having support from the present and past members of the Green Electronic Nanomaterials group at the School of Materials and Mineral Resources Engineering, Universiti Sains Malaysia (SMMRE, USM), especially Nurulhuda Bashirom, Monna Rozana and Subagja Toto Rahmat for the chapters they wrote as well as for their preparing figures and helping to sort out references. I would also like to express my deepest gratitude to my colleagues at SMMRE, USM, for their encouragement and moral support (especially over coffee and cake).

Acknowledgments

I am very grateful for the endless motivation and enthusiasm of my parents, Mr. Lockman Salleh and Mrs. Amiah Daud. Among other things, my mother helped in deciding the book's cover and has provided emotional support throughout. My husband, Amir Hamzah Ghazali, deserves a special thanks for constantly asking about the progress of this book, as do my children, who let me stay up late sometimes to read and reread.

Editor

Zainovia Lockman is a lecturer of Materials Engineering at the School of Materials and Mineral Resources Engineering (SMMRE), Universiti Sains Malaysia (USM). She graduated from Imperial College London in 1999 with a first-class honors degree in Materials Science and Engineering. She received her PhD in 2003 from Imperial College London as well as majoring in electronics material (superconductors). After receiving her PhD, Dr. Lockman worked as a postdoctoral research fellow at Imperial College before moving to University of Cambridge, UK, working on thin film and nanostructured oxides. Her main interest has since revolved around thin film oxide fabrication via oxidation method for various electronic devices. Upon joining USM in 2004, Dr. Lockman and her research team at SMMRE have focused on electronic semiconducting oxide. In 2006, she went for a long research attachment at University of Cambridge to start on anodic oxidation process for photocatalytic TiO_2 nanotubes formation. Dr. Lockman is now leading the Nanomaterials Niche Area at SMMRE. Her research team, Green Electronic Nanomaterials Group, has been working on synthesis of oxide nanomaterials (nanotubular, nanowires, nanoparticles, and nanopores) for environment protection, energy generation, saving, and transfer (i.e., nanomaterials for green technology). The team has many outstanding achievements portrayed by the vast numbers of research publications in notable, international journals, chapter in books, and research grants awarded. Dr. Lockman is a recipient of the Nippon Sheet Glass Foundation Japan award; Malaysian Solid State Science and Technology Society (MASS) Young Researcher award; Young Scientist Award, Springer; L'Oréal-UNESCO for Women in Science Award; and United Kingdom Prime Minister's Initiative 2 for International Education (PMI 2) award through Imperial College London. Her research group has also done various outreach programs to secondary schools and has helped in community projects for enhancing scientific interest among locals. Dr. Lockman is an active member of Young Scientist Network–Academy of Sciences Malaysia, treasurer for the Microscopy Society Malaysia, and a committee member for the Malaysia Nanotechnology Association. She is supervising 13 postgraduate students and has graduated 23 since 2006.

Contributors

Haslinda Abdul Hamid
School of Materials and Mineral
    Resources Engineering
Universiti Sains Malaysia
and
Faculty of Applied Sciences
Universiti Teknologi Mara
Pulau Pinang, Malaysia

Khairunisak Abdul Razak
Green Electronics Nanomaterials
    Group
Science and Engineering of
    Nanomaterials Team
School of Materials and Mineral
    Resources Engineering
and
NanoBiotechnology Research and
    Innovation (NanoBRI)
Institute for Research in Molecular
    Medicine
Universiti Sains Malaysia
Pulau Pinang, Malaysia

Zaid Aws Ali Ghaleb
School of Materials and Mineral
    Resources Engineering
Universiti Sains Malaysia
Pulau Pinang, Malaysia

Nurain Najihah Alias
School of Materials and Mineral
    Resources Engineering
Universiti Sains Malaysia
Pulau Pinang, Malaysia

Christian Laurence E. Aquino
Sustainable Electronic Materials Group
Department of Mining, Metallurgical
    and Materials Engineering
University of the Philippines
Diliman, Quezon City, Philippines

Mary Donnabelle L. Balela
Sustainable Electronic Materials Group
Department of Mining, Metallurgical
    and Materials Engineering
University of the Philippines
Diliman, Quezon City, Philippines

Nurulhuda Bashirom
Green Electronics nanoMaterials Group
Science and Engineering of
    Nanomaterials Team
School of Materials and Mineral
    Resources Engineering
Universiti Sains Malaysia
Pulau Pinang, Malaysia

Andrey Berenov
Department of Materials
Imperial College London
London, UK

Supab Choopun
Department of Physics and Materials
    Science
Faculty of Science
Chiang Mai University
Chiang Mai, Thailand

Luigi A. Dahonog
Sustainable Electronic Materials Group
Department of Mining, Metallurgical
    and Materials Engineering
University of the Philippines
Diliman, Quezon City, Philippines

Niyom Hongsith
School of Science
University of Phayao
Phayao, Thailand

Syahriza Ismail
Carbon Research Technology Group
Faculty of Manufacturing Engineering
Universiti Teknikal Malaysia Melaka
Durian Tunggal, Malaysia

Nurul Izza Soaid
Green Electronics nanoMaterials Group
Science and Engineering of
    Nanomaterials Team
School of Materials and Mineral
    Resources Engineering
Universiti Sains Malaysia
Pulau Penang, Malaysia

Mariatti Jaafar
School of Materials and Mineral
    Resources Engineering
Universiti Sains Malaysia
Pulau Pinang, Malaysia

Ahalapitiya H. Jayatissa
Nanotechnology and MEMS
    Laboratory
Mechanical, Industrial and
    Manufacturing Engineering (MIME)
    Department
The University of Toledo
Toledo, Ohio

Go Kawamura
Department of Electrical and Electronic
    Information Engineering
Toyohashi University of Technology
Toyohashi, Japan

Zainovia Lockman
Green Electronics nanoMaterials Group
Science and Engineering of
    Nanomaterials Team
School of Materials and Mineral
    Resources Engineering
Universiti Sains Malaysia
Pulau Pinang, Malaysia

Ahmad Makarimi Abdullah
School of Materials and Mineral
    Resources Engineering
Universiti Sains Malaysia
Pulau Pinang, Malaysia

Atsunori Matsuda
Department of Electrical and Electronic
    Information Engineering
Toyohashi University of Technology
Toyohashi, Japan

Noorhashimah Mohamad Nor
School of Materials and Mineral
    Resources Engineering
Universiti Sains Malaysia
Pulau Pinang, Malaysia

Hiroyuki Muto
Institute of Liberal Arts and Sciences
Department of Electrical and Electronic
    Information Engineering
Toyohashi University of Technology
Toyohashi, Japan

Nyein Nyein
Department of Physics
University of Meiktila
Mandalay, Myanmar

Aian B. Ontoria
Sustainable Electronic Materials Group
    Department of Mining,
Metallurgical and Materials
    Engineering
University of the Philippines
Diliman, Quezon City, Philippines

Bharat R. Pant
Nanotechnology and MEMS Laboratory
Mechanical, Industrial, and
    Manufacturing Engineering (MIME)
    Department
The University of Toledo
Toledo, Ohio

Subagja Toto Rahmat
Green Electronics nanoMaterials Group
Science and Engineering of
    Nanomaterials Team
School of Materials and Mineral
    Resources Engineering
Universiti Sains Malaysia
Penang, Malaysia

Nur Syafinaz Ridhuan
School of Materials and Mineral
    Resources Engineering
Universiti Sains Malaysia
Pulau Pinang, Malaysia

Monna Rozana
Energy System Engineering
Institut Teknologi Sumatera
Lampung, Selatan, Indonesia

Khairul Arifah Saharudin
School of Materials and Mineral
    Resources Engineering
Universiti Sains Malaysia
Pulau Pinang, Malaysia

Srimala Sreekantan
School of Materials and Mineral
    Resources Engineering
Universiti Sains Malaysia
Pulau Penang, Malaysia

Tan Wai Kian
Institute of Liberal Arts and
    Sciences
Toyohashi University of
    Technology
Toyohashi, Japan

Ekasiddh Wongrat
School of Science
University of Phayao
Phayao, Thailand

Khatijah Aisha Yaacob
School of Materials and Mineral
    Resources Engineering
Universiti Sains Malaysia
Pulau Pinang, Malaysia

Siti Noorhaniah Yusoh
School of Materials and
    Mineral Resources Engineering
Universiti Sains Malaysia
Pulau Pinang, Malaysia

1 Surface Oxidation of Metal for Metal Oxide Nanowires Formation

Zainovia Lockman, Subagja Toto Rahmat, Nurulhuda Bashirom, and Monna Rozana

CONTENTS

1.1 INTRODUCTION

Transition metal oxide (TMO) is a special class of materials with unusual yet useful electronic properties. Many of these properties strongly depend on the nature of the metal (M)–oxygen (O) bonding, as well as on material defects such as vacancies, dislocations, and grain boundaries (Rao, 1989). In some cases, derivation from point defects has led to variation in stoichiometries of the oxide resulting in the emergence of new properties (Rao, 1989, Lany, 2015). MO, M_2O_3, MO_2, M_2O_5, and MO_3 are several typical binary TMOs known to exist with bonding vary from purely ionic to purely covalent. The crystal bonding obviously influences the electrical properties of TMOs (Lany, 2015). There are several binary TMOs that exhibit metal-like properties whereby they have small energy gap or occupied conduction band rendering their applications as conductors for various electronic devices (Rao, 1989, Munoz-Paez, 1994, Guo et al., 2015, Lany, 2015). CrO_2, PdO, TiO, and NbO are examples of high conductivity binary TMOs. Some TMOs exhibit purely insulating behavior with

1

energy gap of as large as 5 eV; ZrO_2 is an example of such oxide. In between the two extremes are TMOs with energy gap of typically between 1 and 3 eV: semiconductors. For this class of TMOs, the presence of defect states within the band gap contributes to the semiconducting nature of the oxide (Lany, 2015). For wide band gap TMOs, band gap narrowing can be done by doping and/or annealing at reduced condition as to induce anionic defects. Conduction in TMOs can either be electronic or ionic. For example, TiO_{2-x} and lithium-doped NiO are pure electronic conductors while Y_2O_3 doped ZrO_2 is an ionic conductor. Some TMOs have both ionic and electronic conduction (Rao, 1989, Lany, 2015). Binary TMO can be in a form simple oxide (MO), typically possessing rock-salt-structure like NiO and dioxides (MO_2) with fluorite or rutile structure like CeO_2 or TiO_2. On the other hand, sesquioxides, M_2O_3, like Cr_2O_3 and Al_2O_3 have corundum structure (Rao, 1989, Munoz-Paez, 1994, Guo et al., 2015, Lany, 2015). TMOs also form as ternary oxides like spinels, perovskites, bronzes, and garnets. Ternary oxides have exhibited a wide range of applications as they undergo phase transformation from one crystal phase to another accompanied by changes in electrical, mechanical, magnetic and optical properties. As this book is focusing only on binary TMOs, readers are encouraged to read from interesting reviews on these oxides for more detailed information (Greenblatt, 1988, Kanhere and Chen, 2014, Duan et al., 2017, Liu et al., 2017). Needless to say, ternary and quaternary TMOs are becoming crucial ingredients for many emerging and important applications especially in energy devices.

Binary TMOs fabricated in 1-D nanostructures have attracted significant attention due to their shown interesting size-dependent and structure-related properties. In particular, the novel properties of 1-D nanostructures of TMOs have rendered them as prime candidates for energy devices: solar cell, hydrogen cell, batteries, fuel cells, energy storage devices (Cheng and Fan, 2012) and thermoelectric devices (Hochbaum and Yang, 2009, Gadea et al., 2018). 1-D TMOs have also been applied in medical diagnosis, detection of environmental pollutants and sensors (Shen et al., 2009, Comini and Sberveglieri, 2010, Comini, 2016, Hung et al., 2017, Dey, 2018). Another important application of TMO 1-D nanomaterials is in the area of environmental protection either as catalyst (Akbari et al., 2018) or photocatalyst to remove toxic contaminants. 1-D TMOs can also be used as adsorbents to remove pollutants from air and water. Accordingly, it has becoming necessary and critical to create reliable and reproducible synthetic protocols for 1-D TMOs fabrication which can generate pure-phased (or well-controlled, doped) TMOs in a desired 1-D morphology and architecture. Among synthetic protocols available, thermal oxidation of metal has demonstrated some obvious advantages as the process seems to be simpler, cheaper and can be considered "greener" if compares with other physical or chemical-based processes as it does not require complex apparatus nor extensive use of energy (Yan et al., 2011).

Oxidation reaction occurs when a metal combines with an atom or with a molecular group and loses electrons (Scully, 1975). In the presence of oxygen gas, a stable solid oxide, MO will cover the surface of metal. This solid reaction product separates metal from the oxygen gas. The growth of MO can proceed through solid-state diffusion of the reactants (oxygen and metal ions) through the oxide film. For a thin film MO, the driving force for reactants transport can be electric field developed across the film whereas for thick film (or termed scale), chemical potential across

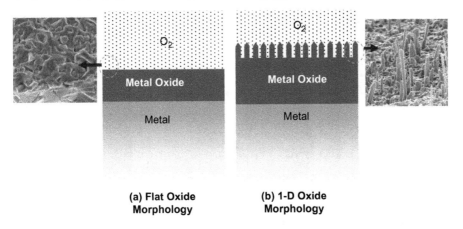

(a) Flat Oxide
Morphology

(b) 1-D Oxide
Morphology

FIGURE 1.1 Illustration of metal oxide formed by thermal oxidation with (a) flat and (b) 1-D surface morphologies as shown from SEM micrographs of the oxides.

the MO will determine the transport process. There exist several mechanisms that explain the transport of ions or electrons through MO. Diffusions of anions and cations across the oxide scale require diffusion paths: lattice, grain boundaries and other easy diffusion paths. The choice of path thus the mechanism of oxidation is largely dependent on the oxidation temperature, defects structure, and the partial pressure of oxygen surrounding the oxide. It has been widely accepted that growth of scale by lattice diffusion via lattice defects occurs at high temperature and results in dense and continuous film which adheres well to the metal substrate (Figure 1.1a). The surface oxide is also rather a flat, planar scale as seen in the micrograph insets in this figure. On the other hand, oxidation at intermediate and low temperature results in non-planar oxide with the surface oxide consists of elongated, fibrous oxides (or termed whiskers in early literatures and 1-D NWs [nanowires] in newer literatures) as illustrated in Figure 1.1b and shown in the micrograph. The micrographs shown are scanning electron microscope (SEM) images of oxidized copper. Therefore, in order to fabricate 1-D NWs on metal by thermal oxidation, several factors need to be explored; transport mechanism of the reactants which is temperature dependent and also the environment at which the oxide is grown. In here, several parameters for NWs formation are summarized, nevertheless before proceeding to this topic, it is best to provide a proper definition of 1-D nanomaterials.

1.2   1-D TMOs

Nanomaterial is a material with any external dimension in the nanoscale or having internal structure or surface structure in the nanoscale. Nanomaterials can be classified as zero-dimensional, one-dimensional, two-dimensional and three-dimensional based on the number of dimensions which are not confined to the nanoscale rage which is <100 nm. Nanoscale is length range approximately from 1 to 100 nm. 1-D nanomaterials have one dimension that is outside the nanoscale. They can have length in micron or even millimeter scale but the diameter of less than 100 nm. Figure 1.2

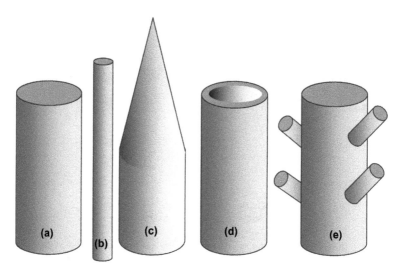

FIGURE 1.2 Schematics of typical 1-D nanostructures: (a) nanorod, (b) nanowire, (c) nanoneedle, (d) nanotube, and (e) branched formed by oxidation of metal.

shows simplified schematics of several forms of 1-D nanostructures. These sketches are based on typical morphologies of 1-D nanostructures reported especially for TMOs fabricated by oxidation process. As mentioned, thermal oxidation results in surface oxide with thickness and surface properties vary depending on the oxidation conditions. As oxide film grows in thickness, a non-planar growth of the oxide may develop with outer surface in a form of 1-D structures. 1-D structures with diameters ranging from several nm to sub-μm were referred to as oxide whiskers in the early literature; whereas the term nanowires and nanorods have been used in recently especially in describing 1-D nanostructures with diameters not exceeding 100 nm. In addition, nanorod (Figure 1.2a) is often considered shorter (sometimes with larger diameter) than nanowire, though the definition is rather arbitrary. Nanoneedle, on the other hand, refers to sharp tip 1-D nanostructure (Figure 1.2c). Nanotube is not a very common structure found on thermally oxidized metal (Figure 1.2d) but is a common structure of anodically oxidized metal (in fluoride electrolyte). Branched nanorod (Figure 1.2e) on the other hand can sometimes be seen especially after long duration oxidation process of metal with low vapor pressure.

As applications of 1-D nanomaterials may pervade all areas of our life, a standardized nomenclature and proper definition are needed not only among academics, for the purpose of reporting, but more importantly for regulators such as health and environmental protection agencies to monitor and protect the environment from 1-D nanomaterials pollution. These are essential as to ensure responsible production, dissemination, applications, and disposal of 1-D nanomaterials. Technical Committee (TC) 229 of the International Organisation for Standardisation (ISO) has published documents ISO/TS 80004 part 1–part 4 (2015), defining nanomaterial; nano-object and nanostructured material. Figure 1.3 illustrates the relationships between these terms (ISO/TS 80004-1:2015). Nano-object is defined as discrete piece of material with one, two or three external dimensions in the nanoscale. The size and shape are

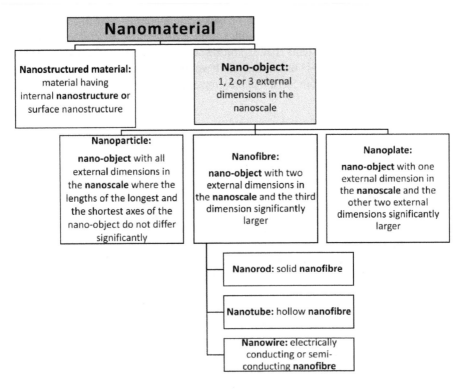

FIGURE 1.3   Definitions of nanomaterial and nano-object.

often intrinsic to their function, so the description and measurement of their size and shape are important and must be considered carefully. The three most basic shapes of nano-objects are summarized in Figure 1.3 as well: nanoparticles, nanofiber, and nanoplate. They represent the most important classes of structural dimensionality to help categorize nano-objects further. As can be seen, 1-D nanomaterials can be classified under nanofibre which is defined as an object with two external dimensions in the nanoscale and the third dimension significantly larger.

1-D nanostructures have been explored and reported since 1950s, around the same time the electron microscope was invented, but only recently this class of nanomaterials has gained enormous attention from researches and industries alike. Among all the 1-D nanostructures reported in literature, nanowires, nanorods, and nanotubes are the most common, particularly due to the improved performance of devices utilizing them over the bulk form. According to ISO/TS 80004-4:2015, nanorod is solid nanofiber, nanotube is hollow nanofibre and nanowire is an electrically conducting or semiconducting nanofiber.

1.3   GENERAL PROPERTIES OF NWs

TC 229 defines nanotechnology as the understanding and control of matter and processes at the nanoscale where the onset of size-dependent phenomena usually enables novel applications. Nanotechnology can also be regarded as a group of emerging technologies that utilize the properties of nanoscale materials that differ from the properties of individual atoms, molecules, and bulk matter, to create improved materials, devices, and systems that exploit these new properties (Ramsden, 2009). Changes in properties of 1-D nanomaterials have made them useful in nanotechnology and these changes in comparison to bulk materials can be categorized as follows:

1.3.1 INTERNAL EFFECT

Internal effect can be generally defined as changes of properties related to the material itself when made in nanoscale. Nanomaterials including NWs have an extraordinary ratio of surface atoms per unit volume. The ratio increases even more with the reduction of the diameter of the NWs. It is known that atoms at the outermost layer of a solid surface have less nearest neighbors; therefore they are comprised of unsatisfied bonds (dangling bonds) and they have more freedom to move (higher entropy) as they are not exactly fixed in place. Because of this they possess excess internal energy which is reduced when they will undergo surface relaxation whereby they shift inwardly making the bond distance between the surface atoms and sub-surface atoms smaller, reducing lattice constant significantly (Cao and Wang, 2011). It is true that surface atoms of a bulk material will also have an asymmetric interaction at the outermost layer but for a nanomaterial, as the surface-to-volume ratio is much higher, the material can be thought to consist of only surface atoms and hence entire solid particles show an appreciable decrease of lattice constant. This will affect several physical properties of the material including crystallization, phase transformation, stabilization and thermal properties. Chemical potential is dependent on the radius of curvature of a surface with higher chemical potential the smaller the material is. Nanomaterial has large surface curvature and hence larger chemical potential which can be reflected by high solubility or vapor pressure as well as high reactivity.

Moreover, due to an absence of bonds at the surface atoms, there exists an excess internal energy of surface atoms compares with atoms in the bulk. This and the higher entropy the surface atoms possess will create a system with high surface free energy; thus, nanomaterials are thermodynamically unstable or metastable. Because of this, nanomaterials have extremely high reactivity and hence can be expected to participate in various surface reactions. Furthermore, with larger surface areas, they possess more sites for reaction to occur in comparison to bulk material. This effect has major consequences in technologies where reactivity and surface area are important like catalysis, sensor, and adsorption.

Another property that can change when TMO is made in 1-D nanostructure is the mechanical properties of the oxide. Growth mechanisms of some NWs would favor the formation of single-crystalline structures and hence can be considered to have a much lower density of line defects than is typically found in bulk material. Therefore, 1-D TMOs can feature mechanical strength and toughness values approaching the theoretical limits of perfect crystals. Moreover, due to the small diameter, some TMO NWs exhibits much high flexibility. This would increase on the robustness of the material making them suitable for applications in devices like actuator and sensors.

1.3.2 EXTERNAL EFFECT

External effect can be defined as changes a material experiences when it is imposed to external physical fields. When 1-D TMO NWs are made to have a diameter less or comparable with a length of any physical phenomena for example, mean-free-path of carriers (electrons or holes), exciton Bohr radius, coherent length, screening length, phonon mean free path, critical size of magnetic domains, and light absorption coefficient then, the materials will no longer behave like their bulk counterpart due to confinement effect. It is best to explain by giving examples of typical devices that utilizes 1-D TMOs like solar cell (especially dye-sensitized solar cell) and photocatalytic-based hydrogen cell via water splitting process. Both of these devices can only function under light illumination and often the main aim of such energy devices is high conversion efficiency. For this, losses in all steps of the conversion processes must be identified and eliminated. For instance, steps required for NWs used to convert sunlight-to-electricity (solar cell) or sunlight-to-hydrogen (hydrogen cell) are photon absorption, electron-hole pairs formation, and electron-hole pairs separation and carriers collection or transfer. The use of TMO NWs arranged in periodic arrays as oppose to planar surfaces like thin film or wafer can help in minimizing losses in each step. For example, absorption losses can be reduced when high aspect-ratio NWs are used as the long axis of the NW is expected to enable adequate area for light absorption, while the radical axis provides short electron-hole separation distance and shorter diffusion path for charge transfer. With shorter distance to diffuse, recombination can be minimized.

NWs also provide a direct pathway for electron transport along their long axis to the back contact reducing recombination. Moreover, as mentioned, the crystallinity of TMOs in 1-D structure is often rather perfect hence recombination due to lattice defects can be much reduced. It is therefore clear that in order to capture much of the benefit of the 1-D nanostructures, TMOs synthesized by thermal oxidation must be phase pure, well crystalline, and perfectly aligned normal to the surface of the substrate.

1.4 SYNTHESIS PROCESS OF TMO NWs

Generally, the synthesis routes of nanomaterials can be divided into top down and bottom up methods. In bottom-up processes, structures are assembled from their subcomponents in an additive fashion. Top-down fabrication strategies use sculpting or etching to carve structures from a larger piece of material in a subtractive fashion. There are a large number of synthesis techniques to produce TMO NWs. Much of these processes have been reviewed in great detail (Fan et al., 2007, Rao and Govindaraj, 2011, Nikoobakht et al., 2013, Dasgupta et al., 2014). Based on process and mechanism of formation, bottom-up techniques can be broadly classified into (a) process and (b) mechanisms for NWs formation as shown in Figure 1.4.

TMOs NWs can be synthesized on a large scale through wet chemical processes (sol-gel, hydrothermal and solvothermal processes), vapor processes (chemical vapor deposition and physical vapor deposition) and thermal processes like decomposition and oxidation method. Thermal oxidation method as mentioned is a process that would allow for direct growth of TMO NWs at elevated temperature typically between 400°C and 800°C in an oxidizing environment. Thermal oxidation requires a bulk metal like foil, thin film, wire or powder and a method to heat it up. As shown in Figure 1.4a, heating method includes furnace, resistive, flame and plasma heating. A flame synthesis differs from furnace heating that the flame heats up a metal substrate at extremely high rate and depending on the nature of the flame used, the flame composition may influence the growth behavior of the oxide. Resistive heating on the other hand requires an electric current to be supplied to the metal substrate. A metal wire, for instance, is connected at both ends to a power source and electric current supplied to the wire will heat it up from the inside. Rate of heating is also high by this heating method.

The time taken for NWs to grow on metal surface by thermal oxidation depends obviously on the selection of the heating method used; time taken to reach a certain desired temperature and can be further elaborated by the choice of heating profile; heating rate, quenching or furnace cool. Time for NWs growth can also vary depending on the nature of the substrate selected; the metal itself; its melting point and vapor pressure, its shape, dimension, purity, grain size and surface cleanliness. The applied atmosphere can be laboratory air, which is the easiest whereby a furnace is left open during the course of oxidation process. For a well-controlled experiment, however, pure oxygen or a mixture of oxygen (with argon, hydrogen or nitrogen) can

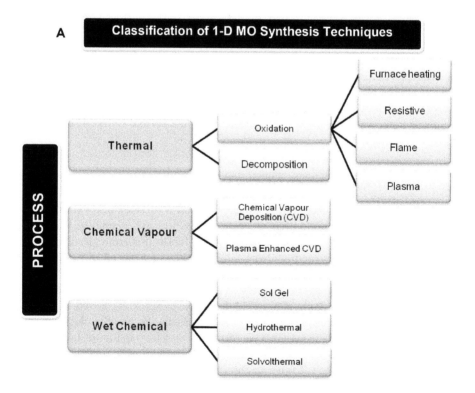

FIGURE 1.4   Classification of 1-D NWs synthesis based on (a) processes.   (*Continued*)

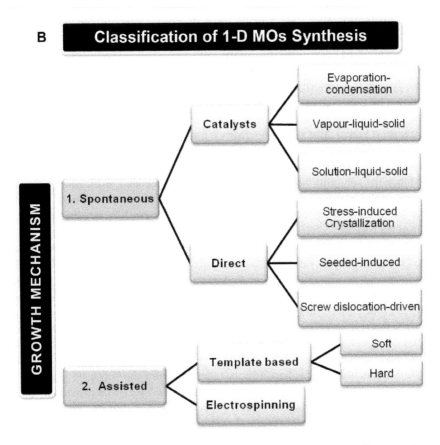

FIGURE 1.4 (Continued) Classification of 1-D NWs synthesis based on (b) mechanism.

also be used. The choice of gases used can be correlated to the dissociation pressure of the TMOs and hence, knowledge on the basic of thermodynamic is required. The oxidizing atmosphere can be static or gas can be flown through the furnace during the course of oxidation process. The rate of gas flown as well as the pressure used can be controlled as to suit the best condition for NWs formation. Wet oxidation can be done by purging the furnace with water vapor and this can induce faster NWs formation on several metal surfaces like iron and copper. There have also been cases where acetone or ethanol is injected in the oxidizing furnace for NWs formation or oxidation in the presence halide salts or potassium hydroxide. Some examples are given in the subsequent sub-heading (and in other chapters in this book as well).

Obviously anisotropic growth is required to produce NWs on metal surfaces, whereby during the oxidation process, crystal grows must be directed (confined) along a certain orientation faster than other directions. Preferably, the NWs formed must be uniform in size with diameter consistent along the length of the nanowires. This can be obtained by confining growth along one direction, with no growth allowed along other directions. In chemical vapor processes, confining growth can be done by catalyst-mediated processes described well by vapor-solid-liquid (VSL) mechanism or a direct growth (or sometimes called vapor-solid (VS)) whereby the presence of a metal catalyst is not required for the confinement of 1-D NW growth (Figure 1.4b).

VSL is the most used mechanism to account for growth of NWs in vapor or gas phase. Typically, a vaporized source material is introduced into a reaction chamber and flown by a carrier gas towards a substrate. Liquid catalysts are present on the surface of the substrate as means to confine the growth of the vaporized constituent. Otherwise, a planar film would develop. With VSL, liquid droplets are formed by eutectic reactions between the precursor and the catalyst. The droplets have a finite solubility of the vaporized precursor thus providing favorable sites for the adsorption. Once the solubility limit within the droplet is exceeded, nucleation will occur underneath the liquid droplet. The growth of the NWs occurs from base to tip and at the tip of the NW; a "cap" is often observed when the NW is viewed under an electron microscope.

In thermal oxidation, this mechanism of growth is often disregarded even though there have been cases when catalysts are required to confine growth of TMOs for 1-D NWs formation or when the metal or its oxide can vaporized during the growth process. The VS mechanism has been used to explain the growth of 1-D NWs without catalysts present, nevertheless the description of NWs formed via VS process is not exactly very clear and not completely accurate for the case of thermal oxidation process unless again, vapor is generated and participated in the growth process during the oxidation process.

The VS mechanism involves defects inherited from the substrate surface (can be the outermost layer of oxide scale before the NWs layer), that provide a "mold" confining growth of 1-D nanostructure. Process without catalysts can therefore be termed as a "direct" process. In thermal oxidation for NWs formation, stress-induced and screw dislocation are typically used to describe the formation of anisotropic growth for direct process. This will be explained later.

From a morphology point of view, direct growth of NWs on metal by thermal oxidation can have two main limitations; alignment and uniformity. The former relates to the growth direction of the NWs which depends largely on the substrate onto which they are growing and the latter is more on the variation in dimension of one NW to another. Surface preparation of the metal surfaces perhaps is worth mentioning here whereby roughens surface can give more dense NWs as oppose to smooth surface. Roughening of the surface can be done by grinding or by using sandpaper. Nevertheless, maintaining uniform diameter along the length of an NW can be a challenge in TMO NWs grown by thermal oxidation as well as maintaining aligned NWs. The physical method can also be chosen to confine growth anisotropically with an added benefit of having control over length and diameter of the NWs formed. As shown in Figure 1.4b, this includes electrospinning and templated techniques. The former is on generating fiber of TMOs by using electrical forces whereas templated technique is a process that requires a template to direct the formation of NWs. Templated technique is perhaps the most common method to produce TMO NWs with good dimension control. This process requires precursor solution of the desired TMO to be deposited into a template with 1-D morphology like elongated porous templates. Anodic aluminum oxide (AAO) and polycarbonate membrane are two examples of porous templates. The inner hole of the AAO template is therefore important as it controls the final diameter of the TMOs NWs. The dimensions of the NWs formed are similar to that of the template. Once solidified, the TMO NWs will be removed from the template often by chemical etching of the template will be subjected to thermal annealing for phase formation and crystallization to occur. Unlike hard templates method, soft template methods do not require a rigid structure as a mold for TMO NWs to grow. Typical soft templates are surfactant, polymer and biopolymer. Surfactants are amphiphilic molecules, including ammonium salts, heterocyclic, carboxylic acid salts, sulfonate salts, and other ionic or nonionic surfactants. It is easy for the amphiphilic molecule groups to form a variety of ordered polymers in a solution, such as liquid crystals, vesicles, micelles, microemulsion, and self-assembled film (Rao, 1989, Hobbs et al., 2012). Using these aggregates as a template, TMO precursor is deposited on the surface or in the interior of the template by solution chemical methods. When the template is removed, powdery TMO NWs will be produced. Therefore, post-process for NWs alignment requires adding in steps to the fabrication procedure of a NW-based device.

1.5   CHARACTERIZATIONS OF NWs

The discovery of TMO NWs by thermal oxidation was spurred not only by the excitement over applications of nanomaterials in nanotechnology, hyped since the 2000s, but by advances in various characterization instruments with electron microscopy as the most important. Electron microscopy techniques, specifically SEM and transmission electron microscopy (TEM) are two important techniques for structural determination of TMO NWs formed by thermal oxidation. The development of field emission emitters in SEM is crucial for the field of nanomaterials characterization and hence field emission SEM (FESEM) is primarily used for imaging the surface and cross section of oxide formation on metal surface. Oxidized metal can be viewed in secondary or backscattered mode with secondary mode for giving shape and topographic identification of the nanostructures whereas backscattered electrons for chemical contrast. This can give an indication of different oxide phases that may form on the oxide surface. In-situ observation of oxide morphology evolution can also be done in SEM (or TEM) and hence the growth process of NWs can be viewed and recorded at real-time. A proposal on the mechanism of formation can then be made based on the moving images of the growing NWs.

For elemental identification, an energy dispersed X-ray method (attached to SEM system) can be used and ratio between metal and oxygen can give an early indication of phases that may exist within the oxide layer. X-ray diffraction (XRD) on the other hand is a more accurate technique for phase determination with glancing angle XRD providing information of the surface oxide and to a lesser extends the inner layer. Apart from crystals studies, XRD can be used to investigate strain and shape factor. The shape factor is related to the size of the crystal and by using Scherrer expression, crystallite size can be deduced. Raman spectroscopy has also gaining much attention in recent years for phase identification, strain measurement, and crystallinity determination. For much detailed analysis of crystallinity high-resolution TEM (HRTEM) can be used whereby atomic resolution lattice images can be obtained.

In addition, a diffraction mode in HRTEM can reveal whether the oxide NW formed is amorphous, single crystalline or polycrystalline. HRTEM also can be used to support a proposed mechanism of thermally grown NWs on metal substrate by providing for instance information on NWs growth direction. Other techniques to characterize NWs would be specific to the properties desired from the oxide material. For example, surface analysis such as x-ray photoelectron spectroscopy can be used to identify the oxidation state of the oxide formed as well as on defects analysis within the oxide. Photoluminescence spectroscopy is a technique to investigate band structure of TMO NWs with semiconducting properties. Photocurrent measurement is often used for optical and electronic properties of the NWs.

1.6   THERMAL OXIDATION PROCESS FOR 1-D TMOs FORMATION

As established so far, thermal oxidation requires a furnace (heating system) and the reactants: metal and oxygen. An oxidation reaction is said to occur when a metal, M (with valency, z) combines with an atom or with a molecular group and loses electrons (Eq. 1.1) in this case oxygen, O. The term oxidation is therefore describing transfer of electrons to adsorbed oxygen molecules forming O^{2-} (Eq. 1.2) and the combination of metal with oxygen to form $MO_{z/2}$ (Eq. 1.3).

$$\text{Oxidation: } M \rightarrow M^{z+} + ze^- \tag{1.1}$$

$$\text{Reduction: } O_2 + ze^- \rightarrow z/2O^{2-} \tag{1.2}$$

$$\text{Combined: } M + O_2 \rightarrow MO_{z/2} \tag{1.3}$$

It is reasonable to assume that a metal would produce only one oxide; for example oxidation of titanium would produce titanium dioxide, TiO_2, and oxidation of copper results in a scale of cupric oxide, CuO. During oxidation process, however, it is not always the case, as oxidation may result in the formation of an oxide with more than one structure (or polymorphs) or oxide with different oxidation states. Moreover, not all metals can be oxidized to produce NWs and they are those that require extremely long duration of oxidation for NWs formation that it becomes unpractical for large-scale production. In addition, not all metals transform to metal oxide when heated at high temperature and hence the thermodynamic of the metal-oxygen system must be consulted. The overall driving energy of metal-oxygen reaction is the free energy associated with the formation of MO. MO will be formed thermodynamically when the ambient oxygen pressure is larger than the dissociation pressure of the MO in equilibrium with the M. The standard free energy of formation of oxides as a function of temperature and corresponding dissociation pressure of the MO are summarized in the form of Ellingham/Richardson diagram. On the other hand, understanding metal-oxygen equilibria and phase diagram is essential for determining all possible oxide phases that may form on the metal surfaces. For a metal that can form several oxides, phase diagram of the metal-oxygen system can provide information on the most probable sequence of the oxides. Often, oxide scale will have the most oxygen-deficient oxide nearer to the metal surface and oxygen-rich oxide next to the gas phase. Therefore, for oxidation of metal for NWs formation, one can predict the phase of the surface NWs by referring to the phase diagram of the system.

Indeed, TMOs synthesis via oxidation can be considered as relatively low temperature process as well as catalyst and surfactant-free process. Nevertheless, catalyst addition is required for oxidation of some metals as to reduce the oxidation temperature and to induce much more uniform NWs formation. Understanding the mechanism of NWs formation is therefore important in ensuring the success of NWs formation on metal surfaces.

1.6.1   TMO Nanowires Formation Mechanism

During thermal oxidation process, the oxide scale on metal surface can thicken and the grains can exhibit grain growth, recrystallization, and plastic deformation depending on the composition of the scale and the oxidation conditions (temperature, time, and partial pressure of oxygen). Oxidation is a thermally induced process; therefore growth of TMOs on the metal surface is determined by cationic and anionic diffusion across the TMO layer. At low and intermediate temperature, diffusion rate is expected to be higher along oxide grain boundaries as opposed to bulk or lattice diffusion. At the oxide/air interface, scale growth can be irregular and give rise to a wide variety of outer oxide surface morphology including 1-D NWs. The formation of NWs never occurs during the initial growth process when the oxide is thin, but is always observed when a certain thickness is achieved. For a TMO exhibiting various phases, the NWs are always consisted of the higher valance oxide, which is suggesting that the growth is independent on the metal but is reliant on the nature of the oxide onto which the NWs are growing, which is often oxide with lower oxidation state. From Figure 1.4b, the growth mechanism of NWs by oxidation can be proposed to follow a plastic flow mechanism or catalysts assisted (base growth), or via a short-circuit diffusion path at the center of the whiskers (tip growth). The following describes these processes further:

1. **Base Growth**
   In this process, extrusion of metal atoms from the substrate occurs whereby outwards diffusion of cations is driven by compressive stress. In this case, NWs will grow from their bases outwards forming elongated structure as illustrated in Figure 1.5. Catalyst-aided process can be thought as base growth as well, when the NWs are formed underneath catalysts droplets diffuse outward. Diffusion of metal cations is required as well as constant supply of oxygen around the growing oxide. Knowledge of the catalyst (for catalyst process) is required.
2. **Tip Growth**
   For this growth mode, the growth front is at the tip of the growing NWs, whereby continuous flow of metal cations from base outwards and constant supply of oxygen are required for NWs formation. There are several possible processes of supplying cations to induce tip growth:
   a.   Evaporation and condensation method
   b.   Surface diffusion
   c.   Internal diffusion

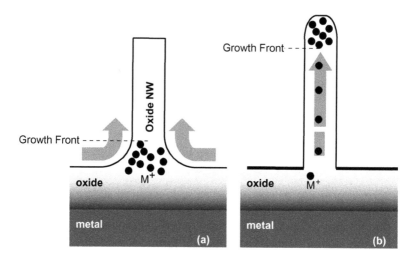

FIGURE 1.5   Schematic of (a) base growth and (b) tip growth process during thermal oxidation of metal for NW formation.

Process (a) would require a metal with low vapor pressure. Zinc is an example whereby in the formation of ZnO NWs, diffusion of Zn^{2+} may result in the formation of Zn vapor that would condense back on the tip forming an elongated 1-D structure. Metals with higher vapor pressure would follow (b) and (c) for their oxide NWs formation. For surface diffusion process to occur high mobility cations with high vapor pressure are required. Growth happens at the surface of the growing NWs forming an elongated structure. Internal diffusion can be regarded as oxidation via an easy diffusion path at the center of the growing NWs. To understand these mechanisms further, it is best to use copper and titanium as examples of metals that can be oxidized to produce 1-D NWs.

1.6.2   CuO Nanowires

CuO and Cu_2O are well-known copper oxides with semiconducting properties (Rakhshani, 1986). They are p-type semiconductors with applications ranging from photovoltaic to photocatalyst. The performance of devices utilizing CuO or Cu_2O is often morphology related whereby they appear to be more efficient when made with 1-D NWs structure. 1-D NWs CuO can be produced by direct oxidation of copper substrate. Figure 1.6 shows photographs of copper foil (a) before and (b) after oxidation from our laboratory. The foil turned darker after oxidation indicating a formation of oxide layer covering its surface. Under SEM, the oxide scale with thickness of ~5 μm is seen to comprise of multi-layered oxides (Figure 1.6c) with utmost surface comprising of either nanorods or NWs depending on oxidation temperature (Figure 1.6d and e). Control over the length and size of the surface 1-D nanostructures formed is a challenge as well as on the coverage (or sometimes defined as density) of NWs on the substrate surface.

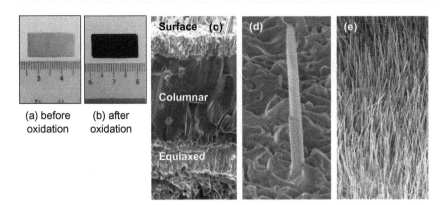

FIGURE 1.6   Appearance of copper foil: (a) before and (b) after oxidation (600°C). (c) Cross section SEM image of (b). SEM image of oxidized copper at (d) 400°C showing a single nanorod and (e) 600°C, showing CuO nanowires.

The importance of temperature on whiskers (or NWs) growth on copper has been reported since 1950s as seen in Table 1.1. Cowley, for instance, observed whiskers with diameter of less than 100 nm on copper substrate oxidized at 600°C for ~25 hours (Cowley, 1954). Gulbrensen et al. on the other hand concluded that oxidation at 250°C–450°C resulted in CuO whiskers formation and with aspect ratio depending on the growth temperature. At 450°C two sizes of whiskers were observed; small whiskers of ~50 nm in diameter and larger whiskers of ~130–150 nm in diameter. The smaller whiskers were reported to be longer than the thicker whiskers. Their electron microscope studies also showed embryonic CuO whiskers to form after 6 hours of reaction at 250°C (Gulbransen et al., 1961). Ali and Wood investigated on the effect of temperature and oxygen partial pressure on the phases of oxide formed (Ali and Wood, 1968). At 200°C and in 1 torr of oxygen, Cu_2O was the only oxide formed but at 600°C, multilayered oxide was observed with CuO (whiskers) as the outer layer. The appearance of CuO whiskers was proposed to be related to the thickness of Cu_2O and localized stress in the Cu_2O layer was responsible for the formation of CuO whiskers. In the 1980s, works on CuO whiskers (or sometimes termed as needles) formation was continued by various groups all over the world. Figure 1.7 shows TEM micrograph of CuO whiskers (Kaito et al., 1986) formed by the oxidation of a copper grid at 400°C in air. Each needle is seen to be in the range of few microns in length and a few tens of nm in width. It was not until early 2000s that the term NW was used instead of whisker. Jiang et al. (2002) concluded on the formation of CuO NWs on copper grids, foils and wires at 400°C–700°C. Electron microscopic studies indicated that these NWs had a controllable diameter in the range of 30–100 nm with lengths of up to 15 μm by varying the temperature and growth time (Jiang et al., 2002).

Apart from temperature, atmospheric condition was also studied. Xu et al. (2004) reported on the morphologies of oxidized copper powders in wet air at 300°C and 800°C with NWs to form at 400°C–600°C. The use of wet oxidation was preferred for faster growth of NWs and thicker diameter NWs. N_2/O_2 gas flow has

TABLE 1.1

Literature Survey on the Formation of CuO NWs under Thermal Oxidation of Copper

Author/Year	Thermal Oxidation Conditions			Morphology	Application
	Atmosphere	Temperature (°C)	Time (hour)		
Cowley (1954)	Air	600	25	CuO' crystals	—
Gulbransen et al. (1961)	Hg oxygen pressure	250–450	6	CuO whiskers	—
Ali and Wood (1968)	Oxygen pressure	200 & 600	3	Cu_2O whiskers, CuO and Cu_2O nanowires	—
Appleby & Tylecote (1970)	Oxygen	900	2 days	CuO	—
Taniguchi (1985)	Oxygen pressure	600–700	Not stated	CuO whiskers	—
Kaito et al. (1986)	Air	450	10 minutes	Copper oxide whiskers	—
Jiang et al. (2002)	Air	400–700	4	CuO nanowires	Catalyst to convert hydrocarbons completely into carbon dioxide and water
Xu et al. (2004)	Wet air	800	4	CuO nanowires	—
Huang et al. (2004)	Water vapor	400–750	0.5, 1, 2, 10, 20	CuO nanowires	—
Kaur et al. (2006)	Air	675	4	CuO nanowires	—
Zhang et al. (2007)	Static air with N_2/O_2 flow	450	4	CuO nanowires	—
Kim et al. (2008)	Dry air	400	12	CuO nanowires	Gas sensors for air quality control in automotive cabin
Chen et al. (2008)	Air	400	8	CuO nanowires	—
Hansen et al. (2008)	Mixture of pure O_2 or $Ar–O_2$	500	2.5	Cupric oxide nanowires	—
Gonçalves et al. (2009)	Air	400–600	3	CuO nanowires	—
Liang et al. (2010)	Air	500	2	Cupric oxide nanowires	—
Vila et al. (2010)	Air	380	3, 14	CuO nanowires	—

(Continued)

TABLE 1.1 (*Continued*)
Literature Survey on the Formation of CuO NWs under Thermal Oxidation of Copper

Author/Year	Thermal Oxidation Conditions			Morphology	Application
	Atmosphere	Temperature (°C)	Time (hour)		
Li et al. (2010)	Static air	600	6	CuO nanowires	Chemical sensor
Choopun et al. (2010)	Air	600	24	CuO nanowires	—
Zhong et al. (2010)	Ambient air	400–500	24–96	CuO nanowires	Gas-sensors
Hsueh et al. (2011)	Air	450	5	CuO nanowires	Humidity sensors
Mumm & Sikorski (2011)	Air	500	2	CuO nanowires	—
Zappa et al. (2011)	Oxygen and Argon	300	15	CuO nanowires	Chemical sensing
Yuan et al. (2011)	Air	450	2	CuO nanowires	—
Mema et al. (2011)	Static oxygen	450	0.5	CuO nanowires	—
Love et al. (2011)	Air	600	0.5	CuO nanowires	Heat pipes, boilers, electronics cooling
Steinhauer et al. (2011)	Ambient air	225 & 500	3.3	CuO nanowires	Conductometric gas sensors
Cheng & Chen (2012)	Air oven	400–500	0.5–7	CuO nanowires	Sensors
Lamberti et al. (2012)	Static air	400	0.5	Cuprous oxide	Li-ion batteries
Filipič & Cvelbar (2012)	Air	400, 500, 600	4	CuO and Cu$_2$O nanowires	Energy conversion devices
Yuan & Zhou (2012)	Nitrogen, Vacuum, Oxygen	450	1	CuO nanowires	—
Li et al. (2013b)	Air	300–700	2, 5	Cupric oxide nanowires	—
Li et al. (2013a)	Air	300–750	6	CuO nanowires	—
Steinhauer et al. (2013b)	Pure oxygen	400	3	CuO nanowires	Sensor
Steinhauer et al. (2013a)	Ambient air	500	3.3	CuO nanowires	Sensor

(*Continued*)

TABLE 1.1 (Continued)
Literature Survey on the Formation of CuO NWs under Thermal Oxidation of Copper

Author/Year	Thermal Oxidation Conditions				Morphology	Application
	Atmosphere	Temperature (°C)	Time (hour)			
Hoa et al. (2014)	Air	400, 500, 600	2		Cupric oxide nanowires	Sensors
Zhang et al. (2014)	Static air	400	12		CuO nanowires	–
Lee & Tuan (2014)	Flowing nitrogen, Air	300	4		CuO nanowires	–
Li et al. (2014)	Air	500	6		Copper oxide nanowires	Lithium-ion batteries
Wu et al. (2014)	Air	500, 600, 700	1, 5, 10		CuO nanowires	Photocatalyst and energy harvesting materials
Yang et al. (2015)	Dry air	300–500	1		CuO nanowires	–
Dorogov et al. (2015)	Air	400	6		CuO whiskers	–
Kim et al. (2015)	Air	500	4		CuO nanowires	Sensor
Zhao et al. (2015)	Air	400–700	2–5		Cupric oxide nanowires	Photoelectric devices
Tang et al. (2016)	Air	400	24		CuO nanowires	–
Scuderi et al. (2016)	Dye degradation in water	550	3		CuO and Cu₂O nanowires	Photocatalyst in degrading organic pollutants
Zheng et al. (2017)	In situ	500	5		CuO nanowires	Monolithic catalyst for plasma-catalytic oxidation of toluene
Hilman et al. (2017)	Dry air	250	2		CuO nanowires	–
Mahmoodi et al. (2018)	Air	400	3, 5, 7, 9		Copper oxide nanowires	Antibacterial

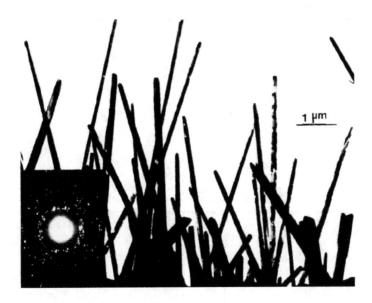

FIGURE 1.7 Microscopy image of a copper oxide whisker. (Reprinted from *J. Cryst. Growth.*, 74, Kaito, C. et al., Electron microscopic studies on structures and reduction process of copper oxide whiskers, 469–479, Copyright 1986, 2018, with permission from Elsevier.)

also been used for oxidation of copper. Oxidation of copper thin film on silicon under a N_2/O_2 gas flow was done at temperature range from 400°C to 500°C produced vertically aligned despite not very uniform NWs on large areas (Zhang et al., 2007). Oxidation in oxygen also resulted in NWs formation at similar temperature window of 500°C–700°C on copper strips as shown in Figure 1.8a–c (Kaur et al., 2006). No NWs can be seen for copper strip oxidized at 800°C (Figure 1.8d). It is evidence from literature that oxidation temperature must be within 200°C–700°C for CuO whiskers formation (Table 1.1). From, Table 1.1, oxidation in argon-oxygen mixture has been also studied but not much improvement in term of CuO NWs formation is achieved in such condition.

Figure 1.9 shows the cross sections of samples prepared at 500°C for (a) 45 minutes and (b) 90 minutes in our laboratories on copper wires in static air. As can be seen, with the increase of oxidation time, CuO NWs become longer and the distribution of the NWs on the substrate is denser. Dense, long NWs are preferred for various applications and hence, time variation is an important parameter in order to determine length and coverage of NWs on the substrate.

It can be surmised that the formation of CuO NWs by direct oxidation of copper can be done at certain temperature, duration, and atmospheric windows. The windows are however; rather wide and hence, forming NWs on copper is not very complicated. Despite that, it appears that the interpretation of the growth mechanism is somewhat complex with no specific conclusion gathered from older and newer literatures alike. The growth mechanisms have however been centered on answering the following questions: What promotes the anisotropic growth during thermal

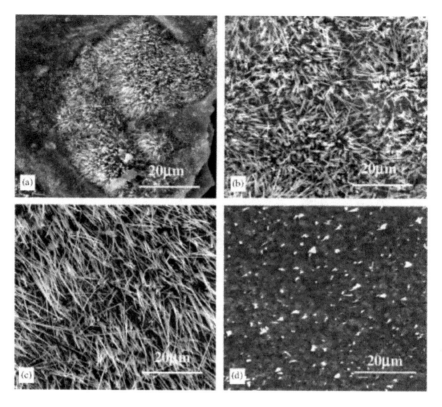

FIGURE 1.8 SEM micrographs of CuO NWs prepared by annealing copper strips for 4 hours under oxygen at different temperatures: (a) 500°C, (b) 600°C, (c) 700°C, and (d) 800°C. (Reprinted with permission from Kaur et al., 2006, Copyright 2018, Elsevier.)

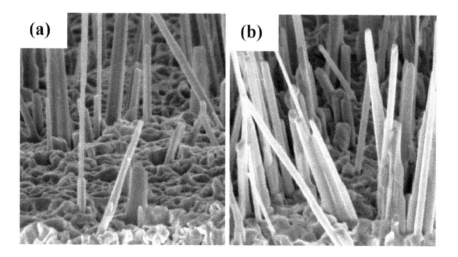

FIGURE 1.9 SEM micrographs of CuO NWs prepared by oxidation of copper at 500°C for (a) 45 minutes and (b) 90 minutes.

oxidation of copper? What is the transport mechanism of the reactants throughout the process and does the growth occurs at the tip or at the base of the NW? The answers vary from one work to another, despite there existing some consistency in explanation, but to date there is no exact mechanism concluded to account for the 1-D growth of CuO NWs during the thermal oxidation process.

One of the first mechanisms proposed is via VS as put forward by Jiang et al. (2002). It is known that oxidation of copper resulted first in Cu_2O formation (Eq. 1.4) and conversion of Cu_2O to CuO (Eq. 1.5). The latter process is slower and due to this there exists significant CuO vapor in the reaction chamber that results in continuous growth of CuO NWs (Jiang et al., 2002).

$$4Cu + O_2 \rightarrow 2Cu_2O \qquad (1.4)$$

$$2Cu_2O + O_2 \rightarrow 4CuO \qquad (1.5)$$

However, VS process for NWs growth has been dismissed by various authors due to the lack of evidence on the vapor formation during thermal oxidation of Cu. During oxidation, diffusion of Cu^+ occurs across the growing scale due to differences in chemical potential. Diffusion is a thermally activated process whereby diffusion of cations will occur along certain diffusion paths; lattice or defects depending on temperature of oxidation. Xu et al. (2004) suggested on the formation of NWs originating from screw dislocations intersecting the surface of the foil and such defects are active only at specific temperature range: 400°C–700°C. However, there was no further elaboration on why the defects are active at that specific temperature ranges or any micrographs that support such growth mechanism.

A rather clearer description was actually given by Rapp in 1984 after observing CuO whiskers on thermally oxidized copper. A screw dislocation contains a hollow core and diffusion along this core will result in NW formation. The core provides a rapid diffusion path by surface diffusion of Cu^{2+} from the Cu_2O layer to the tip forming elongated 1-D nanostructure provided the temperature used for oxidation will allow this to happen. Whisker formation is favored at low temperatures, and the presence of water vapor (Rapp, 1984). The temperature often quoted is either "low temperature" defined as $T < 0.5T_m$ or "intermediate temperature" which is $0.5T_m < T < 0.75T_m$ where T_m is the melting point of metal. In this case, as melting point of copper is 1085°C, then oxidation between 300°C and 800°C should result in the formation of NWs. Dislocations could be inherited from growth stresses due to volume differences between oxide and metal substrate or between oxides of different phases. Screw-dislocation-driven 1-D growth mechanism is somewhat different than what is proposed by Rapp in which upon intersecting the crystal surface, an axial screw dislocation creates self-perpetuating growth steps enabling anisotropic crystal growth under low supersaturation conditions. Screw dislocations can act as deposition sinks under low supersaturation and with constant exposure of the precursor vapor, growth will occur at the dislocation steps.

Kumar et al. (2004) proposed on the growth of CuO NWs by stress-driven mechanism due to the difference in the crystallographic structures, molar volumes and densities of CuO and Cu_2O. CuO has monoclinic structure whereas Cu_2O has a simple cubic structure. Since CuO formed on Cu_2O (as seen from Eq. 1.5), then there must exist substantial stress developed at the interface between the two oxide layers. One way to release stress is by the formation of triangular pyramids at the surface of the oxide scale. As the oxidation proceeds, stress is accumulated and once a critical limit has been reached, the oxide layer relaxes itself by spontaneous growth of NWs from the triangular pyramids. Kaur et al. (2006) also attributed the growth of CuO NWs to a stress-driven process whereby high rate of oxidation (fast growth of CuO) resulted in small crystallites of CuO where NWs were thought to have occurred from. Similarly, Conçalves et al. (2009) on the formation of equiaxed CuO grains on the Cu_2O layer. Short-circuit diffusion of Cu^{2+} ions across grain boundaries and/or defects of the underlying Cu_2O layer resulted in NWs formation on the CuO grains. More recent works also concluded on the growth of NWs directly on the surface CuO grains. In 2011, Yuan et al. performed series of experiment and calculation to account for this. Agreeing with previous works, the formation of CuO NWs is said to be due compressive stresses that occurs because of volume change associated with solid-state transformation at the $CuO|Cu_2O$ interface. Stress developed can induce outward grain-boundary diffusion of Cu^+ ions. They also observed the formation of twin boundaries in NWs. Such morphology is said to serve as the template for initiating the nucleation and growth of CuO NWs. Growth occurs at the tips of the NWs, where atomic steps around the tip provide sites for rapid surface diffusion of cations. A kinetic model based on the stress-driven grain-boundary diffusion has also been proposed.

When titanium is oxidized, titanium oxides such as TiO, Ti_2O_3, Ti_3O_5, and TiO_2 can be formed depending on the duration and temperature of oxidation. Often, TiO_2 is the surface oxide and the lower oxidation state oxides will form layers underneath. The scale formation on titanium is much more complex if compared to copper. The surface oxide TiO_2 is however always in rutile phase (Motte et al., 1976) and hence, the NWs formed are also consisted of rutile TiO_2. Unlike oxidation of copper where CuO NWs are easily formed, oxidation of titanium for TiO_2 NWs formation is not as straightforward. This is reflected by the lack of publications on the formation of NWs on titanium by thermal oxidation (Table 1.2). The mechanism of TiO_2 NWs formation is also not very conclusive. Vapor-phase-assisted growth via VSL or VS can however be ruled out since all possible oxides that grow on titanium have low vapor pressure. At 600°C–800°C, oxidation of titanium occurs by inward diffusion of oxygen but diffusion of titanium ions from the substrate to the tip of the NW is required for 1-D growth. Controlling oxygen content in the atmosphere during oxidation can help in reducing oxygen diffusion in the scale thus inducing NW growth. It has been established by several authors that both the oxygen concentration and the oxidation temperature are known to affect the formation of TiO_2 NWs and hence, strict control of them is required in ensuring the success of NWs growth.

One of the earliest reports on the formation of TiO_2 NWs (whiskers) is by Motte et al. in 1975 whereby the formation of whiskers with high length-to-diameter ratio is said to be achieved by oxidizing pure titanium at 650°C for 120 hours in water vapor (Motte et al., 1976). Higher oxidation temperature (850°C) leads to well-developed crystals normal to the surface with smaller length-to-diameter ratio. The formation of NWs is thought to be similar to the CuO NWs whereby growth is induced by stress developing within the oxide scale. Nevertheless, systematic studies to further verify this are still lacking.

TABLE 1.2
Literature Survey on the Formation of TiO$_2$ NWs under Thermal Oxidation of Titanium

Authors/Year	Substrates	Oxidation Process	Surface Morphology
Motte et al. (1976)	Ti sheet 99.5% purity	Temperature: 650°C–850C; Atmosphere: Water vapor; Oxidation time: 120 hours	TiO$_2$ Whiskers
Peng & Chen (2004)	Ti plates grade 1	Temperature: 850°C; Atmosphere: mixed Ar (200 sccm) & O$_2$ (1 sccm); Oxidation time: 1.5 hours;	Nanoribbons
		Temperature: 850°C; Atmosphere: mixed Ar & Acetone; Oxidation time: 1.5 hours	Nanorods
Daothong et al. (2007)	Ti wires, diameter of 0.25 mm, 99.8% purity	Temperature: 650°C–850C; Atmosphere: Ethanol vapor; Oxidation time: 0.5–3 hours; Post Annealing Temperature: 450°C; Atmosphere: Air; Oxidation time: 0.5–2 hours	TiO$_2$ NWs
Cheung et al. (2007)	Ti foils, Thickness of 0.25 mm, 99.6% purity	1 wt% KF was dropped on the substrate prior to oxidation process 1st stage heating process Temperature: 150°C; Atmosphere: Water vapor; Oxidation time: 20 minutes; Pressure: 7.5 Torr;	TiO$_2$ NWs

(Continued)

TABLE 1.2 (*Continued*)
Literature Survey on the Formation of TiO$_2$ NWs under Thermal Oxidation of Titanium

Authors/Year	Substrates	Oxidation Process	Surface Morphology
Huo et al. (2008)	Ti foils, 99.6% purity	2nd stage heating process Temperature: 450°C–650C; Atmosphere: Water vapor; Oxidation time: 120 minutes; Pressure: 70–80 Torr	TiO$_2$ NWs
Lee et al. (2010)	Ti bar (grade 2), Ti64, β-Ti	Temperature: 850C; Atmosphere: mixed Ar (150 sccm) & Acetone; Oxidation time: 1.5 hours	TiO$_2$ NWs
Lee (2016)	TiO powder, particle size of <325 mesh, 99.9% purity	Temperature: 700°C–900C; Atmosphere: mixed Ar (200–1000 mL/min) & O$_2$ (10–100 ppm); Oxidation time: 7–9 hours	TiO$_2$ NWs
		Temperature: 1000C; Atmosphere: Air; Oxidation time: 2 hours	TiO$_2$ NWs

Several works have then focused on the use of atmosphere with very low concentration of oxygen in order to produce TiO_2 NWs by thermal oxidation. Peng and Chen (2004) reported on the synthesis of aligned TiO_2 nanorod arrays by oxidizing titanium at 850°C in acetone or in a very low concentration of oxygen gas. Following this work, fabrication of TiO_2 NWs or nanorods has been done in a furnace loaded with low-oxygen-content organic compounds such as ethanol, and acetone (Table 1.2). For instance Daothong et al. in 2007 published on TiO_2 NWs by thermal oxidation in ethanol vapor (Daothong et al., 2007) at 650°C–850°C in the presence of ethanol vapor for 30–180 minutes at the pressure of 10 Torr. Huo et al. (2008) used acetone vapor as oxygen source at 850°C for TiO_2 NWs formation as well. Indeed, highly dense, rather well-aligned TiO_2 NWs were produced under these conditions. It is suffice to conclude at such high temperature, the growth of NWs on titanium is only possible when the furnace has less oxygen in it. For the case of NWs grown in acetone or ethanol, mechanism of formation is thought to be due to absorption of acetone or ethanol the surface of the growing NWs whereby inhibition of growth in a-b axes is impeded allowing for uniaxial growth direction.

The synthesis of TiO_2 NWs by thermal oxidation with ethanol or acetone has one drawback which is the presence of amorphous carbon shell surrounding the TiO_2 NWs. Amorphous carbon results in a scattering effect leading to decreased electrical properties of the NWs. In order to remove this shell, post-annealing is required. Post-annealing can however lead to the formation of anatase TiO_2 NWs (Daothong et al., 2007) which can be beneficial for certain applications.

It is well established that TiO_2 grown on Ti is a result of mostly oxygen diffusion inwards (Kofstad et al., 1958) whereas for NWs structure to form outward diffusion of metal cations is required. As mentioned, hollow dislocation core provides rapid surface diffusion to the tip of the growing NWs but not many works concluded on the surface diffusion via dislocation cure for TiO_2 NWs growth. Oxide on titanium also suffers from compressive stress since the Pilling–Bedworth ratio (PBR) value of TiO_2 is 1.7. As the PBR is higher than unity therefore the oxide is under compressive stress. Since the TiO_2 scale is in compression during oxidation process at high temperature, stress induced diffusion is also possible to occur similar to the case of CuO NWs. In the case of titanium oxidation with low concentration of oxygen, ethanol, or acetone vapor, outward diffusion of Ti ions was induced by compressive stress via grain boundary in TiO_2 scale.

There have been reports as seen from Table 1.2 on an effort to reduce oxidation temperature for TiO_2 formation with the use of catalysts. Potassium halide (KF and KI) and potassium hydroxide (KOH) are the catalysts used to assist on growth of TiO_2 NWs on titanium foil substrate and significantly reduced oxidation temperature to less than 650°C. We have also produced TiO_2 NWs by thermal oxidation of KOH deposited titanium foil. Surface oxidation after approximately 500°C is shown in Figure 1.10a. Increasing the temperature to about 600°C resulted in more coverage of the oxide NWs on titanium foil (Figure 1.10b).

The NWs growth mechanism in the catalyst-assisted thermal oxidation method is rather different than NWs formation with low oxygen concentration or with water vapor. Vapor-liquid-solid mechanism is proposed for this growth of NWs with KF catalyst as reported by Cheung et al. (2007). KF catalyst in liquid form can absorb OH^- from water vapor and diffusion of Ti^{4+} from titanium substrate at elevated temperature will result in

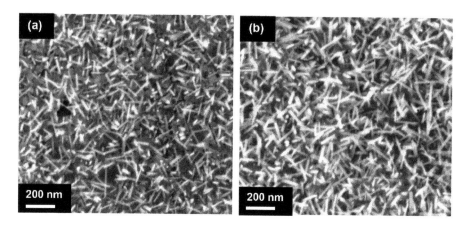

FIGURE 1.10 SEM micrographs of TiO_2 NWs prepared by oxidation of titanium coated with KOH (as catalysts) at (a) ~500°C and (b) ~600°C.

the formation of liquid alloy droplets. The droplets act as seed for NWs form. Continuous feeding of Ti^{4+} and OH^- into the droplet will result in a supersaturated droplet allowing for TiO_2 to nucleate and grow. Nevertheless, more systematic studies are required in order to conclude on the growth mechanism of TiO_2 NWs in various conditions.

1.7   CONCLUSION

TMO is a special class of material with interesting properties, especially when made with 1-D nanostructures. In this chapter, the definitions of 1-D nanostructures are given along with overview of some TMOs. Then we presented on the possibility of thermal oxidation process of metal for TMO NWs formation with brief discussion on CuO and TiO_2 NWs formation. Advantages and disadvantages of the oxidation process are discussed and mechanisms of formation of NWs on both copper and titanium are presented.

ACKNOWLEDGMENT

1-D nanostructures study supported by USM-Research University Grant for Toyohashi University of Technology, Japan-USM collaboration; 1001/PBAHAN/870048.

REFERENCES

Akbari, A., Amini, M., Tarassoli, A., Eftekhari-Sis, B., Ghasemian, N. & Jabbari, E. 2018. Transition metal oxide nanoparticles as efficient catalysts in oxidation reactions. *Nano-Structures & Nano-Objects*, 14, 19–48.
Ali, S. & Wood, G. 1968. The influence of crystallographic orientation on the oxidation of Cu. *Corrosion Science*, 8, 413–422.
Appleby, W. K., & Tylecote, R. F. 1970. Stresses during the gaseous oxidation of metals. *Corrosion Science*, 10, 5, 325–341.
Cao, G. & Wang, Y. 2011. World scientific series in nanoscience and nanotechnology. *Nanostructures and Nanomaterials Synthesis, Properties, and Applications,* 2nd ed., Scientific Publishing Co. PTE LTD, Singapore.

Chen, J., Zhang, F., Wang, J., Zhang, G., Miao, B., Fan, X., Yan, D. & Yan, P. 2008. CuO nanowires synthesized by thermal oxidation route. *Journal of Alloys and Compounds*, 454, 268–273.

Cheng, C. & Fan, H. J. 2012. Branched nanowires: Synthesis and energy applications. *Nano Today*, 7, 327–343.

Cheng, S.-L. & Chen, M.-F. 2012. Fabrication, characterization, and kinetic study of vertical single-crystalline CuO nanowires on Si substrates. *Nanoscale Research Letters*, 7, 119.

Cheung, K. Y., Yip, C. T., Djurišić, A., Leung, Y. H. & Chan, W. K. 2007. Long K-doped titania and titanate nanowires on Ti foil and FTO/quartz substrates for solar-cell applications. *Advanced Functional Materials*, 17, 555–562.

Choopun, S., Hongsith, N. & Wongrat, E. 2010. *Metal-Oxide Nanowires by Thermal Oxidation Reaction Technique*, IntechOpen. DOI: 10.5772/39506. Available from: https://www.intechopen.com/books/nanowires/metal-oxide-nanowires-by-thermal-oxidation-reaction-technique.

Comini, E. & Sberveglieri, G. 2010. Metal oxide nanowires as chemical sensors. *Materials Today*, 13, 36–44.

Comini, E. 2016. Metal oxide nanowire chemical sensors: Innovation and quality of life. *Materials Today*, 19, 559–567.

Cowley, J. 1954. Intensity anomalies in electron diffraction patterns of CuO. *Journal of the Electrochemical Society*, 101, 277–280.

Daothong, S., Songmee, N., Thongtem, S. & Singjai, P. 2007. Size-controlled growth of TiO_2 nanowires by oxidation of titanium substrates in the presence of ethanol vapor. *Scripta Materialia*, 57, 567–570.

Dasgupta, N. P., Sun, J., Liu, C., Brittman, S., Andrews, S. C., Lim, J., Gao, H., Yan, R. & Yang, P. 2014. 25th anniversary article: Semiconductor nanowires–synthesis, characterization, and applications. *Advanced Materials*, 26, 2137–2184.

Dey, A. 2018. Semiconductor metal oxide gas sensors: A review. *Materials Science and Engineering: B*, 229, 206–217.

Dorogov, M., Priezzheva, A., Vlassov, S., Kink, I., Shulga, E., Dorogin, L., Lohmus, R., Tyurkov, M., Vikarchuk, A. & Romanov, A. 2015. Phase and structural transformations in annealed copper coatings in relation to oxide whisker growth. *Applied Surface Science*, 346, 423–427.

Duan, H., Zheng, H., Zhou, Y., Xu, B. & Liu, H. 2017. Stability of garnet-type Li ion conductors: An overview. *Solid State Ionics*, 318, 45–53.

Fan, H. J., Gösele, U. & Zacharias, M. 2007. Formation of nanotubes and hollow nanoparticles based on Kirkendall and diffusion processes: A review. *Small*, 3, 1660–1671.

Filipič, G. & Cvelbar, U. 2012. Copper oxide nanowires: A review of growth. *Nanotechnology*, 23, 19.

Gadea, G., Morata, A. & Tarancon, A. 2018. Semiconductor nanowires for thermoelectric generation. *Semiconductors and Semimetals*, 98, 321–407.

Gonçalves, A., Campos, L., Ferlauto, A. & Lacerda, R. 2009. On the growth and electrical characterization of CuO nanowires by thermal oxidation. *Journal of Applied Physics*, 106, 034303.

Greenblatt, M. 1988. Molybdenum oxide bronzes with quasi-low-dimensional properties. *Chemical Reviews*, 88, 31–53.

Gulbransen, E., Copan, T. & Andrew, K. 1961. Oxidation of copper between 250°C and 450°C and the growth of CuO "whiskers." *Journal of the Electrochemical Society*, 108, 119–123.

Guo, T., Yao, M.-S., Lin, Y.-H. & Nan, C.-W. 2015. A comprehensive review on synthesis methods for transition-metal oxide nanostructures. *CrystEngComm*, 17, 3551–3585.

Hansen, B. J., Lu, G. & Chen, J. 2008. Direct oxidation growth of CuO nanowires from copper-containing substrates. *Journal of Nanomaterials*, 2008, 48.

Hoa, N. D., Quy, N., Jung, H., Kim, D., Kim, H., & Hong, S. -H. 2010. Synthesis of porous CuO nanowires and its application to hydrogen detection. *Sensors and Actuators B: Chemical*, 146, 1, 266–272.

Hilman, J., Yost, A. J., Tang, J., Leonard, B. & Chien, T. 2017. Low temperature growth of CuO nanowires through direct oxidation. *Nano-Structures & Nano-Objects*, 11, 124–128.

Hobbs, R. G., Petkov, N. & Holmes, J. D. 2012. Semiconductor nanowire fabrication by bottom-up and top-down paradigms. *Chemistry of Materials*, 24, 1975–1991.

Hochbaum, A. I. & Yang, P. 2009. Semiconductor nanowires for energy conversion. *Chemical Reviews*, 110, 527–546.

Hsueh, H., Hsueh, T., Chang, S., Hung, F., Tsai, T., Weng, W., Hsu, C. & Dai, B. 2011. CuO nanowire-based humidity sensors prepared on glass substrate. *Sensors and Actuators B: Chemical*, 156, 906–911.

Huang, L., Yang, S., Li, T., Gu, B., Du, Y., Lu, Y. & Shi, S. 2004. Preparation of large-scale cupric oxide nanowires by thermal evaporation method. *Journal of Crystal Growth*, 260, 130–135.

Hung, C. M., Le, D. T. T. & Van Hieu, N. 2017. On-chip growth of semiconductor metal oxide nanowires for gas sensors: A review. *Journal of Science: Advanced Materials and Devices*, 2(3), 263–285.

Huo, K., Zhang, X., Hu, L., Sun, X., Fu, J. & Chu, P. K. 2008. One-step growth and field emission properties of quasialigned TiO_2 nanowire/carbon nanocone core-shell nanostructure arrays on Ti substrates. *Applied Physics Letters*, 93, 013105.

ISO/TS 80004-1:2015, *Nanotechnologies—Vocabulary, Part 2: Nano-Objects Properties* (ISO/TS 80004-1:2015). Retrieved from https://www.iso.org/standard/54440.html.

ISO/TS 80004-4:2015, *Nanotechnologies—Vocabulary, Part 4: Nanostructured Materials Properties* (ISO/TS 80004-4:2015). Retrieved from *https://www.iso.org/standard/52195.html.*

Jiang, X., Herricks, T. & Xia, Y. 2002. CuO nanowires can be synthesized by heating copper substrates in air. *Nano Letters*, 2, 1333–1338.

Kaito, C., Nakata, Y., Saito, Y., Naiki, T. & Fujita, K. 1986. Electron microscopic studies on structures and reduction process of copper oxide whiskers. *Journal of Crystal Growth*, 74, 469–479.

Kanhere, P. & Chen, Z. 2014. A review on visible light active perovskite-based photocatalysts. *Molecules*, 19, 19995–20022.

Kaur, M., Muthe, K., Despande, S., Choudhury, S., Singh, J., Verma, N., Gupta, S. & Yakhmi, J. 2006. Growth and branching of CuO nanowires by thermal oxidation of copper. *Journal of Crystal Growth*, 289, 670–675.

Kim, J.-H., Katoch, A., Choi, S.-W. & Kim, S. S. 2015. Growth and sensing properties of networked p-CuO nanowires. *Sensors and Actuators B: Chemical*, 212, 190–195.

Kim, Y.-S., Hwang, I.-S., Kim, S.-J., Lee, C.-Y. & Lee, J.-H. 2008. CuO nanowire gas sensors for air quality control in automotive cabin. *Sensors and Actuators B: Chemical*, 135, 298–303.

Kofstad, P., Hauffe, K. & Kjollesdal, H. 1958. Investigation on the oxidation mechanism of titanium. *Acta Chemica Scandinavica*, 12, 239–266.

Kumar, A., Srivastava, A. K., Tiwari, P., & Nandedkar, N. V., 2004. The effect of growth parameters on the aspect ratio and number density of CuO nanorods. *Journal of Physics: Condensed Matter*, 16, 8531–8543.

Lamberti, A., Destro, M., Bianco, S., Quaglio, M., Chiodoni, A., Pirri, C. & Gerbaldi, C. 2012. Facile fabrication of cuprous oxide nanocomposite anode films for flexible Li-ion batteries via thermal oxidation. *Electrochimica Acta*, 86, 323–329.

Lany, S. 2015. Semiconducting transition metal oxides. *Journal of Physics: Condensed Matter*, 27, 283203.

Lee, G. H. 2016. Synthesis of TiO_2 nanowires via thermal oxidation process in air. *Materials Research Innovations*, 20(6), 421–424.

Lee, H., Dregia, S., Akbar, S., & Alhoshan, M. 2010. Growth of 1-D TiO_2 nanowires on Ti and Ti alloys by oxidation. *Journal of Nanomaterials*.

Lee, S.-K. & Tuan, W.-H. 2014. Scalable process to produce CuO nanowires and their formation mechanism. *Materials Letters*, 117, 101–103.

Li, A., Song, H., Wan, W., Zhou, J. & Chen, X. 2014. Copper oxide nanowire arrays synthesized by in-situ thermal oxidation as an anode material for lithium-ion batteries. *Electrochimica Acta*, 132, 42–48.

Li, A., Song, H., Zhou, J., Chen, X. & Liu, S. 2013a. CuO nanowire growth on Cu_2O by in situ thermal oxidation in air. *CrystEngComm*, 15, 8559–8564.

Li, D., Hu, J., Wu, R. & Lu, J. G. 2010. Conductometric chemical sensor based on individual CuO nanowires. *Nanotechnology*, 21, 485502.

Li, X., Liang, J., Kishi, N. & Soga, T. 2013b. Synthesis of cupric oxide nanowires on spherical surface by thermal oxidation method. *Materials Letters*, 96, 192–194.

Liang, J., Kishi, N., Soga, T. & Jimbo, T. 2010. Cross-sectional characterization of cupric oxide nanowires grown by thermal oxidation of copper foils. *Applied Surface Science*, 257, 62–66.

Liu, J., Xu, C., Chen, Z., Ni, S. & Shen, Z. X. 2017. Progress in aqueous rechargeable batteries. *Green Energy & Environment* 3, 1, 20–41.

Love, C. J., Smith, J. D., Cui, Y. & Varanasi, K. K. 2011. Size-dependent thermal oxidation of copper: Single-step synthesis of hierarchical nanostructures. *Nanoscale*, 3, 4972–4976.

Mahmoodi, A., Solaymani, S., Amini, M., Nezafat, N. B. & Ghoranneviss, M. 2018. Structural, morphological and antibacterial characterization of CuO nanowires. *Silicon*, 10, 1427–1431.

Mema, R., Yuan, L., Du, Q., Wang, Y. & Zhou, G. 2011. Effect of surface stresses on CuO nanowire growth in the thermal oxidation of copper. *Chemical Physics Letters*, 512, 87–91.

Motte, F., Coddet, C., Sarrazin, P., Azzopardi, M. & Besson, J. 1976. A comparative study of the oxidation with water vapor of pure titanium and of Ti-6Al-4V. *Oxidation of Metals*, 10, 113–126.

Mumm, F. & Sikorski, P. 2011. Oxidative fabrication of patterned, large, non-flaking CuO nanowire arrays. *Nanotechnology*, 22, 105605.

Munoz-Paez, A. 1994. Transition metal oxides: Geometric and electronic structures: Introducing solid state topics in inorganic chemistry courses. *Journal of Chemical Education*, 71, 381.

Nikoobakht, B., Wang, X., Herzing, A. & Shi, J. 2013. Scalable synthesis and device integration of self-registered one-dimensional zinc oxide nanostructures and related materials. *Chemical Society Reviews*, 42, 342–365.

Peng, X. & Chen, A. 2004. Aligned TiO_2 nanorod arrays synthesized by oxidizing titanium with acetone. *Journal of Materials Chemistry*, 14, 2542–2548.

Rakhshani, A. 1986. Preparation, characteristics and photovoltaic properties of cuprous oxide—A review. *Solid-State Electronics*, 29, 7–17.

Ramsden, J. J. 2009. *Applied Nanotechnology, The Conversion of Research Results to Products*, Elsevier, Oxford, UK.

Rao, C. N. R. 1989. Transition metal oxides. *Annual Review of Physical Chemistry*, 40, 291–326.

Rao, C. R. & Govindaraj, A. 2011. *Nanotubes and Nanowires*, 2nd ed., Royal Society of Chemistry, London, UK.

Rapp, R. 1984. The high temperature oxidation of metals forming cation-diffusing scales. *Metallurgical Transactions B*, 15 B, 195–212.

Scuderi, V., Amiard, G., Boninelli, S., Scalese, S., Miritello, M., Sberna, P. M., Impellizzeri, G. & Privitera, V. 2016. Photocatalytic activity of CuO and Cu_2O nanowires. *Materials Science in Semiconductor Processing*, 42, 89–93.

Scully, J. C. 1975. *Chapter 1. The Fundamentals of Corrosion*, 2nd ed., Pergamon Press, Great Britain, UK.

Shen, G., Chen, P.-C., Ryu, K. & Zhou, C. 2009. Devices and chemical sensing applications of metal oxide nanowires. *Journal of Materials Chemistry*, 19, 828–839.

Steinhauer, S., Brunet, E., Maier, T., Mutinati, G. C. & Köck, A. 2013a. Suspended CuO nanowires for ppb level H$_2$S sensing in dry and humid atmosphere. *Sensors and Actuators B: Chemical*, 186, 550–556.

Steinhauer, S., Brunet, E., Maier, T., Mutinati, G. C., Köck, A., Freudenberg, O., Gspan, C., Grogger, W., Neuhold, A. & Resel, R. 2013b. Gas sensing properties of novel CuO nanowire devices. *Sensors and Actuators B: Chemical*, 187, 50–57.

Steinhauer, S., Brunet, E., Maier, T., Mutinati, G. C., Köck, A., Schubert, W. D., Edtmaier, C., Gspan, C. & Grogger, W. 2011. Synthesis of high-aspect-ratio CuO nanowires for conductometric gas sensing. *Procedia Engineering*, 25, 1477–1480.

Tang, C. M., Wang, Y. B., Yao, R. H., Ning, H. L., Qiu, W. Q. & Liu, Z. W. 2016. Enhanced adhesion and field emission of CuO nanowires synthesized by simply modified thermal oxidation technique. *Nanotechnology*, 27, 395605.

Taniguchi, S. 1985. Stresses developed during the oxidation of metals and alloys. *Transactions of the Iron and Steel Institute of Japan*, 25, 3–13.

Vila, M., Díaz-Guerra, C. & Piqueras, J. 2010. Optical and magnetic properties of CuO nanowires grown by thermal oxidation. *Journal of Physics D: Applied Physics*, 43, 135403.

Wu, F., Myung, Y. & Banerjee, P. 2014. Unravelling transient phases during thermal oxidation of copper for dense CuO nanowire growth. *CrystEngComm*, 16, 3264–3267.

Xu, C. H., Woo, C. H. & Shi, S. Q. 2004. Formation of CuO nanowires on Cu foil. *Chemical Physics Letters*, 399, 62–66.

Yang, Q., Guo, Z., Zhou, X., Zou, J. & Liang, S. 2015. Ultrathin CuO nanowires grown by thermal oxidation of copper powders in air. *Materials Letters*, 153, 128–131.

Yuan, L. & Zhou, G. 2012. Enhanced CuO nanowire formation by thermal oxidation of roughened copper. *Journal of Electrochemical Society*, 159, C205–C209.

Yuan, L., Wang, Y., Mema, R. & Zhou, G. 2011. Driving force and growth mechanism for spontaneous oxide nanowire formation during the thermal oxidation of metals. *Acta Materialia*, 59, 2491–2500.

Zappa, D., Comini, E., Zamani, R., Arbiol, J., Morante, J. R. & Sberveglieri, G. 2011. Copper oxide nanowires prepared by thermal oxidation for chemical sensing. *Procedia Engineering*, 25, 753–756.

Zhang, K., Rossi, C., Tenailleau, C., Alphonse, P. & Chane-Ching, J.-Y. 2007. Synthesis of large-area and aligned copper oxide nanowires from copper thin film on silicon substrate. *Nanotechnology*, 18, 275607.

Zhang, Q., Wang, J., Xu, D., Wang, Z., Li, X. & Zhang, K. 2014. Facile large-scale synthesis of vertically aligned CuO nanowires on nickel foam: Growth mechanism and remarkable electrochemical performance. *Journal of Materials Chemistry A*, 2, 3865–3874.

Zhao, X., Wang, P., Yan, Z. & Ren, N. 2015. Room temperature photoluminescence properties of CuO nanowire arrays. *Optical Materials*, 42, 544–547.

Zheng, M., Yu, D., Duan, L., Yu, W. & Huang, L. 2017. In-situ fabricated CuO nanowires/Cu foam as a monolithic catalyst for plasma-catalytic oxidation of toluene. *Catalysis Communications*, 100, 187–190.

Zhong, M. L., Zeng, D. C., Liu, Z. W., Yu, H. Y., Zhong, X. C. & Qiu, W. Q. 2010. Synthesis, growth mechanism and gas-sensing properties of large-scale CuO nanowires. *Acta Materialia*, 58, 5926–5932.

2 Formation of 1-D Metal Oxide Nanostructures via Thermal Oxidation

*Christian Laurence E. Aquino, Luigi A. Dahonog,
Aian B. Ontoria, and Mary Donnabelle L. Balela*

CONTENTS

2.1 INTRODUCTION

One-dimensional (1-D) nanostructured metal oxides offer a wide array of applications ranging from supercapacitors (Khandare and Santosh, 2017, Mei et al., 2017, Qiu et al., 2018), energy conversion (Major et al., 2017, Mo et al., 2017), gas sensors (Hsu et al., 2017, Li, 2017, Tan et al., 2017), environmental cleaning (Babu et al., 2016, Dastkhoon et al., 2017, Wang et al., 2018), photodetectors (Weng et al., 2009, Wang et al., 2011, Zhang et al., 2017) and a lot more. The wide application of 1-D metal oxide nanostructures calls for the need to establish facile methods that will enable cost-effective and controllable upscaling of the fabrication of these versatile materials. Among the non-templated methods for the synthesis of 1-D nanostructures, thermal oxidation is undeniably the simplest to perform as it only involves heating a parent material of pure composition, usually in the form of metal foils, sheets or pre-synthesized nanoparticles, at intermediate temperatures relative to the metal's melting temperature. The process may be done in open conditions or in closed systems where purging of gases is possible (Fu et al., 2001, Hiralal et al., 2008, Yuan et al., 2012a, Budiman et al., 2016). The oxidizing environment may be in ambient conditions as in air or may be varied by purging of different gases such as pure oxygen or water vapor (Budiman et al., 2016). Other key factors which are also often varied include the time and temperature of oxidation, surface roughness, and gas partial pressure. Combinations of these operating parameters will affect the morphology of the formed nanostructures, which is not only limited to 1-D nanostructures such as nanorods, nanobelts, nanowires, etc. but may also extend to other more complicated morphologies as well (Fu et al., 2001, Hiralal et al., 2008, Liu et al., 2014, Budiman et al., 2016).

Thermal oxidation of pure metals has already been widely studied in the past few decades and has involved different metals like iron (Fe) (Fu et al., 2001, Hiralal et al., 2008, Yuan et al., 2012a, Budiman et al., 2016), copper (Cu) (Kaur et al., 2006, Chen et al., 2008), zinc (Zn) (Yuan et al., 2014), tungsten (W) (Florica et al., 2016), and titanium (Ti) (Behera and Chandra, 2016), and the less common thermally oxidized metals like cobalt (Co) (Lee, 2016), manganese (Mn) (Yu et al., 2005), nickel (Ni) (Abubakar et al., 2017) among others. The simplicity of the process and the ease of which operating parameters can be monitored and controlled compared to other solution-based synthesis methods make thermal oxidation suitable for large-scale fabrication of nanostructured metal oxides. Although some researchers have proposed mechanisms involved in the growth of 1-D oxide nanostructures synthesized by this method, a clearer and overarching science is yet to be established and may differ depending on the metal (Hiralal et al., 2008, Yuan et al., 2012, Budiman et al., 2016). Understanding the mechanisms involved in the growth process will enable the design of appropriate oxidizing conditions and thus allow for the tuning of the nanostructures to a desired morphology for a specific application.

2.2 IRON OXIDE NANOSTRUCTURES

Fe is certainly one of the most versatile and most abundant metals from among the metallic elements. Fe oxide nanostructures have already been established to have applications in electronics, photocatalysis (Huang et al., 2011, Cao et al., 2015), biological systems (Rajendran et al., 2017, Rufus et al., 2017), supercapacitors (Xu et al., 2015, Nathan and Boby, 2017) among others. Generally, Fe oxide exists in three forms depending on the environmental conditions where it was formed. These are FeO (wustite), Fe_3O_4 (magnetite), and α-Fe_2O_3 (hematite). α-Fe_2O_3 is known to be the most thermodynamically stable oxide of Fe under ambient conditions (Yuan et al., 2012a, Budiman et al., 2016). Of all the Fe oxides synthesized by thermal oxidation, α-Fe_2O_3 can be directly derived from pure Fe to form 1-D nanostructure.

Thermal oxidation of Fe is often at a wide range of temperatures around 200°C to 800°C at varying ramp rates. In this process, the growth of Fe oxide nanostructures is generally a function of different operating parameters such as oxidation temperature, time and atmosphere.

2.2.1 EFFECT OF TEMPERATURE AND GROWTH TIME

Oxidation of Fe is actually spontaneous at ambient conditions. However, the oxide layer formed at this condition does not exhibit distinct nanostructured morphology (Hiralal et al., 2008). Depending on the oxidation temperature, different Fe oxide layers may form on the metal substrate. At temperatures below 570°C, a multilayer of $Fe_3O_4|\alpha$-Fe_2O_3 film is formed. On the other hand, at higher oxidation temperatures, $Fe|FeO|Fe_3O_4|\alpha$-Fe_2O_3 layers are made. These layers have already been characterized in a number of studies (Hiralal et al., 2008, Yuan et al., 2012b, Budiman et al., 2016). A sample cross section of an oxidized Fe sheet from a study by Yuan et al. (2012b) is shown in Figure 2.1.

FeO is thermodynamically unstable at temperatures below 570°C explaining its non-formation at 400°C as shown in Figure 2.1a. The Fe_3O_4 and FeO grains are visually coarse and columnar while the α-Fe_2O_3 grains directly in contact with the nanowires are composed of considerably finer grains. The Fe_3O_4 and α-Fe_2O_3 layers in Figure 2.1a have thicknesses of around 2 μm and 250 nm, respectively (Yuan et al., 2012b). Figure 2.1b shows the three-layered structure of FeO, Fe_3O_4, and α-Fe_2O_3 with thicknesses of around 5, 2, and 0.4 μm, respectively (Yuan et al., 2012b). This suggests

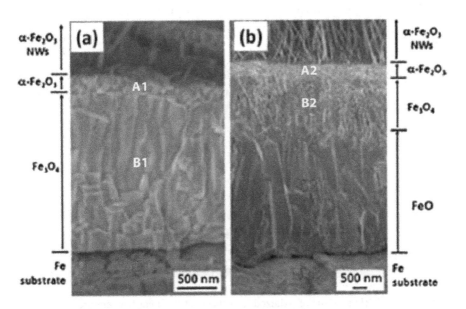

FIGURE 2.1   Iron oxide layers at (a) 400 and at (b) 600°C oxidation temperature. (Reprinted from *Mater. Sci. Eng. B*, 177, Yuan, L. et al., The origin of hematite nanowire growth during the thermal oxidation of iron, 327–336, Copyright (2012), with permission from Elsevier.)

that the FeO layer is not really essential for nanowire formation. It can be noted from the ordering of the layers that α-Fe_2O_3 is always formed at the surface with a distinct nanowire morphology. Oxidation of Fe becomes less favorable at deeper layers since transport of reacting species becomes more difficult. Thus, only the lower oxides, such as FeO and Fe_3O_4 form at these areas. The lower oxide layer forms at the expense of the higher oxide layer above it. This is why only α-Fe_2O_3 can be directly tuned to form nanostructures in thermal oxidation process. The lower oxides will always exist as a sublayer below the α-Fe_2O_3 layer (Hiralal et al., 2008, Pujilaksono et al., 2010, Yuan et al., 2012b). Post-treatment of the synthesized α-Fe_2O_3 nanowires in a reducing atmosphere however, can also alter the stable phases present as explained later in the following section.

Figure 2.2 shows the X-ray diffraction (XRD) pattern of oxidized iron sheets synthesized at different temperatures (Grigorescu et al., 2012). The patterns clearly show a multi-phase composition of α-Fe_2O_3 and Fe_3O_4 which agrees with the cross-sectional images shown earlier. Sharp peaks indicate high degree of crystallinity of both oxides. It can be seen from the patterns that peaks attributed to α-Fe_2O_3 become more pronounced at increasing temperatures particularly the peak at $2\theta = 33.67°$ which is indexed to the 104 plane. The increase in intensities suggests that more α-Fe_2O_3 phase was formed at higher temperatures, which also agrees with the slight thickness increase in the α-Fe_2O_3 layer in Figure 2.1.

Higher oxidation temperature generally increases nanowire density (Hiralal et al., 2008, Bertrand et al., 2009, Yuan et al., 2012b). Nanowire diameter, length, and growth rate also increase with temperature (Hiralal et al., 2008, Yuan et al., 2012b). Oxidation rates are enhanced as temperature is increased hence, the corresponding effects to nanostructure morphology (Bertrand et al., 2009). The enhanced

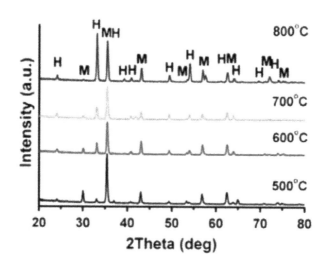

FIGURE 2.2 XRD patterns of oxidized iron sheets at different temperatures. (Reprinted from *Electrochem. Commun.*, 23, Grigorescu, S. et al., Thermal air oxidation of Fe: Rapid hematite nanowire growth and photoelectrochemical water splitting performance, 59–62, Copyright (2012), with permission from Elsevier.)

oxidation rate is attributed to smaller α-Fe_2O_3 grain sizes due to higher temperatures providing more diffusion paths for atomic transport (Bertrand et al., 2009). Smaller α-Fe_2O_3 nanowires are formed at lower oxidizing temperatures and then broadening of the nanowires occur at higher temperatures (Liu et al., 2014, Budiman et al., 2016). Some other researches produced a plate-like morphology for the α-Fe_2O_3 nanostructures upon subjecting to higher oxidation temperatures typically ranging from 600°C to 800°C (Budiman et al., 2016). The length of exposure or time of growth also influences the surface morphology of the α-Fe_2O_3 nanowires. It was shown that oxidation comes at very early stages and then growth proceeds with time (Hiralal et al., 2008, Yuan et al., 2012b). Initial formation of nanostructures was determined to have started on the grains of α-Fe_2O_3 and later progressed as the reaction proceeded (Hiralal et al., 2008). Generally, longer growth time enhances uniaxial growth thus producing longer, needle-like structures for Fe oxidized in air and oxygen (Hiralal et al., 2008, Liu et al., 2014).

2.2.2 EFFECT OF OXIDATION ATMOSPHERE

Altering the oxidizing atmosphere is one key factor in tuning surface morphology of α-Fe_2O_3 nanostructures. In a study by Hiralal et al. (2008), the morphology of the α-Fe_2O_3 nanostructures on the surface and their corresponding size distribution remained the same upon subjecting to different partial pressures of oxygen during thermal oxidation (Hiralal et al., 2008). However, the thickness of the oxide layers increased. The results of Yuan et al. (2012a) on the other hand showed no significant change in terms of α-Fe_2O_3 nanowire height (Yuan et al., 2012a). But the nanowire diameter was decreased, leading to higher nanowire density. α-Fe_2O_3 nanobelts were found to preferentially form at lower oxygen partial pressures (Yuan et al., 2012a).

Budiman et al. (2016) showed that the addition of water vapor in the oxidation atmosphere could greatly affect the surface morphology and density of the α-Fe_2O_3 phase at significantly shorter periods of time. Figure 2.3 shows the morphology of the synthesized α-Fe_2O_3 in the presence of water vapor. Oxidation for about 1 min at 500°C after ramp-up produced a dense, mixed α-Fe_2O_3 nanowire and nanosheet. The structures generally appear broader at the bottom of the wires and sheets. At longer oxidation time, an apparent lateral growth occurred where the tapered nanowires evolved to become blade-like in appearance as in Figure 2.3d and e in contrast to a needle-like morphology in dry air (Budiman et al., 2016). Consequently, the enhancement of oxidation rates brought by water vapor also promoted the growth of the nanostructures. Thus, in the synthesis of α-Fe_2O_3 1-D nanostructures, oxidation in dry air is more desired since water vapor will enhance growth rates and turn the nanowires to nanosheets.

Since only α-Fe_2O_3 nanostructures can be directly tuned to form a 1-D morphology in an oxidizing atmosphere, subjecting the synthesized α-Fe_2O_3 nanowires in a reducing atmosphere can be done to form Fe_3O_4 nanowires. This was successfully performed by Qin Han et al. (2007) by reducing the α-Fe_2O_3 nanowires in a 10:1 N_2:H_2 reducing atmosphere at 410°C to 430°C. Figure 2.4 shows the SEM (scanning electron microscope) image of the as-synthesized α-Fe_2O_3 nanowires and Fe_3O_4 nanowires after reduction. The size and shape of the resulting Fe_3O_4 nanowires

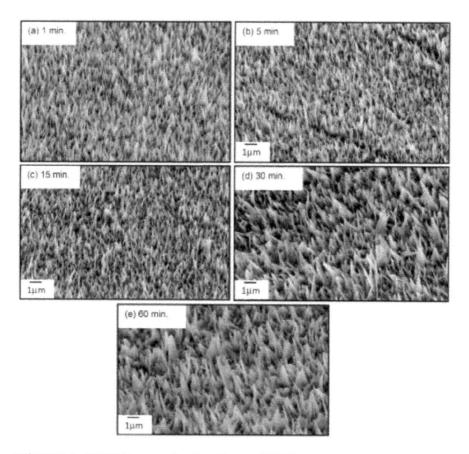

FIGURE 2.3   FESEM images of oxidized iron at 500°C in water vapor at (a) 1 (b) 5 (c) 15 (d) 30 and (e) 60 min. (Reprinted from *Appl. Surf. Sci.*, 380, Budiman, F. et al., Rapid nanosheets and nanowires formation by thermal oxidation of iron in water vapor and their applications as Cr(VI) adsorbent, 172–177, Copyright (2016), with permission from Elsevier.)

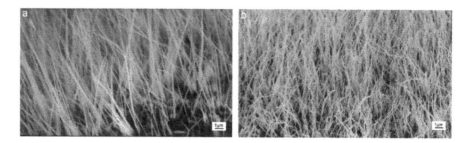

FIGURE 2.4   SEM images of (a) as-synthesized α-Fe_2O_3 nanowires and (b) Fe_3O_4 nanowires after gas reduction. (Reprinted from *J. Cryst. Growth*, 307, Han, Q. et al., Synthesis and magnetic properties of single-crystalline magnetite nanowires, 483–489, Copyright (2007), with permission from Elsevier.)

is close to that of the α-Fe$_2$O$_3$ nanowires. The α-Fe$_2$O$_3$ nanowires not only served as the precursor material for the Fe$_3$O$_4$ nanowires but as the 1-D template as well (Han et al., 2007).

2.2.3 GROWTH MECHANISM

The growth mechanism of 1-D nanostructures is commonly described using vapor-solid (VS) or vapor-liquid-solid (VLS) approach. The former requires evaporation and consequent condensation of precursor material to form the structures while the latter necessitates a catalyst (usually a droplet) for uniaxial growth to transpire. However, the growth of 1-D nanostructures via thermal oxidation does not follow VS or VLS mechanisms since usual oxidation temperatures do not reach the melting point of Fe. The formation of a vapor or liquid phase through incipient melting is improbable.

Despite not having a universally accepted growth mechanism, studies do agree that the growth of the α-Fe$_2$O$_3$ nanostructures is primarily driven from the 'relief' of compressive stresses formed at the Fe$_3$O$_4$|α-Fe$_2$O$_3$ interface due to solid-state transformations (Hiralal et al., 2008, Yuan et al., 2012b, Budiman et al., 2016). The compressive stresses are incurred due to the differences in specific volumes of the oxide layers. The Pilling–Bedworth ratio (ratio of volume of oxide to that of the consumed metal) of Fe$_3$O$_4$ and α-Fe$_2$O$_3$ were reported to be 2.10 and 2.14, respectively (Yuan et al., 2012b). There is a slight volume mismatch at Fe$_3$O$_4$/α-Fe$_2$O$_3$ interface, but this is said to be already enough to commence nanowire growth at the surface of the α-Fe$_2$O$_3$ layer. What is seemingly being debated to date is the transport mechanism of the reacting species during nanowire growth. Some researchers suggested that Fe atoms from the parent substrate diffuse along a tunnel found on the core of a screw dislocation of the nanowire (Hiralal et al., 2008, Budiman et al., 2016) and react consequently with the oxygen atoms in the atmosphere. Experimental observations, which suggest the bricrystalline nature of the oxide nanowires (absence of a tunnel inside the nanowire) do not strongly support this claim (Yuan et al., 2012b). An extensive study by Yuan et al. (2012a,b) has provided a more comprehensive approach in describing the growth mechanism of the α-Fe$_2$O$_3$ nanowires and nanobelts formed during thermal oxidation. Figure 2.5 shows the growth mechanism for nanowires and nanobelts as elucidated by Yuan et al. (2012a). The interfacial reaction at the Fe$_3$O$_4$|α-Fe$_2$O$_3$ interface stimulates the growth of the nanostructures. The Fe atoms normally move towards the substrate surface via grain boundary (GB) diffusion since this is the path that would necessitate less energy. However, at more enhanced interfacial reaction rates, like increasing O$_2$ partial pressures, compressive stresses on the Fe$_3$O$_4$|α-Fe$_2$O$_3$ interface also increase thereby enabling movement of Fe atoms not only through GB diffusion but through lattice diffusion as well. The formation of both nanobelts and nanowires suggest a combined diffusion mechanism of the Fe atoms from the parent substrate to the surface as shown in Figure 2.5a and b. Growth of the nanostructures then proceeds via surface diffusion on the metal oxide nanowire/nanobelt driven by concentration gradient (Hiralal et al., 2008, Yuan et al., 2012a).

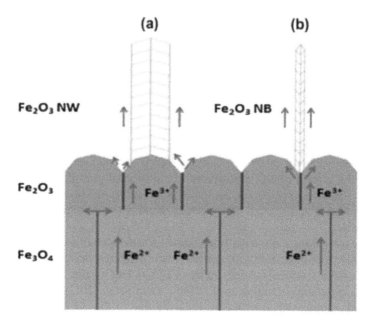

FIGURE 2.5 Mass transport mechanism of Fe ions for oxide (a) nanowire and (b) nanobelt growth. (Reprinted from *J. Mater. Res.*, 27, Yuan, L. et al., The growth of hematite nanobelts and nanowires—tune the shape via oxygen gas pressure, 1014–1021, Copyright (2012), with permission from Cambridge University Press.)

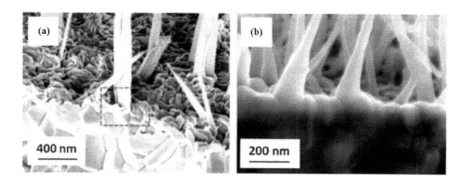

FIGURE 2.6 Root regions of α-Fe$_2$O$_3$ nanostructures formed at oxygen pressures (a) 0.1 and (b) 200 Torr. (Reprinted from *J. Mater. Res.*, 27, Yuan, L. et al., The growth of hematite nanobelts and nanowires—tune the shape via oxygen gas pressure, 1014–1021, Copyright (2012), with permission from Cambridge University Press.)

It was shown that nanobelts are associated with the geometry of the grain boundary where it originated. Figure 2.6a shows that the base of the nanobelts, which is formed at low O$_2$ partial pressures, clearly follows the grain boundary geometry (Yuan et al., 2012a). On the other hand, nanowires grow directly on the grains where it serves as its structure template as in Figure 2.6b. The bicrystallinity or polycrystallinity of the nanowires is attributed to the faceting of the grains at the surface of the material as shown in Figure 2.5a. Although this mechanism provides insight on the basis of the growth of 1-D α-Fe$_2$O$_3$ nanostructures, quantifying at great accuracy the appropriate parameters for the synthesis of a desired morphology still remains a challenge due to the convolution of the different thermodynamic and kinetic factors involved in the process (Yuan et al., 2012a).

2.3   ZINC OXIDE NANOSTRUCTURES

Zinc oxide is a semiconductor with a direct wide band gap energy of 3.37 eV with a high exciton binding energy of 60 meV at room temperature (Zhang et al., 2012). ZnO is a polymorphic material existing as three different crystal structures, namely, rock-salt, which can be obtained only under high pressure conditions; zinc-blende, which has been grown on cubic substrates; and the most thermodynamically favored phase, wurtzite (Escobedo-Morales et al., 2016). Among the nanostructures formed by ZnO, one-dimensional (1-D) nanostructures have several advantages. 1-D ZnO include nanobelts (Kong et al., 2003, Fan et al., 2006, Cui et al., 2008), nanohelices (Telford, 2005, Xiang et al., 2009), nanorods (Yi et al., 2005, Dhara et al., 2012, Oleg et al., 2012), nanowhiskers and nanowires (Xu, 2003, Gong et al., 2012, Zhang et al., 2012). 1-D ZnO nanostructures, particularly nanowire arrays, may be grown or embedded on a steady substrate, providing higher surface area compared to plain thin film. ZnO nanowires may be synthesized using several routes, including chemical vapor deposition (CVD) (Wan and Ruda, 2010, Zhan et al., 2017), physical vapor deposition (PVD) (Wang et al., 2005, Tigli and Juhala, 2011), pulsed laser deposition (PLD) (Nishimura et al., 2007, Dietrich and Grundmann, 2014), metal organic chemical vapor deposition (MOCVD) (Park et al., 2008, Nath et al., 2012, Sallet, 2014), thermal evaporation (Yao et al., 2002, Lee et al., 2003, Umar et al., 2006), hydrothermal synthesis (Sun et al., 2006, Yun et al., 2012, Han et al., 2013), and thermal oxidation (Xu et al., 2011, Yuan et al., 2014, Florica et al., 2016, Wu et al., 2017). Most of these processes either require special substrate or catalysts to assist growth. Others require high processing temperature as in the case of thermal evaporation and PVD. In thermal evaporation, adhesion of the nanostructure on the substrate is an issue. A more straightforward synthesis method is thermal oxidation, which only involves a selection of zinc and oxygen sources and carried out in temperatures between 400°C and 1000°C (Khanlary et al., 2012, Yuan et al., 2014, Florica et al., 2016, Wu et al., 2017). Thermal oxidation of Zn foil or powder in an oxygen-rich environment results in the growth of a variety of nanostructures, such as nanorods (Liang et al., 2008, Somvanshi and Jit, 2013, Kim and Leem, 2017), nanowhiskers (Liu et al., 2007, Noothongkaew et al., 2013), and tetrapods (Lee, 2011, Bae and Lee, 2012, Yuan et al., 2014). Direct oxidation of metallic zinc is a simple and cost-effective method for producing high purity ZnO nanostructures (Khanlary et al., 2012, Yuan et al., 2014).

2.3.1   EFFECT OF OXIDATION TEMPERATURE

Temperature serves as the most important factor in the formation of ZnO nanostructures (Colorado and Colorado, 2017). The oxidation temperatures for the growth of

ZnO nanostructures range from 400°C to 1000°C which covers both the melting (420°C) and boiling point (907°C) of Zn. No ZnO nanostructures were formed on surface of the Zn foil at 200°C–300°C (Dhara et al., 2012). On the other hand, ZnO nanowires with diameter of around 20 nm and length of about 500 nm appeared on the Zn substrates at 400°C. At higher oxidation temperatures, the dimensions of the as-prepared 1-D nanostructures increased. When the Zn foil was oxidized at 600°C, tapered nanowires with a diameter of 500 nm at the base and a length of up to 2 μm grew on the foil surface. Large rods were then formed on the surface of the Zn foil at 800°C. These rods are tapered with a diameter of 1.8 μm at the base and lengths of several microns.

In an earlier work (Khanlary et al., 2012), Zn thin films deposited on glass substrates were oxidized in ambient atmosphere for 1 h at 450°C to 650°C with 50°C step size for each sample. At 450°C and 500°C, the ZnO thin films were not completely oxidized and metallic Zn was still present below the ZnO surface. Zn peaks found at $2\theta = 36.53°$ and $77.43°$ are assigned to the reflections of the (002) and (004) planes. Only a small peak at about $2\theta = 34.5°$ indicates the formation of ZnO, which increases in intensity as the temperature was raised from 450°C to 500°C. Complete oxidation to ZnO occurred at temperatures of 550°C and higher. Figure 2.7 shows the SEM micrographs of the Zn thin films after oxidation in air at 450°C–650°C for 1 h. The ZnO nanowires formed at 450°C have a diameter of around 38 nm and length of about 280 nm as seen in Figure 2.7a. As the oxidation temperature was increased, the resulting nanowires became thicker and longer until the oxidation temperature reached 600°C. This can be attributed to the larger Zn droplets aggregated on the surface of the glass substrate. These bigger droplets act as nucleation sites where thicker nanowires would grow (Khanlary et al., 2012). The samples formed at 500°C to 550°C have diameters and lengths of about 52–72 nm and 420–770 nm, respectively. At 600°C, more nanowires were produced. These nanowires have diameters of about 87 nm and lengths of about 1.4 μm as shown in Figure 2.7c. Figure 2.7e shows that by increasing the oxidation temperature to 650°C, the length and the nanowires concentration decreased.

2.3.2 EFFECT OF OXIDIZING ATMOSPHERE

Xu et al. (2011) demonstrated the influence of the oxidizing atmosphere using a substrate with a composition of $Cu_{0.64}Zn_{0.36}$. The alloy was oxidized at 500°C under air, N_2, and N_2-O_2 for 3 h. Scanning electron microscope images reveal the formation of oxide scale on the surface of the brass substrate in all samples. For sample oxidized in N_2 atmosphere, large equiaxial grains with diameters of about 1 μm developed on planar scales. On the other hand, small equiaxial grains with 50-nm diameters are shown in the high magnification image. Under N_2-5%O_2 atmosphere, heavily convoluted scales were formed. When viewed under high magnification, ZnO nanowires with diameters between 50 and 100 nm and lengths of 5–10 μm were seen together with nanowalls. These nanowalls have thickness and width of about 70–100 nm and 10 μm, respectively. When the concentration of oxygen was increased to N_2-15%O_2,

FIGURE 2.7   SEM images of the ZnO nanowires grown by thermal oxidation of Zn thin films in air for 1 h at (a) 450°C, (b) 500°C, (c) 550°C, (d) 600°C, (e) 650°C. (Reprinted from Khanlary, M.R. et al., *Molecules*, 17, 5021–5029, 2012.)

non-uniform nanostructures was obtained. Fewer nanowires and nanowalls, which typically grow on the rough surfaces of the substrates, were produced.

The thermodynamic conditions for the formation of ZnO in oxidizing environment are governed by the following equations:

$$2Zn + O_2 \leftrightarrow 2ZnO;$$

(2.1)

$$\Delta G^0 = -701{,}200 + 200.98T\,(\text{J});\tag{2.2}$$

$$P_{O_2}^{eq}(ZnO) = \frac{(a_{ZnO})^2}{(a_{Zn})^2} \times \exp\left(\frac{\Delta G^0}{RT}\right)\tag{2.3}$$

where ΔG^0 is the change in the Gibbs energy for the oxidation when all the different species are present in their standard states; $P_{O_2}^{eq}$ is the equilibrium pressure of oxygen for the formation of ZnO; a_{ZnO} and; a_{Zn} are the thermodynamic activities of ZnO and Zn, respectively; R is the gas constant (8.31 J/mol K); and T is the absolute temperature. The thermodynamic activity for a pure phase, such as ZnO is equal to 1. For a pure metallic precursor such as Zn foil or Zn powder, the activity is treated as unity as well. So for a given temperature, and Zn metal concentration, the required partial pressure of oxygen may be calculated for the oxidation process to occur.

2.3.3 Effect of Oxidation Time

The work performed by Chao et al. (2013) shows the time dependence of ZnO nanowire lengths during thermal oxidation. In their work, metallic Zn films were deposited on Si (100) by RF magnetron sputtering with discharge power of 120 W and thermally oxidized at 430°C for 10–500 min. The variation in nanowire length obeys a parabolic law and may be fitted with an equation:

$$\text{Length} = 0.27 \times t^{0.5}\tag{2.4}$$

where t is the reaction duration in min. This suggests that the growth mechanism is due to Zn ion diffusion driven by stress (Chao et al., 2013, Yuan et al., 2014).

2.3.4 ZnO Growth Mechanism

Figure 2.8 shows the most widely accepted growth mechanism for the oxidation of Zn to form ZnO at different temperature ranges. For samples synthesized below the melting point of Zn (420°C), the well-accepted mechanism for the catalyst-free ZnO nanostructure growth is the solid-solid mechanism, where solid Zn oxidizes into solid ZnO. Figure 2.8a shows the axial movement of the Zn^{+2} ions towards the ZnO surface during growth below the melting point of Zn. The Zn atoms on the surface become reactive at a sufficient temperature and react with the gaseous oxygen species present in the atmosphere forming a ZnO layer on top of the surface. Zn^{+2} ions then diffuse from the metal towards the ZnO surface due to the gradient in concentration and a localized electric field resulting from the polar crystal structure and the solid/gas boundary where O^{-2} is abundant (Ren et al., 2007). It may also be due to a stress driven mechanism owing to the large molar volume difference between the ZnO and Zn layer. Due to the absence of stress at the outer surface of ZnO, Zn ions from the high-stress ZnO/Zn interface are driven by the stress gradient towards the solid/gas interface. This promotes nucleation and growth on the adjacent facets

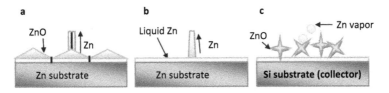

FIGURE 2.8 Different transformation processes occurring at different oxidation temperature: (a) Solid-solid transformation at temperatures below the melting point of Zn (less than 420°C), (b) Liquid-solid transformation at temperatures between the melting and boiling temperature of Zn, (c) Vapor-solid transformation at temperatures above the boiling point of Zn (above 907°C). (Redrawn from *J. Cryst. Growth*, 390, Yuan, L. et al., Temperature-dependent growth mechanism and microstructure of ZnO nanostructures grown from the thermal oxidation of zinc, 101–108, Copyright (2014), with permission from Elsevier.)

of ZnO grains (Yuan et al., 2014). At temperatures between the melting and boiling point of Zn, however, a different mechanism is proposed as seen in Figure 2.8b. At this temperature range, Zn droplets form on the surface acting as liquid catalysts that quickly reacts with oxygen present in the atmosphere, resulting in the formation ZnO, which would elongate to form 1-D nanostructures (Khanlary et al., 2012, Escobedo-Morales et al., 2016). For oxidation temperatures above the boiling point of Zn, tetrapods tend to be the preferred ZnO nanostructure morphology that is formed. The vapor-solid transformation dominates. Zn evaporates and reacts with oxygen in the atmosphere and forms ZnO nuclei acting seed adopting an octahedral-twinned shape. The high growth rate of the octahedral multiple twins results to the cracking of the twin boundaries. ZnO particles continue to grow preferentially along the cracks in the octahedral-twinned nucleus which result to the formation of tetrahedral geometry. These nanostructures are typically deposited at the vicinity around the Zn source (Yuan et al., 2014).

2.4   TUNGSTEN OXIDE NANOSTRUCTURES

Much attention has been given to the synthesis of 1-D nanostructures of the different oxides of tungsten (W) due to its excellent electrochromic, photochromic, and gaschromic properties. Oxides of W can exist in several W:O ratios such as tungsten trioxide (WO_3), $W_{18}O_{49}$, W_3O_8, W_5O_{14}, $WO_{2.9}$ which offers a variety of interesting properties for different applications (Yu et al., 2009). Synthesis of these different oxides are commonly done through solvo/hydrothermal routes (Song et al., 2006), thermal evaporation (Kim et al., 2010), chemical vapor deposition (Kim et al., 2010), and thermal oxidation (Gu et al., 2002, Yu et al., 2009, Houweling et al., 2011, Zappa et al., 2015).

Starting materials in the synthesis of 1-D W oxide nanostructures via thermal oxidation are commonly in the form of filaments, wires, and films. However, most researches commonly involve fabrication of a thin film of W prepared via RF magnetron sputtering (Zappa et al., 2015, Behera and Chandra, 2016), or DC sputtering (Su et al., 2015) on an appropriate substrate before subjecting to oxidation in a furnace. Compared to Fe, Cu, Zn, and Ti, thermal oxidation of W to form 1-D nanostructures is not much explored. As such, a universally accepted mechanism for the growth of 1-D WO_x nanostructures are yet to be established and may differ depending on the approach.

2.4.1 EFFECT OF OXIDATION ATMOSPHERE/TEMPERATURE

The effect of varying oxidation parameters during the thermal oxidation of W is not much explored in literature, unlike Fe and Cu. However, this factor still provides significant effects on the growth of W oxide nanowires from its parent material. Behera and Chandra (2016) investigated the effect of oxidation temperature on the growth of WO_3 nanowires from a RF sputtered W thin film of thickness of around 100 nm. The films were then oxidized on a tube furnace at varying temperatures for 4 h in an ambient atmosphere. As-deposited W film is made up of particles with diameters around 100 nm. Nanorods were formed at 300°C with diameter and lengths of around 80 and 200 nm. Increasing oxidation temperature to 400°C–500°C resulted to an increase in both length and density of the nanorods (Behera and Chandra, 2016). Nanorod diameter and length was about 100 nm and 1 μm at this temperature range.

Kojima et al. (2007) varied the H_2 to O_2 ratio during the annealing of RF deposited W thin films by controlling gas flow rates. The films were annealed in a furnace at 800°C for 10 min at varying ratio of H_2 and O_2 (RHO), by allowing some air leakage of about 2 sccm. Belt-like structures with widths of about 100–500 nm were formed in 0 RHO atmosphere. At increasing RHO, however, the number of nanobelts started to decrease and high-aspect ratio nanowires were favorably formed. These nanowires have diameters and lengths of about 10–100 nm and 1 μm, respectively. Further increasing the RHO to more than 0.8 resulted in the decrease in the number of nanowires. These findings show that the density of nanowires decreases to a certain degree with increasing RHO. On the other hand, for the samples with RHO more than 0.8, no significant differences were observed in terms of length and width.

2.4.2 OXIDATION IN VARIOUS SUBSTRATE/SOURCE MATERIALS

Various source materials for 1-D WO_x nanostructures have been explored by a number of researches. One of the earliest work was done by Gu et al. (2002) where W tips was first electrochemically etched by 1 M KOH with 2 V applied voltage before immediately annealing in a furnace. Argon (Ar) gas was pumped in the furnace at a constant flow rate of 300 sccm for 10 min. Figure 2.9 shows the SEM and TEM (transmission electron microscopy) images of the grown W oxide nanowires on etched W tip. Synthesized WO_x nanowires had lengths of around 300 nm with uniform diameters ranging from 10 to 30 nm. Nanowires grew considerably erect and perpendicular to the surface of the W tip as shown in Figure 2.9a. The nanowires were covered with a thin amorphous layer of W oxide of around 1 nm in thickness as seen in Figure 2.9b. The electron diffraction pattern in the inset image of Figure 2.9b shows a lattice spacing of 3.8 Å. The two vertical planes are indexed to the {010} and {103} peaks of monoclinic $W_{18}O_{49}$ crystal.

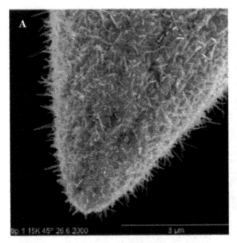
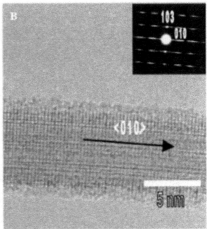

FIGURE 2.9 (a) SEM image of WO_x nanowires grown on an electrochemically etched W tip; (b) TEM and electron diffraction pattern of WO_x nanowire. (Reprinted from with permission from Gu, G. et al., *Nano Lett.*, 2, 2203–2209, Copyright (2002), American Chemical Society.)

Furthermore, Gu et al. (2002) were also able to demonstrate growth of $W_{18}O_{49}$ nanowires on a flat W substrate. The W substrates were first reduced in a H_2 gas at 700°C for 5 min to remove native oxides present in the surface before heating in an Ar atmosphere at 300 sccm flowrate for different oxidation periods. $W_{18}O_{49}$ nanowire lengths were found to increase from 500 nm to 1.6 µm for the plates annealed for 1 and 3 h, respectively. This agrees well with the case of Cu and Fe, where nanowire lengths increase at longer oxidation time.

2.4.3 Effect of Sputtering Conditions

Metallic W thin films are usually prepared via RF or DC magnetron sputtering. Varying sputtering parameters like temperature, film thickness, and sputtering gas has been shown to affect the characteristics of the nanowires grown after thermal oxidation of the films. Zappa et al. (2015) prepared W thin films via RF magnetron sputtering and investigated the effect of sputtering temperature and film thickness on the morphology of the thermally oxidized nanowires. Sputtering power was kept at 100 W with a chamber pressure of 5.5×10^{-3} mbar. Thermal oxidation of the films was done at 550°C for 1 h at a constant O_2 flowrate of 2 sccm. Annealing of the resulting nanowires was then carried out for the fabrication of a gas-sensing device.

Figure 2.10 shows the SEM images of the nanowires at varying sputtering temperature and film thickness. Sputtering temperature was found to have a strong influence on both diameter and length of the nanowires. WO_3 nanowires from film sputtered at 300°C are larger than the film produced at 200°C. However, nanowire diameter remained almost constant at about 40 nm. On the other hand, the nanowires

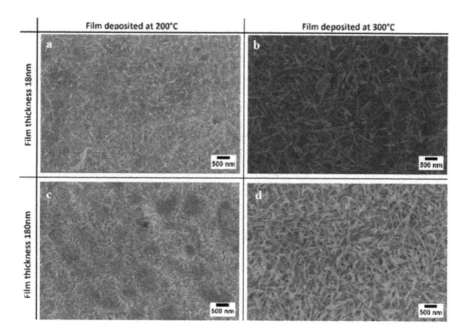

FIGURE 2.10 SEM image of WO₃ nanowires at varying sputtering temperature and film thicknesses. (a) film thickness of 18 nm deposited at 200°C; (b) film thickness of 18 nm deposited at 300°C; (c) film thickness of 180 nm deposited at 200°C; (d) film thickness of 180 nm deposited at 300°C. (From Zappa, D. et al., *Anal. Methods*, 7, 2203–2209, 2015, Reprinted by permission of The Royal Society of Chemistry.)

from the film sputtered at 200°C showed more uniformity in terms of nanowire dimensions. For the samples sputtered at 200°C, the film thickness (18 and 180 nm) did not strongly influence the diameter of the nanowires. However, a slight increase in nanowire diameter was observed for the nanowires prepared from the film produced at 300°C. The average diameter slightly increased from 31 to 38 nm.

Initial film thickness also affects the resulting nanowire density. A higher nanowire density is apparent for films with thickness of 180 nm as seen in Figure 2.10 (Zappa et al., 2015). This was attributed to the abundance of bulk W as compared to the 18 nm film. Some nanowires would be expected to grow later than the others during oxidation. Having a thinner film means that W is more easily consumed during oxidation leading to less nucleation sites for new nanowires. Conversely, more W atoms are available in a thicker film for the nucleation of new nanowires and for the growth of the existing nanowires.

2.5 COPPER OXIDE NANOSTRUCTURES

Copper (Cu) oxides are p-type semiconducting materials which have been known to have excellent optical, electrical, and physical properties. It has two phases, namely cupric oxide (CuO) and cuprous oxide (Cu₂O) which has a band gap of 1.08–2.08 eV and 2.0 eV, respectively (Phiwdang et al., 2013, Mayekar et al., 2014, Zayyoun et al., 2016).

The favorable band gaps of these materials allow them to operate at higher voltages, temperatures, and frequencies which make them useful in numerous applications, such as gas sensors (Cerqui et al., 2014, Duc et al., 2014), lithium-ion batteries (Liu et al., 2013), optoelectronics (Yu et al., 2016), and high temperature superconductors (Uchida, 2014). Additionally, the narrow band gap of these materials makes them useful in catalysis (Poreddy et al., 2015).

Cu oxide nanowires have been synthesized by several methods such as hydrothermal route (Filipič and Cvelbar, 2012), electrospinning (Khalil et al., 2014), and self-catalytic growth process (Hsieh et al., 2003). On the other hand, thermal oxidation (Chen et al., 2008) has been reported to be effective and simple for the growth of these nanowires in a Cu substrate. Different factors influence the growth of these nanowires, including reaction time and temperature. These factors can control the length distribution (Kaur et al., 2006) and diameter of the nanowires (Liu et al., 2014). The oxidative environment (Xu et al., 2004) and the condition (effect of surface stresses) of the Cu substrate (Mema et al., 2011) were also considered to be factors affecting the growth of nanowires.

2.5.1 EFFECT OF TEMPERATURE

Cu oxide nanowires with diameters of 30–250 nm have been fabricated on Cu foil at a wide temperature range of 400°C–800°C (Gonçalves et al., 2009). Nanowires were mostly observed at temperatures 400°C–700°C. No nanowires were observed for temperatures less than 300°C and above 800°C there is an increase in CuO nanowire density after heating at 400°C to 600°C. The formation of CuO nanowires at different temperatures is determined by the microstructure of the underlying Cu_2O layer where Cu diffusion occurs. At lower temperatures (<300°C), the growth of nanowires is attributed to the grain boundary diffusion (Gonçalves et al., 2009). At these temperatures, the Cu_2O layer has small and defective grains, which result to a decrease in growth rate (Gonçalves et al., 2009, Yuan et al., 2011, Li et al., 2013a). Interface reaction at lower temperatures is also slow due to the low mobility of Cu cations and vacancies (Yuan et al., 2011). Thus, the rate of stress generation by the solid-state phase transformation is not sufficient to activate effective grain boundary diffusion for the oxide growth (Gonçalves et al., 2009, Yuan et al., 2011). This explains the absence of nanowires at 300°C. As the temperature is increased, the thermally induced growth in the Cu_2O layer results to a decrease in the diffusion paths (Mimura et al., 2006). Thus, the density of the nanowires decreases. For temperatures greater than 800°C, the oxidation of Cu is due to the lattice diffusion of Cu ions rather than grain boundary diffusion. The interface strain associated with the phase transformation at the Cu_2O/CuO interface can be released quickly by fast lattice diffusion. This leads to uniform growth of CuO grains. Thus, oxide growth by the lattice diffusion causes the flatter morphologies at higher temperatures. Thus, the formation of nanowires is mostly observed in the mid-oxidation temperatures which promote the growth via effective grain boundary diffusion through the thin CuO layer (Li et al., 2013).

Other than Cu sheets or foils as substrates, Cu powders were also employed for the growth of oxide layers (Li et al., 2013b, Yang et al., 2015). Cu_2O/CuO and CuO nanowires were formed on the surface of Cu powders after heating at 300°C–600°C.

Cu_2O/CuO nanowires with diameters of 40 nm and lengths of about 0.1–1.4 μm were observed on the surface of the Cu powder after annealing at 300°C. The nanowires were more apparent when heated at 400°C. The Cu_2O/CuO nanowires have diameter between 50–100 nm and lengths of 0.5–3.0 μm.

2.5.2 EFFECT OF TIME

The effect of time on the formation of CuO nanowires has also been studied by different researchers. Chen et al. (2008) studied the effect of time on the dimension of the produced CuO nanowires annealed at a constant temperature of 400°C. After 0.5 h, CuO nanowires with diameters of 500 nm were produced on the surface of the foil. Extending the time to 2 h resulted to CuO nanowires with a length of 2–3 μm. This was further improved to 6–10 μm as the time was increased to 4 and 8 h. In addition to the increase in nanowire length, more CuO nanowires were present at longer oxidation time. The increase in length with respect to time can be attributed to the parabolic diffusion law.

On the other hand, Gonçalves et al. (2009) observed the dependence of nanowire length to the growth time when the Cu foil annealed at different temperatures (400°C and 600°C). Figure 2.11 is the CuO nanowires produced at 400°C at different time intervals. As seen from the images, the average diameter of the CuO nanowires remains roughly constant while the average length increases with time. The change in length indicates that the growth of CuO nanowires is a thermally activated process.

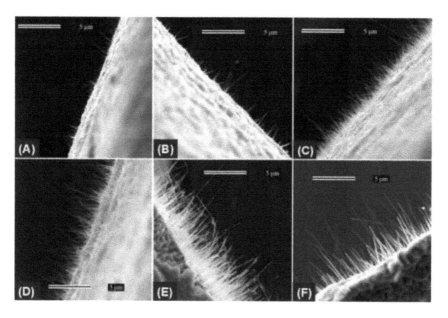

FIGURE 2.11   SEM images of CuO nanowires obtained at different oxidation time (a) 10 min, (b) 30 min, (c) 1 h, (d) 3 h, (e) 10 h and (f) 18 h at a temperature of 400°C. (Reprinted with permission from Gonçalves, A.M. et al., *J. Appl. Phys.*, 106, 034303. Copyright (2009), with permission from AIP Publishing.)

2.5.3 Effect of Atmosphere

The environment during thermal oxidation also significantly influences both the morphology and density of the CuO nanowires formed. CuO nanowires were produced at different temperatures in a mixture of oxygen and argon as the atmosphere. At 400°C, the CuO nanowires have an average diameter of 170 nm. On the other hand, CuO nanowires with diameters of about 50–80 nm were observed at 250°C to 300°C. At a temperature range of 250°C–400°C with both oxygen and argon in the environment, a decrease in the oxygen concentration resulted in an increase in the diameter of CuO nanowires.

On other hand, Xu et al. (2004) studied the effect of various environment on the formation of CuO nanowires. CuO nanowires were not formed in nitrogen. However, small amounts of CuO nanowires were produced when oxidation was performed in air. The CuO nanowires have an average diameter of 100 nm and length up to 5 μm. In highly oxidizing environment, the CuO nanowire density was significantly increased. The CuO nanowires have diameters up to 100 nm and length of about 15 μm.

Meanwhile, the addition of water vapor was also studied (Xu et al., 2004). The SEM images of the Cu foil after annealing at 400°C in wet gases are shown in Figure 2.12. Similar to dry nitrogen, CuO nanowires were not formed in a wet nitrogen environment as in Figure 2.12a. However, large and sharp oxide grains grew on the surface instead of scales. In air, the addition of water vapor resulted in a higher

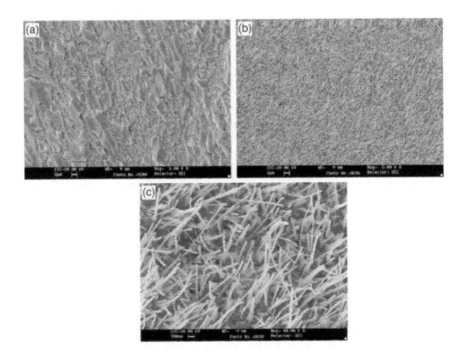

FIGURE 2.12 Morphologies formed at 400°C: (a) no nanowires were formed in wet nitrogen and CuO nanowires formed in (b, c) wet air. (Reprinted from *Superlattices Microstruct.*, 36, Xu, C. et al., The effect of oxidative environments on the synthesis of CuO nanowires on Cu substrates, 31–38, Copyright (2004), with permission from Elsevier.)

nanowire density. The CuO nanowires have diameters of 80 nm and lengths of 3 μm. In contrast, the addition of water vapor in the oxygen atmosphere causes the nanowire density to relatively decrease (SEM not shown).

Thermodynamically, the oxide nanowire will only form if the partial pressure of oxygen (P_{O_2}) is larger than the dissociation pressure of the oxide ($P_{O_2\text{-dis}}$) in equilibrium with Cu (Xu et al., 2004). If traces of oxygen are present with nitrogen, with the partial pressure of oxygen above the dissociation pressure of Cu_2O but below that of CuO, formation of Cu_2O is more favorable. When heated in air or oxygen, nanowires in both phases were formed. Higher oxygen pressure resulted in the growth of CuO nanowire alone (Xu et al., 2004). Oxidation in pure oxygen yielded longer nanowires with higher density. This suggests that high partial pressure of oxygen causes higher nucleation and growth rate for the formation of nanowire (Xu et al., 2004). The addition of water vapor did not affect the length of the nanowires but the density of the nanowires as evident in Figure 2.12. The density of the nanowires produced in wet air is relatively higher than in dry air. This suggests that the water vapor affects the nucleation rate of the nanowires.

2.5.4   Growth Mechanism

It has been reported that the growth of CuO nanowires occurs due to the grain boundary diffusion of Cu ions through the Cu_2O layer and the oxygen ions through the outmost CuO layer (Li et al., 2013). It was also proposed that the difference in the molar volumes of Cu and its oxide produces high compressive stresses which then serve as the driving force for the growth of CuO nanowires (Yuan et al., 2011). Being cation-deficient oxides, the growth of Cu_2O and CuO is shown to be controlled via outward diffusion of cations during the oxidation of Cu (Liang et al., 2011). Oxygen was then adsorbed on the surface of Cu lattice site forming an electron hole. The adsorbed oxygen was ionized forming another hole and Cu^{2+} ion enter the surface to partner with oxygen ions. Consequently, the process leads to another hole and vacancy in the cation sub-lattice (Yuan et al., 2011). Formation of Cu cation vacancies and electron holes were observed at the CuO/oxygen interface (Yuan et al., 2011). This occurs when the oxygen gas at the CuO surface utilizes Cu^{2+} ions from the outer CuO lattice to form new CuO products. On the other hand, the growth of Cu_2O layer requires decomposition of an oxygen-rich layer, in this case, the CuO layer (Gonçalves et al., 2009). There is a phase-boundary reaction at the Cu_2O/CuO interface, which also generates new cation vacancies and electron holes. These vacancies and holes then migrate through the Cu_2O layer (Yuan et al., 2011, Li et al., 2013).

Based from the parabolic diffusion law, the Cu cations are incorporated into the substrate before reaching the base of the nanowire. For prolonged oxidation as well as higher oxidation temperatures, the growing CuO substrate may gradually bury CuO nanowires from their bases and up as shown in Figure 2.13. Since the Cu_2O layer grows via decomposing the CuO layer at the Cu_2O/CuO interface, continued growth of the Cu_2O layer can consume all the CuO phase underneath the CuO nanowires leading to a direct contact of CuO nanowires root with the Cu_2O as shown in Figure 2.13. This mechanism involves the two simultaneous processes

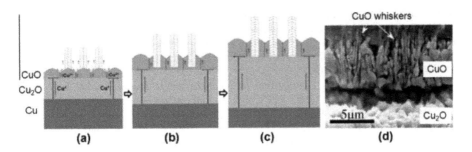

FIGURE 2.13 Schematic of the formation mechanism of nanowires (a) Initial growth of CuO nanowires on the outer surface of the CuO grains, (b) growth of the CuO substrate buries the roots of nanowires, (c) decomposition of the CuO layer at the Cu_2O/CuO interface leading to a direct contact between the roots of the nanowires and the Cu_2O layer and (d) cross-sectional image for the oxidation of Cu showing the direct contact of CuO nanowires on the Cu_2O layer. (Reprinted from *Acta Materiala*, 59, Yuan, L. et al., Driving force and growth mechanism for spontaneous oxide nanowire formation during the thermal oxidation of metals, 2491–2500, Copyright (2011), with permission from Elsevier.)

such as burying oxide nanowires by the growing CuO substrate and the decomposition of the CuO substrate at the Cu_2O/CuO interface. Figure 2.13d shows the experimental result depicting the CuO whiskers being buried by the CuO layer and their direct contact with the Cu_2O phase. It can be seen that the CuO nanowires are buried by the CuO layer and the nanowire roots are close to the Cu_2O/CuO (Yuan et al., 2011).

2.6   CONCLUSION

The simplicity of thermal oxidation as a process to synthesize 1-D metal oxide nanostructures makes it a good candidate for large-scale production. Thermal oxidation does not require complicated equipment and potentially harmful chemical substances. This would then enable practical application of 1-D metal oxide nanostructures for industrial processes, such as environmental purification, photodegradation, among others. It was shown that tuning of the dimensions of the nanostructures can be done to address specific functions by controlling surface characteristics of parent material and oxidation parameters such as time, temperature, and atmosphere.

REFERENCES

Abubakar, D., N. Ahmed, S. Mahmud, and N. Algadri. 2017. Properties of NiO nanostructured growth using thermal dry oxidation of nickel metal thin film for hydrogen gas sensing at room temperature. *Materials Research Express* 4, no. 7: 075009
Babu, B., K. Mallikarjuna, C. Reddy, and J. Park. 2016. Facile synthesis of Cu@TiO$_2$ core shell nanowires for efficient photocatalysis. *Materials Letters* 176: 265–269.

Bae, D., and G. Lee. 2012. Characterization of ZnO tetrapods prepared by a simple oxidation of Zn plate in air atmosphere. *Japanese Journal of Applied Physics* 51: 06FG01.

Behera, B., and S. Chandra. 2016. Synthesis of nanostructured tungsten oxide by thermal oxidation method and its integration in sensor for VOCs detection. *Advanced Materials Letters* 7, no. 9: 697–701.

Bertrand, N., C. Desgranges, D. Poquillon, M.C. Lafont, and D. Monceau. 2009. Iron oxidation at low temperature (260–500°C) in air and the effect of water vapor. *Oxidation of Metals* 73, no. 1–2: 139–162.

Budiman, F., N. Bashirom, W. Tan, K. Razak, A. Matsuda, Z. Lockman. 2016. Rapid nanosheets and nanowires formation by thermal oxidation of iron in water vapour and their applications as Cr(VI) adsorbent. *Applied Surface Science* 380: 172–177.

Cao, Z., M. Qin, B. Jia et al. 2015. One pot solution combustion synthesis of highly mesoporous hematite for photocatalysis. *Ceramics International* 41, no. 2: 2806–2812.

Cerqui, C., A. Ponzoni, D. Zappa et al. 2014. Copper oxide nanowires for surface ionization based gas sensor. *Procedia Engineering* 87: 1023–1026.

Chao, L., S. Tsai, C. Lin et al. 2013. Vertically aligned ZnO nanowires prepared by thermal oxidation of RF magnetron sputtered metallic zinc films. *Materials Science in Semiconductor Processing* 16, no. 5: 1316–1320.

Chen, J.T., F. Zhang, J. Wang et al. 2008. CuO nanowires synthesized by thermal oxidation route. *Journal of Alloys and Compounds* 454, no. 1–2: 268–273

Colorado, S.A., and H.A. Colorado. 2017. Manufacturing of zinc oxide structures by thermal oxidation processes as scalable methods towards inexpensive electric generators. *Ceramics International* 43, no. 17: 15846–15855.

Cui, Q., K. Yu, N. Zhang, and Z. Zhu. 2008. Porous ZnO nanobelts evolved from layered basic zinc acetate nanobelts. *Applied Surface Science* 254, no. 11: 3517–3521.

Dastkhoon, M., M. Ghaedi, A. Asfaram et al. 2017. Simultaneous removal of dyes onto nanowires adsorbent use of ultrasound assisted adsorption to clean waste water: Chemometrics for modeling and optimization, multicomponent adsorption and kinetic study. *Chemical Engineering Research and Design* 124: 222–237.

Dhara, S., and P.K. Giri. 2012. ZnO nanorods arrays and heterostructures for the high sensitive UV photodetection. In *Nanorods*, O. Yalçin (Ed.), pp. 1–32. InTech, Rijeka, Croatia.

Dietrich, C.P., and M. Grundmann. 2014. Pulsed-Laser Deposition of ZnO Nanowires. *Wide Band Gap Semiconductor Nanowires* 1: 303–323.

Duc, L.D., D.T. Le, N.V. Duy et al. 2014. Single crystal cupric oxide nanowires: Length- and density-controlled growth and gas-sensing characteristics. *Physica E: Low-dimensional Systems and Nanostructures* 58: 16–23.

Escobedo-Morales, A., R. Aranda-García, E. Chigo-Anota et al. 2016. ZnO micro- and nanostructures obtained by thermal oxidation: Microstructure, morphogenesis, optical, and photoluminescence properties. *Crystals* 6, no. 10: 135.

Fan, H.J., B. Fuhrmann, R. Scholz et al. 2006. Vapour-transport-deposition growth of ZnO nanostructures: Switch between c-axial wires and a-axial belts by indium doping. *Nanotechnology* 17, no. 11: S231–S239.

Filipič, G., and U. Cvelbar. 2012. Copper oxide nanowires: A review of growth. *Nanotechnology* 23, no. 19: 194001.

Florica, C., N. Preda, A. Costas et al. 2016. ZnO nanowires grown directly on zinc foils by thermal oxidation in air: Wetting and water adhesion properties. *Materials Letters* 170: 156–159.

Fu, Y., J. Chen, and H. Zhang. Synthesis of Fe_2O_3 nanowires by oxidation of iron. *Chemical Physics Letters* 350, no. 5–6 (2001), 491–494.

Gonçalves, A.M., L.C. Campos, A.S. Ferlauto, and R.G. Lacerda. 2009. On the growth and electrical characterization of CuO nanowires by thermal oxidation. *Journal of Applied Physics* 106, no. 3: 034303.

Gong, M.G., Y.Z. Long, X.L. Xu, H.D. Zhang, and B. Su. 2012. Synthesis, Superhydrophobicity, Enhanced Photoluminescence and Gas Sensing Properties of ZnO Nanowires. In *Nanowires—Recent Advances*, X. Peng (Ed.), pp. 77–100. InTech, Rijeka, Croatia.

Grigorescu, S., C. Lee, K. Lee et al. 2012. Thermal air oxidation of Fe: Rapid hematite nanowire growth and photoelectrochemical water splitting performance. *Electrochemistry Communications* 23: 59–62.

Gu, G., B. Zheng, W.Q. Han et al. 2002. Tungsten oxide nanowires on tungsten substrates. *Nano Letters* 2, no. 8: 849–851.

Han, Q., Z. Liu, Y. Xu, and H. Zhang. 2007. Synthesis and magnetic properties of single-crystalline magnetite nanowires. *Journal of Crystal Growth* 307, no. 2: 483–489.

Han, Z., S. Li, J. Chu, and Y. Chen. 2013. Controlled growth of well-aligned ZnO nanowire arrays using the improved hydrothermal method. *Journal of Semiconductors* 34, no. 6: 063002.

Hiralal, P., H. Unalan, H. Wijayantha et al. 2008. Growth and process conditions of aligned and patternable films of iron(III) oxide nanowires by thermal oxidation of iron. *Nanotechnology* 19, no. 45.

Houweling, Z.S., J.W. Geus, M. Jong et al. 2011. Growth process conditions of tungsten oxide thin films using hot-wire chemical vapor deposition. *Materials Chemistry and Physics* 131, no. 1–2: 375–386.

Hsieh, C., J. Chen, H. Lin, and H. Shih. 2003. Synthesis of well-ordered CuO nanofibers by a self-catalytic growth mechanism. *Applied Physics Letters* 82, no. 19: 3316–3318.

Hsu, C., L. Chang, and T. Hsueh. 2017. Light-activated humidity and gas sensing by ZnO nanowires grown on LED at room temperature. *Sensors and Actuators B: Chemical* 249: 265–277.

Huang, J., M. Yang, C. Gu et al. 2011. Hematite solid and hollow spindles: Selective synthesis and application in gas sensor and photocatalysis. *Materials Research Bulletin* 46, no. 8: 1211–1218.

Kaur, M., K.P. Muthe, S.K. Despande et al. 2006. Growth and branching of CuO nanowires by thermal oxidation of copper. *Journal of Crystal Growth* 289, no. 2: 670–675.

Khalil, A., C. Dimas, and R. Hashaikeh. 2014. Electrospun copper oxide nanofibers as infrared photodetectors. *Applied Physics A* 118, no. 1: 217–224.

Khandare, L. and T. Santosh 2017. Gold nanoparticles decorated MnO_2 nanowires for high performance supercapacitor. *Applied Surface Science* 418: 22–29.

Khanlary, M.R., V. Vahedi, and A. Reyhani. 2012. Synthesis and characterization of ZnO nanowires by thermal oxidation of Zn thin films at various temperatures. *Molecules* 17, no. 12: 5021–5029.

Kim, D., and J. Leem. 2017. Catalyst-free synthesis of ZnO nanorods by thermal oxidation of Zn films at various temperatures and their characterization. *Journal of Nanoscience and Nanotechnology* 17, no. 8: 5826–5829.

Kim, H., K. Senthil, and K. Yong. 2010. Photoelectrochemical and photocatalytic properties of tungsten oxide nanorods grown by thermal evaporation. *Materials Chemistry and Physics* 120, no. 2–3: 452–455.

Kojima, Y., K. Kasuya, T. Ooi et al. 2007. Effects of oxidation during synthesis on structure and field-emission property of tungsten oxide nanowires. *Japanese Journal of Applied Physics* 46, no. 9B: 6250–6253.

Kong, X., X. Sun, and Y. Li. 2003. Synthesis of ZnO nanobelts by carbothermal reduction and their photoluminescence properties. *ChemInform* 34, no. 37.

Lee, G. 2011. Synthesis and cathodoluminescence of ZnO tetrapods prepared by a simple oxidation of Zn powder in air atmosphere. *Ceramics International* 37, no. 1: 189–193.

Lee, G. 2016. Synthesis of TiO_2 nanowires via thermal oxidation process in air. *Materials Research Innovations* 20, no. 6: 421–424.

Lee, J., M. Kang, S. Kim et al. 2003. Growth of zinc oxide nanowires by thermal evaporation on vicinal Si(100) substrate. *Journal of Crystal Growth* 249, no. 1–2: 201–207.

Li, A., H. Song, J. Zhou et al. 2013a. CuO nanowire growth on Cu_2O by in situ thermal oxidation in air. *CrystEngComm* 15, no. 42: 8559.

Li, X., J. Liang, N. Kishi, and T. Soga. 2013b. Synthesis of cupric oxide nanowires on spherical surface by thermal oxidationmethod. *Materials Letters* 96: 192–194.

Li, Z. 2017. Supersensitive and superselective formaldehyde gas sensor based on NiO nanowires. *Vacuum* 143: 50–53.

Liang, H., L. Pan, and Z. Liu. 2008. Synthesis and photoluminescence properties of ZnO nanowires and nanorods by thermal oxidation of Zn precursors. *Materials Letters* 62, no. 12–13: 1797–1800.

Liang, J., N. Kishi, T. Soga, and T. Jimbo. 2011. The synthesis of highly aligned cupric oxide nanowires by heating copper foil. *Journal of Nanomaterials* 2011: 1–8.

Liu, D., Z. Yang, P. Wang et al. 2013. Preparation of 3D nanoporous copper-supported cuprous oxide for high-performance lithium ion battery anodes. *Nanoscale* 5, no. 5: 1917.

Liu, H., R. Wu, Z. Huang et al. 2007. Characteristics and photocatalytic effects of Zn/ZnO nanowhiskers compared with ZnO nanoparticles. *Journal of Wuhan University of Technology-Materials Science* 22, no. 4: 643–648.

Liu, Z.W., M.L. Zhong, and C.M. Tang. 2014. Large-scale oxide nanostructures grown by thermal oxidation. *IOP Conference Series: Materials Science and Engineering* 60: 012022.

M. Kaur, K.P. Muthe, S.K. Despande et al. 2006. Growth and branching of CuO nanowires by thermal oxidation of copper. *Journal of Crystal Growth* 289: 670–675

Major, J., R. Tena-Zaera, E. Azaceta. 2017. Development of ZnO nanowire based CdTe thin film solar cells. *Solar Energy Materials and Solar Cells* 160: 107–115.

Mayekar, J., V. Dhar, and S. Radha. 2014. Synthesis of copper oxide nanoparticles using simple chemical route. *International Journal of Scientific and Engineering Research* 5, no. 10: 928–930.

Mei, J., W. Fu, Z. Zhang et al. 2017. Vertically-aligned Co_3O_4 nanowires interconnected with $Co(OH)_2$ nanosheets as supercapacitor electrode. *Energy* 139: 1153–1158.

Mema, R., L. Yuan, Q. Du et al. 2011. Effect of surface stresses on CuO nanowire growth in the thermal oxidation of copper. *Chemical Physics Letters* 512, no. 1–3: 87–91

Mimura, K., J. Lim, M. Isshiki et al. 2006. Brief review of oxidation kinetics of copper at $350°C$ to $1050°C$. *Metallurgical and Materials Transactions A* 37, no. 4: 1231–1237.

Mo, Z., Y. Huang, S. Lu et al. 2017. Growth of ZnO nanowires and their applications for CdS quantum dots sensitized solar cells. *Optik—International Journal for Light and Electron Optics* 149: 63–68.

Nath, S., J. Prakash, J. Xiong, and J. Myoung. 2012. Synthesis of ZnO Nanowire by MOCVD Technique: Effect of Substrate and Growth Parameter. *Nanowires—Recent Advances.* Intech.

Nathan, D.M., and S.J. Boby. 2017. Hydrothermal preparation of hematite nanotubes/reduced graphene oxide nanocomposites as electrode material for high performance supercapacitors. *Journal of Alloys and Compounds* 700: 67–74.

Nishimura, J., R. Guo, M. Higashihata, and T. Okada. 2007. Optical characteristics of ZnO nanowires synthesized by nano-particles assisted pulsed laser deposition. *2007 Conference on Lasers and Electro-Optics—Pacific Rim.*

Noothongkaew, S., S. Pukird, W. Sukkabot et al. 2013. Zinc oxide nanostructures synthesized by thermal oxidation of zinc powder on Si substrate. *Applied Mechanics and Materials* 328: 710–714.

Oleg, V., N. Arkady, N. Andrey, N. Gennady, A. Artem, and A. Anatoly 2012. ZnO nanorods: Synthesis by catalyst-free CVD and thermal growth from salt composites and application to nanodevices. *Nanorods.* Intech.

Park, W.I., C. Lee, J. Chae et al. 2008. Ultrafine ZnO nanowire electronic device arrays fabricated by selective metal-organic chemical vapor deposition. *Small* 5, no. 2: 181–184.

Phiwdang, K., S. Suphankij, W. Mekprasart, and W. Pecharapa. 2013. Synthesis of CuO nanoparticles by precipitation method using different precursors. *Energy Procedia* 34: 740–745.

Poreddy, R., C. Engelbrekt, and A. Riisager. 2015. Copper oxide as efficient catalyst for oxidative dehydrogenation of alcohols with air. *Catalysis Science & Technology* 5, no. 4: 2467–2477.

Pujilaksono, B., T. Jonsson, M. Halvarsson et al. 2010. Oxidation of iron at 400–600°C in dry and wet O_2. *Corrosion Science* 52, no. 5: 1560–1569.

Qiu, Y., H. Fan, X. Chang, H. Dang et al. 2018. Novel ultrathin Bi_2O_3 nanowires for supercapacitor electrode materials with high performance. *Applied Surface Science* 434: 16–20.

Rajendran, K., G. Suja et al. 2017. Evaluation of cytotoxicity of hematite nanoparticles in bacteria and human cell lines. *Colloids and Surfaces B: Biointerfaces* 157: 101–109.

Ren, S., Y.F. Bai, J. Chen et al. 2007. Catalyst-free synthesis of ZnO nanowire arrays on zinc substrate by low temperature thermal oxidation. *Materials Letters* 61, no. 3: 666–670.

Rufus, A., N. Sreeju, V. Vilas, and D. Philip. 2017. Biosynthesis of hematite (α-Fe_2O_3) nanostructures: Size effects on applications in thermal conductivity, catalysis, and antibacterial activity. *Journal of Molecular Liquids* 242: 537–549.

Sallet, V. 2014. Metal-organic chemical vapor deposition growth of ZnO nanowires. *Wide Band Gap Semiconductor Nanowires* 1: 265–302.

Somvanshi, D., and S. Jit. 2013. Catalyst free growth of ZnO nanorods by thermal evaporation method. *AIP Conference Proceedings* 1536: 125.

Song, X., Y. Zhao, and Y. Zheng. 2006. Hydrothermal synthesis of tungsten oxide nanobelts. *Materials Letters* 60, no. 28: 3405–3408.

Su, C.H, C.Y. Su, and Y.F. Lin. 2015. Microstructural characterization and field emission properties of tungsten oxide and titanium-oxide-doped tungsten oxide nanowires. *Materials Chemistry and Physics* 153: 353–358.

Sun, Y., N.G. Ndifor-Angwafor, J. Riley, and M. Ashfold. 2006. Synthesis and photoluminescence of ultra-thin ZnO nanowire/nanotube arrays formed by hydrothermal growth. *Chemical Physics Letters* 431, no. 4–6: 352–357.

Tan, J., M. Dun, and L. Li. 2017. Self-template derived CuO nanowires assembled microspheres and its gas sensing properties. *Sensors and Actuators B: Chemical* 252: 1–8.

Telford, M. 2005. ZnO nanohelix: A new twist. *Materials Today* 8, no. 11: 10.

Tigli, O., and J. Juhala. 2011. ZnO nanowire growth by physical vapor deposition. *11th IEEE International Conference on Nanotechnology*. Portland, Oregon, USA.

Uchida, S. 2014. Copper oxide superconductors. *High Temperature Superconductivity*: 23–59.

Umar, A., H. Ra, J. Jeong et al. 2006. Synthesis of ZnO nanowires on Si substrate by thermal evaporation method without catalyst: Structural and optical properties. *Korean Journal of Chemical Engineering* 23, no. 3.

Wan, H., and H. Ruda. 2010. A study of the growth mechanism of CVD-grown ZnO nanowires. *Journal of Materials Science: Materials in Electronics* 21, no. 10: 1014–1019.

Wang, G., Z. Li, M. Li et al. 2018. Synthesizing vertical porous ZnO nanowires arrays on Si/ITO substrate for enhanced photocatalysis. *Ceramics International* 44, no. 2: 1291–1295.

Wang, L., X. Zhang, S. Zhao et al. 2005. Synthesis of well-aligned ZnO nanowires by simple physical vapor deposition on c-oriented ZnO thin films without catalysts or additives. *Applied Physics Letters* 86, no. 2: 024108.

Wang, S.B., C.H. Hsiao, S.J. Chang et al. 2011. A CuO nanowire infrared photodetector. *Sensors and Actuators A: Physical* 171, no. 2: 207–211.

Weng, W.Y., T.J. Hsueh, S.J. Chang et al. 2009. Laterally-grown ZnO-nanowire photodetectors on glass substrate. *Superlattices and Microstructures* 46, no. 5: 797–802.

Wu, Z., S. Tyan, C. Lee, and T. Mo. 2017. Bidirectional growth of ZnO nanowires with high optical properties directly on Zn foil. *Thin Solid Films* 621: 102–107.

Xiang, Wu, C. Wei, and Q. Feng-Yu. 2009. Spontaneous formation of single crystal ZnO nanohelices. *Chinese Physics B* 18, no. 4: 1669–1673.

Xu, C. 2003. Growth of hexagonal ZnO nanowires and nanowhiskers. *Scripta Materialia* 48, no. 9: 1367–1371.

Xu, C., C. Woo, and S. Shi. 2004. The effects of oxidative environments on the synthesis of CuO nanowires on Cu substrates. *Superlattices and Microstructures* 36, no. 1–3: 31–38.

Xu, C.H., Z.B. Zhu, H.F. Lui et al. 2011. The effect of oxygen partial pressure on the growth of ZnO nanostructure on $Cu_{0.62}Zn_{0.38}$ brass during thermal oxidation. *Superlattices and Microstructures* 49, no. 4: 408–415.

Xu, X., C. Cao, and Y. Zhu. 2015. Facile synthesis of single crystalline mesoporous hematite nanorods with enhanced supercapacitive performance. *Electrochimica Acta* 155: 257–262.

Yang, Q., Z. Guo, X. Zhou et al. 2015. Ultrathin CuO nanowires grown by thermal oxidation of copper powders in air. *Materials Letters* 153: 128–131.

Yao, B.D., Y.F. Chan, and N. Wang. 2002. Formation of ZnO nanostructures by a simple way of thermal evaporation. *Applied Physics Letters* 81, no. 4: 757–759.

Yi, G.-C., C. Wang, and W.I. Park. 2005. ZnO nanorods: Synthesis, characterization and applications. *Semiconductor Science and Technology* 20, no. 4: S22–S34.

Yu, Q., W. Wu, J. Zhang et al. 2009. Aligned tungsten oxide nanowires on tungsten (100) substrates. *Materials Letters* 63, no. 26: 2267–2269.

Yu, T., Y. Zhu, X. Xu et al. 2005. Controlled growth and field-emission properties of cobalt oxide nanowalls. *Advanced Materials* 17, no. 13: 1595–1599.

Yu, X., T. Marks, and A. Facchetti. 2016. Metal oxides for optoelectronic applications. *Nature Materials* 15, no. 4: 383–396.

Yuan, L., C. Wang, R. Cai et al. 2014. Temperature-dependent growth mechanism and microstructure of ZnO nanostructures grown from the thermal oxidation of zinc. *Journal of Crystal Growth* 390: 101–108.

Yuan, L., Q. Jiang, J. Wang, and G. Zhou. 2012a. The growth of hematite nanobelts and nanowires—tune the shape via oxygen gas pressure. *Journal of Materials Research* 27, no. 7: 1014–1021.

Yuan, L., Y. Wang, R. Cai et al. 2012b. The origin of hematite nanowire growth during the thermal oxidation of iron. *Materials Science and Engineering: B* 177, no. 3: 327–336.

Yuan, L., Y. Wang, R. Mema, and G. Zhou. 2011. Driving force and growth mechanism for spontaneous oxide nanowire formation during the thermal oxidation of metals. *Acta Materialia* 59, no. 6: 2491–2500.

Yun, J., Z. Zhang, and Y. Zhang. 2012. Magnetic and optical properties of ZnO nanowire arrays synthesized by a simple hydrothermal process. *2012 12th IEEE International Conference on Nanotechnology (IEEE-NANO)*.

Zappa, D., A. Bertuna, E. Comini et al. 2015. Tungsten oxide nanowires for chemical detection. *Analytical Methods* 7, no. 5: 2203–2209.

Zayyoun, N., L. Bahmad, L. Laânab, and B. Jaber. 2016. The effect of pH on the synthesis of stable Cu2O/CuO nanoparticles by sol–gel method in a glycolic medium. *Applied Physics A* 122, no. 5.

Zhan, Z., L. Xu, J. An et al. 2017. Direct catalyst-free chemical vapor deposition of ZnO nanowire array UV photodetectors with enhanced photoresponse speed. *Advanced Engineering Materials* 19, no. 8: 1700101.

Zhang, D., Y. Sheng, J. Wang et al. 2017. ZnO nanowire photodetectors based on Schottky contact with surface passivation. *Optics Communications* 395: 72–75.

Zhang, Y., M. Ram, E. Stefanakos, and D.Y. Goswami. 2012. Synthesis, characterization, and applications of ZnO nanowires. *Journal of Nanomaterials* 2012, 1–22.

3 Fabrication of 1-D ZnO by Thermal Oxidation Process

Supab Choopun, Ekasiddh Wongrat, and Niyom Hongsith

CONTENTS

3.1 INTRODUCTION

1-D nanostructures possess a high distinction in nanoscience and nanotechnology due to a fundamental conception of 1-D nanostructures. They have altered many properties of materials which attracted attention in today's technology, including electronic, magnetic, optoelectronic, quantum confinement effect, electrochromic, thermoelectric, superconductivity, and piezoelectronic (Grela and Colussi 1996, Qurashi et al. 2010, Devan et al. 2012, Zhang et al. 2015, Benjwal et al. 2018). Those 1-D nanostructures not only changed their electronic structures to physical/chemical properties, but they also demonstrated the great potentials in a maintaining a charge carrier on their surface to utilize in sensors or solar cells applications. Since the size reduction and dimensionality of 1-D format produces the higher surface area and reduces the recombination rate of charge carrier (Benjwal et al. 2018), they can be further applied to enhance the surface to volume ratio in gas sensor or solar cell applications. Furthermore, the smaller size and larger length of the homogenously synthesized 1-D nanostructures are expected to perform as the building block in nanoelectronic device architectures (Qurashi et al. 2010, Devan et al. 2012, Zhang et al. 2015, Kang et al. 2017).

For 1-D nanostructure materials, the metal oxide semiconductors are considerably the candidate materials because they can be naturally synthesized in diverse nanostructure arrangements, for instance ZnO, CuO, SnO_2, In_2O_3, TiO_2, NiO, and WO_3. Among these metal oxide semiconductors, ZnO nanostructures are energetically influenced with their practical properties in new nanoscience and nanotechnology utilization. There are many techniques to fabricate ZnO 1-D nanostructures: thermal evaporation (Kong et al. 2012, Tu et al. 2017), chemical vapor deposition (Xiang et al. 2007, Hu et al. 2012), wet chemical method (Samanta et al. 2015, Das et al. 2017), microwave-assisted thermal oxidation (Thepnurat et al. 2015), and thermal oxidation (Wongrat et al. 2009, Wongrat et al. 2012, Mihailova et al. 2013, Wongrat et al. 2016, Xu et al. 2017).

A thermal oxidation technique is one of a promising approach for fabricating 1-D metal oxide that has various benefits with less drawbacks. It can effectively arrange the ZnO nanostructure form and its good merits include high performance synthesis, low cost, reproducibility, and high mass production. A thermal oxidation process has an impact on oxidizing metal or corrosion effect in various technologies. This approach strongly occurs from the surface chemical reaction between metallic substance and oxygen molecule which acts as ionic bond of a negative oxygen and a positive metal. When those ions are in thermodynamic equilibrium conditions, the metal oxide is completely created. Normally, most metals can be oxidized to form metal oxides at the atmospheric surrounding. So, thermal oxidation process can be simply utilized to synthesize 1-D nanostructures of metal oxides. Basically, the nanostructures of metal oxides can be naturally formed via the thermal oxidation reaction of metal.

In this chapter, ZnO nanostructures that can be formed from zinc metal will be given as an example for fabrication of 1-D metal oxide by thermal oxidation process. Growth kinetics of this thermal oxidation process are also explored in term of thermodynamic parameters, especially Gibbs free energy in order to explain growth mechanism of 1-D metal oxide from thermal oxidation reaction. Moreover, the 1-D nanostructure formation is significantly described via nucleation formation process.

3.2 PREPARATION TECHNIQUE BY THERMAL OXIDATION PROCESS

There are many previous reports about the 1-D ZnO nanostructures as listed in Table 3.1. It can be seen that 1-D ZnO nanostructures can be simply prepared by thermal oxidation process at various conditions and can be obtained with wide varieties of 1-D ZnO nanostructures and wide range of sizes and growth directions.

TABLE 3.1
Lists of ZnO Nanostructures Synthesized by Thermal Oxidation Technique

Materials	Temperature (°C)	Time	Morphology	Diameter (nm)	Growth Direction	References
ZnO	300	5 min	Nanowire and nanoflake	100–150	—	Hsueh et al. (2008)
ZnO	600	24 h	Nanowire	100–500	—	Wongrat et al. (2009)
ZnO	300–600	1 h	Nanoneedle	20–80	[0001]	Yu and Pan (2009)
ZnO	500	1 h	Nanoplate	200–600	[11$\bar{2}$0]	Kim et al. (2004)
ZnO	200–500	30 min	Nanowire	30–350	[0001]	Schroeder et al. (2009)
ZnO	300–600	1 h	Nanowire	12–52	[11$\bar{2}$0]	Fan et al. (2004)
ZnO	<400	30 min	Nanowire	20–150	[11$\bar{2}$0]	Ren et al. (2007)
ZnO	600	1.5 h	Nanowire	30–60	[0001]	Sekar et al. (2005)
ZnO	400–600	1 h	Nanowire and nanorod	20	[2$\bar{1}$10]	Liang et al. (2008)
ZnO	500	24 h	Nanowire	40–400	—	Wongrat et al. (2016)
ZnO	390	10 h	Nanowire	25–85	—	Xu et al. (2011)
ZnO	530	15 min	Nanoneedles	20–80	—	Mihailova et al. (2013)
ZnO	500	15 min	Nanowire	44.2	[10$\bar{1}$0]	Wu et al. (2017)

The thermal oxidation technique can be roughly classified by the external driving heating energy into three main approaches including: (1) thermal heating oxidation technique in the furnace; (2) current heating oxidation technique; and (3) microwave-assisted thermal oxidation technique. These techniques can be used to prepare 1-D nanostructures of metal oxides or compounds of metal oxides. Each thermal oxidation technique is fully described in the following:

3.2.1 THERMAL HEATING OXIDATION TECHNIQUE IN THE FURNACE

In this technique, the continuous thermal heating generated from the furnace is an external heating energy to drive metal atoms to form metal oxide compounds. The metal atoms can be oxidized during either in (a) solid phase or (b) gas phase.

For (a) metal solid phase, a metal layer is used for thermal heating oxidation and can be prepared as thin films or thick films by several techniques. Here, zinc metal layer is prepared by DC magnetron sputtering technique, and by evaporation technique as a thin film (Yawong et al. 2005), and also by screening technique as a thick film. For DC magnetron sputtering technique, zinc metal thin film is sputtered onto an alumina substrate under argon pressure of 30 mtorr at power of 200 watts. The sputtering time is varied at 30, 60, and 90 minutes for various thicknesses of the zinc metal layer. For the evaporation technique, 0.3 g of zinc powder is put in a tungsten boat and thermally evaporated onto alumina substrate (Wongrat et al. 2012) under pressure of 5×10^{-5} torr and the electrical current is routinely adjusted to achieve the heating with the maximum current of 30 A. For the screening technique, the zinc powder is crushed and then blended with polyvinyl alcohol (PVA) to form a paste (Wongrat et al. 2009) and then screened as the thick film on the alumina substrate.

Next, the zinc thin films and thick films on the alumina substrate are placed in a tube furnace and heated at desire temperatures at normal atmosphere for thermal oxidation, as illustrated in Figure 3.1. The heating temperatures for sputtered zinc thin films are performed at 600°C, 800°C, and 900°C for 6 hr under air atmosphere. It is found that the sputtered zinc thin film can be completely oxidized to form ZnO columnar-like structures (Yawong et al. 2005). The evaporated zinc thin film and screened zinc thick film on alumina substrates are progressively heated for 24 hr

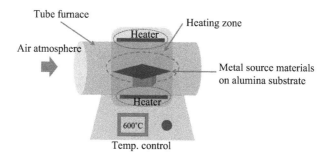

FIGURE 3.1 Schematic diagram of ZnO nanostructures synthesis apparatus from solid phase of zinc metal source performed on alumina substrate by thermal oxidation technique.

under air atmosphere at the heating temperatures of 500°C and 700°C, respectively. After the thermal oxidation process, the color of obtained products change from gray to white, suggesting the occurrence of the chemical oxidation reaction and the transformation of zinc (Zn) metal to zinc oxide (ZnO) (Wongrat et al. 2009, Wongrat et al. 2012).

The ZnO sample transformed from metallic zinc screened on alumina substrate illustrates the wire-like structures in range of 100–500 nm and has the length in micron level with the sharp tip (Wongrat et al. 2009). In addition, the ZnO sample transformed from metallic zinc evaporated on alumina substrate also represents the highly density wire-like structures with diameter in range of 40–200 nm (Wongrat et al. 2012).

For (b) metal gas phase, the metal atoms are oxidized via chemical vapor deposition (CVD) technique as shown in Figure 3.2. The metallic zinc powder is put in a ceramic boat and placed in a small horizontal quartz tube. The alumina substrates are intentionally stuck onto the top and bottom of a small quartz tube to encounter the downstream vapor from metal source. As a beginning process, a rotary pump was turned on until it pumped down to the pressure of 60 torr. The temperature controller was set up to the target temperature which boosted up to 500°C and 600°C for our experiment. The Ar gas constantly flowed at a rate of 500 ml/min for 5 min, then the O_2 gas directly flowed onto the system at a rate of 50 ml/min for 20 min. At the end of the process, the heater of tube furnace was switched off to cool down to the room temperature.

From surface morphology results, the growth of ZnO from zinc powder shows that the morphologies depend on the substrates position which placed away from source material as seen in Figure 3.3a–d. The representative of ZnO at the top of a small quartz tube, sintered at a temperature of 500°C, exhibit the uniform well-align nanowires morphology as displayed in Figure 3.3a. It can be seen that the average diameters are 150 nm. However, when ZnO, grown at the bottom of a small quartz

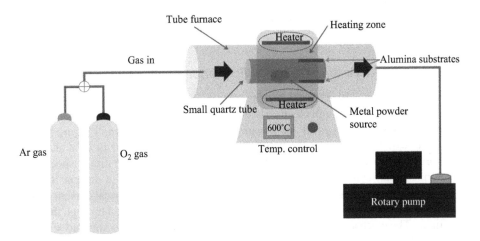

FIGURE 3.2 Schematic diagram of ZnO nanostructures synthesis apparatus from zinc powder vaporized by thermal oxidation technique.

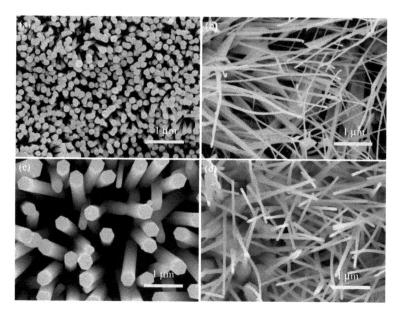

FIGURE 3.3 FE-SEM images of ZnO grown on alumina substrates by evaporation from zinc powder at the temperatures of (a), (b) 500°C and (c), (d) 600°C.

tube, shows the longer nanowires and expresses the connection between nanowires to the others as shown in Figure 3.3b. In addition, the average diameters are 43 nm. While the ZnO grown at the top of a small quartz tube with a sintering temperature of 600°C reveals the apparently distinct morphology. Their structure features consist of the larger well-align nanowires with a hexagonal shape along the longitudinal direction and the larger nanowires connected with the others as visibly appeared in Figure 3.3c and d. The average diameters of two structure features are 376 and 76 nm for FESEM (field emission scanning electron microscope) images of Figure 3.3c and d, respectively.

3.2.2 CURRENT HEATING OXIDATION TECHNIQUE

For this technique, the external heating energy is a DC electrical current passing through the metal. The high electrical current energizes the metal resulting in transformation to the excited gases and plasma transformation as shown in Figure 3.4.

In our experiment, the zinc powder, purity 99.99%, is mixed with PVA and then compressed into a cylinder with height and diameter of 1 cm and 6 mm, respectively. The zinc metal cylinder with electrodes is put on the stand of a current heating system as shown in Figures 3.4 and 3.5a and b. When the DC electrical current power supply is switched on, the glowing plasma is immediately generated through the zinc cylinder. The electrical currents are adjusted to be 5 and 10 A for comparative study. The oxidation reaction spontaneously takes place during plasma generation in air atmosphere.

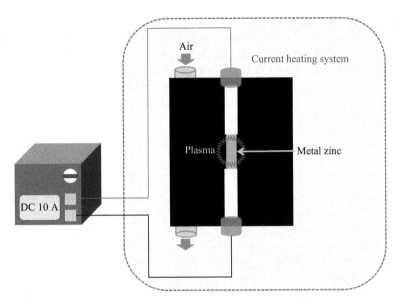

FIGURE 3.4   Schematic diagram of current heating apparatus for thermal oxidation process of a cylindrical zinc metal.

FIGURE 3.5   (a) and (b) Current heating setup for thermal oxidation process of zinc metal cylinder. (c) and (d) FE-SEM images of tetrapod-like ZnO structures heated with DC electrical current of 5 A and 10 A, respectively.

The morphologies of ZnO products from the current heating oxidation technique are comparatively observed in Figure 3.5c and d for the condition with DC electrical currents of 5 and 10 A, respectively. For 5 A condition, the obtained ZnO structures illustrate tetrapod-like products with an average diameter at the middle of the legs of 85 nm and length in the order of several microns as seen in Figure 3.5c. However, for 10 A condition, the interconnected tetrapod-like ZnO structures can be partially discovered with an average diameter at the middle of legs of 225 nm and length in the order of several microns as shown in Figure 3.5d.

3.2.3 MICROWAVE-ASSISTED THERMAL OXIDATION TECHNIQUE

This technique utilizes microwave radiation as a heating energy to metal. Metals can absorb microwave and get heated very fast to form plasma. Since the metal plasma is very active due to very high energy, the oxidation reaction can rapidly occur during plasma formation resulting in metal oxide nanostructures. For ZnO nanostructure preparation, zinc (Zn) powder (2 g; 99.99%, less than 50 μm particle size) is used as a precursor and put on a quartz substrate that is placed in a quartz tube having a diameter of 2.8 cm and length of 10 cm in a household microwave oven (SHARP model), as shown in Figure 3.6a. The Zn metal powder was then irradiated with microwave power of 700 W at a frequency of 2.45 GHz (λ = 12 cm) for 60 s under atmospheric

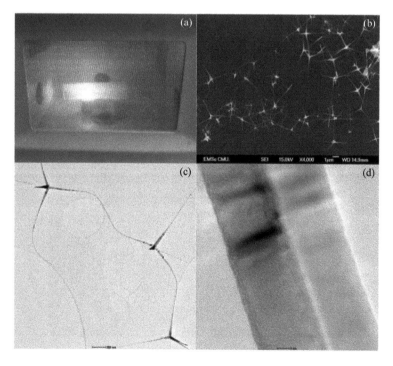

FIGURE 3.6 Microwave assisted thermal oxidation technique (a) the plasma in microwave oven (b) FE-SEM image of ITN-ZnO nanostructures (c) and (d) TEM bright field image and high-resolution image of ITN-ZnO nanostructures, respectively.

conditions. After cooling down to room temperature, the wool-like ZnO structures were observed in the quartz tube (Thepnurat et al. 2015).

The obtained ZnO products can be distinguished into two areas with two different morphologies. In the bottom area, the products with opaque white color are obtained and exhibit tetrapod-like ZnO structures (T-ZnO) about 1 μm in diameter at the middle of their legs and 10–30 μm in length similarly to the nanostructures obtained from the current heating oxidation technique as discussed earlier. In the top area, the products with transparent white color are obtained and exhibit the tetrapod-like structure with leg-to-leg linking (ITN-ZnO) as seen in Figure 3.6b–d. The legs of ZnO tetrapods are clearly interlinked with the legs of neighbor tetrapods at the end of the legs. The diameter at the middle of each leg is in the order of 50 nm and is much smaller than that of T-ZnO in the bottom area (Thepnurat et al. 2015).

3.3 THERMODYNAMIC TREATMENT OF THERMAL OXIDATION REACTION

The theoretical approach for the explanation of chemical reaction occurring in the oxidation process of metal is a thermodynamic based on Gibbs free energy. Gibbs free energy is a powerful parameter for the identification of a stable system. Generally, the basic thermodynamic parameter as Gibbs free energy change (per mole) is definitely expressed and at a constant temperature. Gibbs free energy change can be written as the following equation:

$$dG = VdP \qquad (3.1)$$

where G, V, and P variables are defined as Gibbs free energy, system volume, and pressure, respectively. During the thermal oxidation process of metal, the oxygen molecules in air play an important role. Therefore, the ideal gas approximation of oxygen molecule is simply considered by taking a specific volume (per mole) as:

$$V = \frac{RT}{P} \qquad (3.2)$$

where R and T are defined as a gas constant and absolute temperature, respectively. According to equations (3.1) and (3.2), a specific volume from equation (3.2) is represented in equation (3.1) and the integration is undertaken in the interval of state 1 (P_1) and state 2 (P_2). The definite integral result is delivered by:

$$\Delta G = RT \ln \frac{P_2}{P_1} \qquad (3.3)$$

From equation (3.3), the pressure P_1 corresponds to the standard state pressure of 1 atm and the Gibbs free energy change is defined as $\Delta G = G - G^0$, where G^0 is the Gibbs free energy at a standard state. Typically, for most systems this is not ideal, leading to a new function called activity (a). In addition, the pressure is directly proportional to the activity (a) of substance. Therefore, the equation (3.3) is replaced as (Gaskell 2003):

$$G = G^0 + RT \ln a \tag{3.4}$$

To thoroughly explain the oxidation reaction of a metallurgical substance in metal-gas interaction, the chemical reaction scheme is specified with one mole of oxygen molecule as (Lad 1995, Fu and Wagner 2007):

$$\frac{2}{n}M + O_2 \rightarrow \frac{2}{n}MO_n \tag{3.5}$$

In equation (3.5), it can be seen that the Gibbs free energy at a temperature change through the chemical reaction. The Gibbs free energy change at a non-equilibrium condition is not zero. When it carries out into an equilibrium condition, the Gibbs free energy is zero. The Gibbs free energy change of this chemical reaction according to equation (3.5) is written as:

$$\Delta G_T = \sum G_{T,\text{products}} - \sum G_{T,\text{reactants}} \tag{3.6}$$

So, by applying equation (3.6) for chemical reaction according to equation (3.5), the Gibbs free energy change is newly represented by:

$$\Delta G_T = \frac{2}{n}G_{MO_n} - \left(\frac{2}{n}G_M + G_{O_2}\right) \tag{3.7}$$

By substituting the Gibbs free energy of each substance from equation (3.4) into equation (3.7), it becomes:

$$\Delta G_T = \frac{2}{n}\left(G_{MO_n}^0 + RT \ln a_{MO_n}\right) - \left(\frac{2}{n}\left(G_M^0 + RT \ln a_M\right) + \left(G_{O_2}^0 + RT \ln a_{O_2}\right)\right) \tag{3.8}$$

or

$$\Delta G_T = \frac{2}{n}G_{MO_n}^0 - \frac{2}{n}G_M^0 - G_{O_2}^0 + RT \ln\left(\frac{(a_{MO_n})^{2/n}}{(a_M)^{2/n}a_{O_2}}\right) = \Delta G^0 + RT \ln(K) \tag{3.9}$$

when

$$\Delta G^0 = \frac{2}{n}G_{MO_n}^0 - \frac{2}{n}G_M^0 - G_{O_2}^0 \tag{3.10}$$

and the equilibrium constant (K) is given by:

$$K = \frac{\left(a_{MO_n}\right)^{2/n}}{\left(a_M\right)^{2/n} a_{O_2}}$$

(3.11)

From equation (3.9), $\Delta G_T = 0$ when the obtained system carries out until it turns into an equilibrium condition. Moreover, the activity of a solid is cancelled because it is a constant of 1. However, the activity of the gas phase in an equilibrium condition is close to its partial pressure. So, the equation (3.9) is definitely rewritten as:

$$\Delta G^0 = RT \ln P_{O_2}$$

(3.12)

where P_{O_2} is an oxygen partial pressure.

From equation (3.12), it can be implied that a relation plot between the standard Gibbs free energy change versus temperatures at a constant oxygen partial pressure exhibit a straight line. The standard Gibbs free energy change is zero at the temperature of 0 K and down into negative values with the temperature increasing. Moreover, the standard Gibbs free energy variation for metal oxidation reaction displays the pressure-independent characteristics and it is regularly launched as a linearly distinctive ΔG^0–T dependent. This relation is experimentally obtained by calculating the enthalpy contribution change (ΔH^0) and entropy contribution change (ΔS^0) (Gaskell 2003). The given ΔG^0–T dependent for oxidation reaction of solid and liquid copper that transform to cuprous oxide are illustrated as an example by (Gaskell 2003):

$$\Delta G^0 = -338,900 - 14.2T \ln T + 247T$$

(3.13)

Equation (3.13) is valid for solid copper in temperature ranges of 298–1356 K and (Gaskell 2003):

$$\Delta G^0 = -390,800 - 14.2T \ln T + 285.3T$$

(3.14)

Equation (3.14) is valid for liquid copper in temperature range 1356–1503 K.

To comparatively investigate ΔG^0–T dependent, the standard Gibbs free energy change of solid nickel is also given as (Johnson and Stracher 1995):

$$\Delta G^0 = -489,160 + 197.08T \qquad \text{J/mol}$$

(3.15)

To substantially describe the thermal oxidation reaction of metal, the Ellingham diagram construction is a useful tool to explain the possibility of metal reacting on

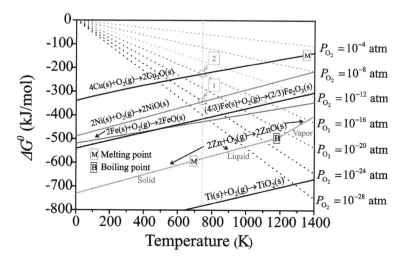

FIGURE 3.7   Ellingham diagram of various metals for metal oxides formation.

the oxygen and becoming a mostly common metal-oxide form. The plotted graph of ΔG^0–T dependent obtained from equations (3.12) to (3.15) is called the Ellingham diagram and the ΔG^0–T dependent of copper oxide and nickel oxide are comparatively illustrated in Figure 3.7. While the other chemical reactions of metal oxides are also plotted in the Ellingham diagram in Figure 3.7. There are many benefits for the chemical reaction explanation of metal to react on oxygen, hydrogen, carbon monoxide-carbon dioxide, especially oxygen affinity of metal. Commonly, more negative Gibbs free energy changes of chemical reaction can predict the oxide stability. For instance (see in Figure 3.7), when solid copper and nickel are collectively placed in the closed systems at the same temperature of 743 K under oxygen partial pressure below 10^{-24} atm, two metals cannot be oxidized to form metal-oxides. However, if the oxygen is continuously flowed into the system until it reaches a constant oxygen partial pressure at 10^{-24} atm (see in grey circle 1), the oxidation reaction of nickel starts to occur and forms nickel oxide in thermodynamic equilibrium state. If the oxygen partial pressure is increased to above 10^{-24} atm (less than 10^{-16} atm), the oxidation reaction of nickel still progressively occurs. At this mechanism, it is clearly seen that the copper oxidation process cannot spontaneously continue. Finally, if the oxygen partial pressure is increased up to 10^{-16} atm, the copper oxidation spontaneously continues into a thermodynamic equilibrium state including copper, oxygen, and copper oxide in the closed system (see in grey circle 2). It is clearly seen that the nickel oxidation continues easier than that of copper oxidation. On the other hand (also see in Figure 3.7), if the initial state of those metals including copper and nickel are placed into the effectively closed system at the oxygen partial pressure above 10^{-16} atm (see in grey circle 2); both metals simultaneously react to oxygen molecule and transform to metal-oxides. Consequently, the oxidation reaction consumes oxygen, resulting in a decreasing of oxygen quantity in the system. The initial oxygen partial pressure is decreased to reach around 10^{-16} atm. This mechanism leads to maintaining an equilibrium condition of copper, oxygen and copper oxide substances. After that, the oxygen amount is moderately consumed until its partial pressure value is below 10^{-16} atm resulting in terminating oxidation reaction of an oxidized copper. Nevertheless, the oxidation reaction occurrence of nickel to form nickel oxide is still unstable. So, it continuously keeps reacting until the oxygen partial pressure is at around 10^{-24} atm (see in grey circle 1). As a result, a chemical reaction of nickel, oxygen, and nickel oxide substances perfectly reaches the equilibrium in the system.

At this oxygen partial pressure, it is the lowest value of oxygen consumption desire to continue a copper oxidation reaction to become copper oxide. Therefore, the formerly chemical reaction of copper oxide creation becomes unstable and it decomposes to become copper and oxygen (Gaskell 2003). This mechanism is very essential for a possible explanation of chemical reactions of a variety substances involved with synthesis, growth, and new substance formation.

3.4 GROWTH KINETICS OF THERMAL OXIDATION PROCESS

In the previous section, we have mainly provided the thermodynamic approach to predict the possibility of metal-oxide creation from metals. The highest possibility of metals for oxidation are the metals having the most negative standard Gibbs free energy.

In this section, growth kinetics of this thermal oxidation process are investigated in terms of thermodynamic parameters in order to explain the growth mechanism of 1-D metal oxide nanostructure formation. The explanation is divided into three parts according to the classification of the growth by the thermal oxidation technique.

Actually, the kinetic theory of oxidation mechanism of pure metals to form metal oxide was originally proposed in a few researches including Wagner (1935, 1936), Wagner and Grünewald (1938), Cabrera and Mott (1949). Wagner proposed that the particles flux variation of reaction at high temperature is further dominant when is compared with the charge species concentration (Wagner 1935, 1936, Wagner and Grünewald 1938, Martin and Fromm 1997). Later, Mott theory (Wagner 1935, 1936, Wagner and Grünewald 1938) was classically introduced by the potential barrier disparity explanation between two oppositely separated electrical charges which produced the strong field to drive either metal ions or oxygen ions migrating across an adherent oxide film (Martin and Fromm 1997). However, there is still no kinetic theory of nanostructure formation of metal oxide grown via the thermal oxidation technique. Therefore, the possible mechanism is proposed in order to evidently explain for nanostructure formation, especially 1-D nanostructures.

3.4.1 GROWTH KINETICS OF ZnO NANOWIRE/NANOROD FOR THERMAL HEATING OXIDATION TECHNIQUE IN THE FURNACE

3.4.1.1 Oxidization of Zinc Metal Layer

For (a) oxidization of zinc metal layer, the growth kinetics has been explained in our previous reports (Choopun et al. 2010, Wongrat et al. 2011) for ZnO nanowire formation. The growth mechanism can be described in four steps as follows:

3.4.1.1.1 Step 1: Oxygen Adsorption

A firstly key factor of thermal oxidation mechanism is an oxygen adsorption process. The initial state of oxidation reaction, the neutrality metal atoms occupy the lattice sites where some adjacent atom site has a vacancy (Chernavskii et al. 2007) as illustrated in Figure 3.8a. After an external activation energy, thermal energy is utilized on metal atoms such as zinc. They loss electrons by migration mechanism up to the surface and form metal ions (M^{++}). While their surface is entirely covered with electron population as shown in Figure 3.8b. For oxygen adsorption mechanism, the physisorption mechanism of weakly bonded oxygen molecules takes place to accept the electron migrated from metal; then this generates the strongly bonded oxygen species of chemisorption mechanism in the variety forms of O_2^-, O^-, O^{2-} (Martin and Fromm 1997, Wongrat et al. 2012, 2016) as further indicated in Figure 3.8c (O^{2-} form is e.g.). Typically, the adsorbed oxygen ion species depend on working temperatures, which is usually O_2^- adsorbed oxygen species at the temperatures below 200°C. When the temperatures are up above 250°C, either the adsorbed oxygen species become O^- or O^{2-} (Martin and Fromm 1997, Hongsith et al. 2010, Wongrat et al. 2012, 2016). The chemical reaction of one mole oxygen molecule to form chemisorption oxygen species is usually written as (Choopun et al. 2010, Wongrat et al. 2011):

$$O_2(gas) + \bar{e} \rightarrow O_2^-(ads)$$

$$O_2(gas) + 4\bar{e} \rightarrow 2O^{2-}(ads) \tag{3.16}$$

$$O_2(gas) + 2\bar{e} \rightarrow 2O^-(ads)$$

3.4.1.1.2 Step 2: Surface Oxidization to Form Nuclei

Since the metal ions (M^{++}) and adsorbed oxygen ions (O^{2-}) are oppositely separated, both metal ions and adsorbed oxygen ions gradually transport via diffusion mechanism through the active layer (Martin and Fromm 1997, Chernavskii et al. 2007). Firstly, the adsorbed oxygen ions transportation mechanism is assumed to be dominant more than that of metal ions transportation. This leads to a larger diffusion path of adsorbed oxygen ions and deep movement of adsorbed oxygen ions into the active layer. However, most metals contain an imperfection which is lattice defects and takes effect on ions movement. The lattice defects are mainly vacancy and interstitial defects. When the adsorbed oxygen ions deeply migrate into the closest neighbors of an interstitial site without expelling authentic atoms, they match to metal ions by interstitial process to form the new metal oxide as displayed in Figure 3.8d (Shewmon 1989). In addition, if the adsorbed oxygen ions are also more mobile than that of metal ions; they migrate by vacancy mechanism into the closest neighbors of an unoccupied site or vacancy lattice of matrix atoms (Shewmon 1989, Chernavskii et al. 2007). The new metal-oxide is also created in the active layer as displayed in Figure 3.8d and the chemical reaction can be written as (Chernavskii et al. 2007):

$$M^{++} + O^{2-} \rightarrow MO \tag{3.17}$$

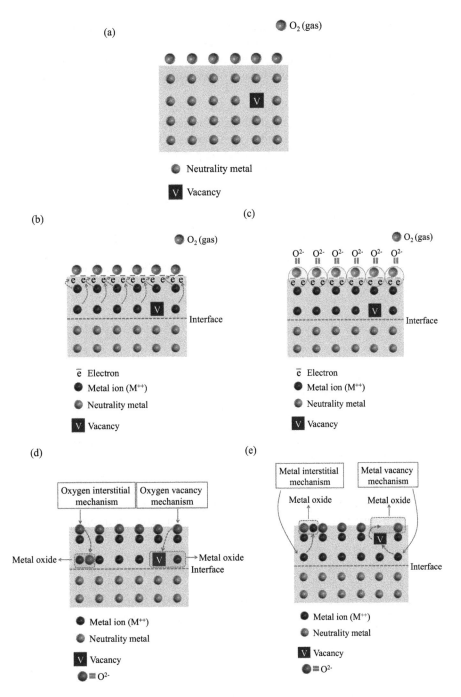

FIGURE 3.8 Schematic diagram of (a) metal and oxygen before interaction (b) metals loss electrons owing to an external activated energy (c) oxygen adsorption phenomenon (d) adsorbed oxygen ions diffusion mechanism (e) metal ions diffusion mechanism.

On the contrary, if the diffusion mechanism of metal ions is more dominant than that of adsorbed oxygen ions, the metal ions migrate out to react with the adsorbed oxygen ions without expelling authentic atoms and create the new metal-oxide on the surface of the active layer by vacancy or interstitial mechanism as represented in Figure 3.8e (Shewmon 1989).

At low temperatures (temperatures below room temperature), the oxygen molecules in air dissociate on the metals surface and capture electrons from those metals and turn to the molecular adsorbed oxygen species as O_2^-. A contact between metals and molecular adsorbed oxygen species often produces the surface potential during oxide formation (Cabrera and Mott 1949, Fehlner and Mott 1970, Martin and Fromm 1997). This surface potential builds the extremely electrical field due to a separated distance between metals and molecular adsorbed oxygen species less than 10 nm (Cabrera and Mott 1949, Martin and Fromm 1997). As a result, the sufficiently driving force enables either metal ions or molecular adsorbed oxygen ions to migrate across the oxide thin film and the metal oxide continuously grows up to a limit thickness (Cabrera and Mott 1949, Martin and Fromm 1997). The thickness of metal oxide thin film is experimentally expressed in an inverse logarithmic law (Cabrera and Mott 1949, Fehlner and Mott 1970, Chernavskii et al. 2007):

$$\frac{1}{x} = A - B \ln t \tag{3.18}$$

where x is a thickness of metal oxide thin film, t is a time duration, A and B are the pressure and temperature dependence parameters (Chernavskii et al. 2007).

From equation (3.18), the metal oxide thickness is moderately formed until it saturates to a stable film. At this stable stage, the oxidation chemical reaction of metal stops and the crystalline formation has barely occurred (Cabrera and Mott 1949). The crystalline formation plays an important role for creation of nuclei which are seedings of nanostructure formation.

Nevertheless, the metal oxide with a dramatically synthesized nanostructure is a crystalline pattern grown at high working temperatures in which an external energy is enough to drive nuclei of those ions in nanostructures arrangement. The ions exhibit the great diffusion length although there is no strong field due to the separated charges (Cabrera and Mott 1949). The oxide layer concentrations are regularly grown via metal or oxygen ions diffusion mechanisms, penetrating into the oxide layer. For instance, metal such as zinc penetrates to the interstitial sites and completely forms a ZnO layer (Samal 2016). The chemical reaction of metal zinc excess in the interstitial site to perform ZnO lattice defect is freshly adapted as (Kofstad 1988, Samal 2016):

$$Zn_i'' + O^{2-} \rightarrow ZnO \tag{3.19}$$

Besides, it is possible to occur a ZnO configuration from an oxygen diffusion mechanism especially at high working temperatures. The oxygen ion species moves into the vacancy site that is surrounded with the positive metal ions. Those ions are then appealed and react to become the ZnO format (Samal 2016).

At high working temperatures in thermal oxidation reaction of metal to form metal oxide format, either metal or oxygen ions diffuse through the oxide layer by interstitial or vacancy mechanism. The oxide film thickness is raised with time duration according to the parabolic scaling law (Tammann 1920, Pilling and Bedworth 1923, Rapp 1984) given by (Mott 1940, Fehlner and Mott 1970, Delalu et al. 2000, Machado et al. 2002, Chernavskii et al. 2007, García et al. 2008):

$$x^2 = kt \tag{3.20}$$

where x is the oxide film thickness, t is time duration, and k is a parabolic rate constant which depends on temperatures (Chernavskii et al. 2007, Young 2008). Interestingly, the parameter k can be experimentally determined at various temperatures (García et al. 2008).

Most nanostructure formation, especially 1-D nanostructure, synthesized via thermal oxidation process takes place at high working temperatures. At the beginning, as in equation (3.20), the metal oxide layer is first formed and meanwhile the recrystallization with a roughness format of oxide layer is established. The growth mechanism begins from the recrystallization phenomenon of an oxide layer, which is a key in nanostructure formation during a heating growth (Shen et al. 2010). The smoothly compressed layer of metal oxide rarely occurs at these high working temperatures. This leads to form a rough oxide layer likely induced from movement of either metal or oxygen ions to produce the metal oxide nuclei by nucleation process. Typically, during the thermal oxidation process, the metal or oxygen ions constantly migrate through the oxide layer depending on temperatures and diffusion length of each substance. The nuclei creation from the smaller ions with more concentrations causes a coalescence behavior of the contacted neighbor ions. Suddenly, those ions are merged together to increase the nuclei size, forming a bigger one. This phenomenon is analogous to the coalescence of two identical water droplets adhered on the smooth surface and then become a new larger size (Choopun et al. 2010, Wongrat et al. 2011). By considering a growth kinetic of 1-D nanostructures in a thermodynamically global approach, the Gibbs free energy is the important parameter to explain the nucleation mechanism. Generally, the oxide nuclei have the associated Gibbs free energy change, depending on nucleation shapes. An ideal shape to form a new phase is a homogeneous nucleation. First, let's consider the homogeneous nucleation of a new phase occurrence of solid, surrounded with old phase of liquid and the solid spherical shape nucleation takes place inside the liquid old phase as illustrated schematically in Figure 3.9a. The Gibbs free energy change associated to the phase transition is given by (Kalyanaraman 2008):

$$\Delta G_N = \frac{4}{3}\pi r^3 \Delta g_v + 4\pi r^2 \gamma_{SL} \tag{3.21}$$

where ΔG_N is the total Gibbs free energy change of a nucleus. Δg_v is Gibbs free energy change per unit volume which is usually a negative value. γ_{SL} is a surface energy of solid-liquid interfacial region and r is a spherical radius.

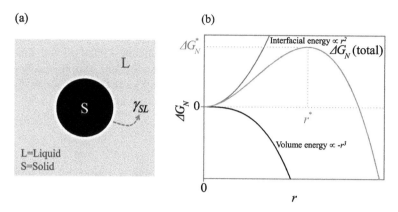

FIGURE 3.9 Schematic diagram of (a) homogeneous nucleation mechanism (b) relation of Gibbs free energy versus radius.

From equation (3.21), it was found that the total Gibbs free energy change depends on two terms, including a volume energy $((4/3)\pi r^3 \Delta g_v)$ and a surface energy $(4\pi r^2 \gamma_{SL})$. Since the magnitude of Δg_v is normally a negative value, the volume energy significantly decreases with an increase of the spherical radius (larger negative energy). Whereas the surface energy increases positively with an increase of the spherical radius as shown in Figure 3.9b. The total Gibbs free energy firstly increases, until it reaches the maximum barrier of the total Gibbs free energy. Later, it declines to a lower value and turns into the negative value. This indicates that the nucleation process spontaneously continues. From Figure 3.9b, it can be mathematically derived that the radius at the maximum barrier of the total Gibbs free energy is the critical radius (r^*) which exists at $\left(d\Delta G_N / dr \right) = 0$ and can obtained as (Kalyanaraman 2008):

$$r^* = -\frac{2\gamma_{SL}}{\Delta g_v} \tag{3.22}$$

In order to obtain the maximum barrier height of the total Gibbs free energy, the critical radius is substituted in equation (3.21) and the critical Gibbs free energy (ΔG_N^*) is written as (Kalyanaraman 2008):

$$\Delta G_N^* = \frac{16\pi \gamma_{SL}^3}{3\Delta g_v^2} \tag{3.23}$$

In thermodynamic meaning, the total Gibbs free energy at $r < r^*$ is lower than that of critical free energy. This phenomenon keeps dissolving particle embryos back to the old phase and the new phase is unfavorable. In contrast, the total Gibbs free energy at $r > r^*$ is higher than that of critical free energy. This can overcome the critical energy barrier height and the nucleation process can spontaneously continue to form nuclei, leading to the stable formation of a new phase.

Interestingly, the homogenous nucleation is a basis of the heterogenous nucleation process, which is utilized to explain the metal oxide nanostructure formation from the thermal oxidation approach of metal substances. Generally, the thermal oxidation process of metals to create the metal oxide comes from diffusivity of either metal or oxygen ions. Those ions cause to the chemical reaction on the underlying substrate and the nucleation process occurs via ionic diffusion mechanism to form oxide nuclei on the substrate. The heterogenous nucleation with cap shape and double cap-shape as displayed in Figure 3.10a and b are mainly considered instead of homogenous nucleation. To compare the thermodynamic parameters, the cluster as cap shape is defined radius r_1, contact angle θ_1 whereas the cluster as double cap-shape is defined as upper cap radius r_1, contact angle θ_1 and bottom cap radius r_2, contact angle θ_2 (Zhou 2009). Likewise, the critical free energy of cap shape and double cap-shape nucleation are derived as:

$$\Delta G_N^* = \frac{16\pi \gamma_{MO}^3}{3\Delta g_v^2} f\left(\theta_1\right)$$

(3.24)

and (Zhou 2009)

$$\Delta G_N^* = \frac{16\pi \gamma_{MO}^3}{3\Delta g_v^2} h\left(\theta_1, \theta_2\right)$$

(3.25)

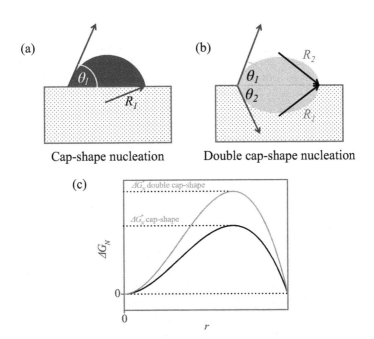

FIGURE 3.10 Schematic diagram of nucleation with (a) cap-shape (b) double cap-shape, and comparatively plotted graph relation of (c) Gibbs free energy versus radius of cap-shape and double cap-shape nucleation.

where Δg_v is Gibbs free energy change relating to the chemical reaction of metal to form metal oxide as illustrated similarly in equation (3.9), γ_{MO} is the surface energy of metal oxide (Zhou 2009). In addition, $f(\theta_1)$ and $h(\theta_1,\theta_2)$ are shape factors for cap-shape and double cap-shape nucleation, respectively (Zhou 2009). They are equal to (Fletcher 1958, Liu 1999, 2000, 2001, 2002a, 2002b, Zhou 2009):

$$f\left(\theta_1\right) = \frac{\left(2+\cos\theta_1\right)\left(1-\cos\theta_1\right)^2}{4}$$ (3.26)

and (Zhou 2009)

$$h\left(\theta_1,\theta_2\right) = \frac{\chi}{\chi-1}\left(\frac{\chi-1}{\chi}+\frac{\sin\theta_2}{\chi\sin\theta_1}\cdot\frac{\gamma_{MOS}}{\gamma_{MO}}\right)^3 f\left(\theta_1\right)$$ (3.27)

where γ_{MOS} is the surface energy of interfacial metal oxide-metal substrate, χ is defined as the Pilling–Bedworth ratio ($\chi = V_{OX}/V_M$). V_{OX} and V_M are specified as molar volume of oxide and metal, respectively (Revie and Uhlig 2008, Zhou 2009). The mostly metal oxide arrangement has regularly this ratio more than 1 ($\chi > 1$) (Revie and Uhlig 2008). To consider the free energy barrier height of heterogenous nucleation with two shapes, the total Gibbs free energy versus radius is comparatively plotted in Figure 3.10c by assuming the arbitrary parameters with $\gamma_{MOS}/\gamma_{MO} = 1/4$ and $\chi = 1.5$ (Zhou 2009). It can be clearly seen that the critical barrier energy of cap-shape nucleation is lower than that of double cap-shape nucleation. This indicates that the heterogenous nucleation with a cap-shape format needs less activation energy to spontaneously grow. This shape supports the surface diffusivity of either metal or oxygen ions to form 2-D nuclei and the metal oxide film are effectively created (Zhou 2009). Usually, 2-D nucleation process cannot grow the nanostructures because 2-D nucleation comes from major metal ions movement which it has a diffusion limit due to ions breaking existence at neighboring bonds. The metal ions cannot move openly on the oxide layer (Zhou 2009). Interestingly, G. Zhou has found that the 3-D nuclei are the dominantly optional configuration forming a Cu_2O island on a consumed Cu metal substrate (Zhou and Yang 2004, Zhou et al. 2008, Zhou 2009). Also, it still corresponds to our previous report with CuO nanowires synthesized by a thermal oxidation technique (Raksa et al. 2009). We have successfully synthesized the copper oxide nanowires from heating a copper plate of a thickness of 0.1 mm in a tube furnace at a working temperature of 600°C for 6 hr. The obtained copper product exhibited the copper metal consumption into the copper substrate, they consisted of layer by layer from bottom to top of the product with an observation of Cu, Cu_2O, and CuO nanowires (Raksa et al. 2009, Choopun et al. 2010). These confirmed results show the dominantly grown 3-D nuclei format, although 2-D nuclei have the critical barrier energy lower than that of 3-D nuclei. This is possibly due to the higher kinetic energy of oxygen ions activated at high working temperatures, migrating as translational motion across the metal layer (Wintterlin et al. 1996, Carley et al. 1999, Schmid et al. 2001, Ciacchi and Payne 2004, Zhou 2009). Moreover, at high working temperatures the activation energy is higher to overcome the critical barrier energy resulting in spontaneous growth of nuclei. Thus, it suggests that the 3-D nuclei from double cap-shape nucleation are the main key in 1-D nanostructure formation.

3.4.1.1.3 Step 3: Nuclei Arrangement

Subsequently, the metal oxide nuclei have already been created on the metal substrate by overcoming the critical energy barrier height. Then, they spontaneously grow to form a metal oxide crystal. Naturally, after the most metal oxide nuclei are formed, they try to consistently adjust the minimize surface energy with nuclei arrangement mechanism (Choopun et al. 2010, Cai et al. 2014). The confirmation is evidently considered from the specific surface energy of selected ZnO given by 1.2, 1.4, and 1.6 J/m² for {0001}, {11$\bar{2}$0}, {10$\bar{1}$0} facets, respectively (Jiang et al. 2002). The formed ZnO nuclei will grow continuously until their length and size are in a stable state. After the growth process was finished, the mostly obtained nanostructures of metal oxide exhibited growth direction corresponding to the lowest miller index (Choopun et al. 2010). Because they have a small specific surface energy. Furthermore, the growth direction also depends on the growth velocity, of which the lower miller index will have a faster growth velocity (Cai et al. 2014). For instance, the fastest growth direction of ZnO nanorods is [0001] direction (Cai et al. 2014). In addition, the density of nuclei is microscopically explained with nuclei probability, which it is used to distinguish the nuclei numbers converting into the stable nuclei. Here, the nuclei probability (P_N) is written by (Choopun et al. 2010):

$$P_N = \frac{N}{N_0} = \exp\left(-\frac{\Delta G_N^*}{k_B T}\right) \qquad (3.28)$$

where N is the surface nuclei concentrations that attained the critical size, N_0 is the total embryos concentrations of atom/ions contacted on the substrate, k_B and T are Boltzmann constant and absolute temperature, respectively.

From equation (3.28), the nuclei concentrations (N) are used to investigate the nuclei numbers, which readily advance to form the metal oxide nanostructures while the nuclei probability is utilized to explain the nuclei density with a temperature effect. It was found that nuclei probability (P_N) increases as temperature increases at the same constant surface energy and Gibbs free energy (Δg_v). The experimental confirmation is obtained from ZnO nanostructures density synthesized at various temperatures by the thermal oxidation approach, and it found that the ZnO nanowire numbers increase with increasing heating temperatures above melting point of zinc substance (in range of 500°C–800°C) (Choopun et al. 2010).

3.4.1.1.4 Step 4: 1-D Nanostructures Formation

The 1-D nanostructures formation spontaneously occur owing to the surface pressure difference between oxide nuclei and metal base substrate. Typically, the surface pressure difference depends on the force; therefore, the driving force is produced through the nuclei surface. As we know, the pressure difference relates to the surface energy and nuclei radius as the Laplace equation (Stolen et al. 2004, Choopun et al. 2010, Wongrat et al. 2011):

$$\Delta P = P_l - P_s = \frac{2\gamma_{sf}}{r_s} \tag{3.29}$$

where P_l and P_s are the pressures of liquid and solid phase, respectively. γ_{sf} is the surface energy of nuclei at the interface of oxide and surrounding oxygen. r_s is the solid radius of oxide nuclei.

From equation (3.29), the surface pressure difference occurs at the interface of oxide nuclei (with a solid phase) and metal base substrate (with a liquid phase). The solid phase comes from the nuclei radius, larger than that of the critical radius, conducting to the spontaneous growth of nuclei, and the stable solid phase of oxide nuclei is completely reached. However, the metal base substrate is still a liquid phase because the heating temperatures are mostly above the melting point of metal. For example, the melting point of zinc is about 419.6°C, but the heating temperatures are routinely obtained above 500°C as investigated in our previous reports (Wongrat et al. 2009, 2012, 2016, Wongrat and Choopun 2011). As mentioned above, this driving force determined from the surface pressure difference can push the oxide nuclei away from the metal base substrate, meanwhile the metal/oxygen ions constantly diffuse into the stable oxide nuclei. This mechanism leads to the nanostructures growth along their length. The driving force due to pressure difference is further dominant at the nuclei radius below 100 nm (Choopun et al. 2010). For summary, the schematic diagram of the growth mechanism of metal oxide nanostructures prepared via the thermal heating oxidation technique in the furnace for oxidization of zinc metal layer is illustrated in Figure 3.11.

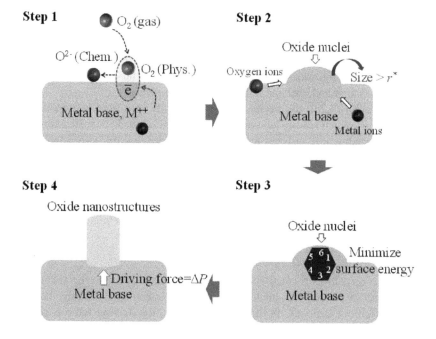

FIGURE 3.11 Schematic diagram of growth mechanism with 4 steps of metal oxide nanostructures synthesized via thermal oxidation of zinc pasted on the substrate.

3.4.1.2 Oxidization of Zinc Metal Gas via CVD Process

For oxidization of zinc metal gas via the CVD process, the growth mechanism is principally explained by self-catalytic vapor-liquid-solid (SC-VLS) mechanism that is based on VLS mechanism. A previously discussed VLS mechanism comes from the metal catalyst as the early standard constituent. But the SC-VLS mechanism is ideally preferred without a catalyst metal. The formerly generated substantial particles migrate along a substrate where the nucleation site is steadily formed, revealing the subsequent nanowire growth (Purushothaman et al. 2012). The detailed growth kinetic of ZnO with 1-D nanostructures are described in the following steps:

3.4.1.2.1 Step 1: Metal Liquid Droplet Formation

Since the 1-D nanostructures were created at the temperatures above a melting point of the metallic zinc, the growth kinetic explanation begins with the temperatures dependence on the vapor pressure of substances, especially the solid phase source. Although the solid phase source has the lower vapor pressure at a normal atmospheric surrounding, when the temperatures are raised, the vapor pressures are increased too. This can drive up the solid phase of metallic zinc to become the vapor phase and this phenomenon is dominant at the lower pressure as illustrated in Figure 3.12 (step 1). Thereby, the metallic vapor is transported across the lower temperatures region to encounter the solid substrate in the direction of a downstream line of carrier gas, where the operation temperatures are lower than that of the metallic source (T of metal source more than T substrate, $T_1 > T_2$, also see in Figure 3.12). At the lower substrate temperature, consequently, the ratio of actual vapor concentration to equilibrium concentration as the supersaturation ratio is further established (Kimura and Maruyama 2002, Choi 2012). As a result, the liquid phase of metallic zinc is immediately condensed on the formerly obtained substrate. The clusters of liquid droplet are constantly produced by the nucleation mechanism. The small clusters will dissolve and cannot grow to become the metallic droplets, while the larger clusters can grow preferentially to become the metallic droplets. Importantly, the preferred shape of nucleation process is heterogeneous nuclei with a cap-shape (see in Figure 3.10a) molding with a metallic liquid precipitation on the solid substrate. Then the created stabile metallic liquid droplet functions as the metal catalyst, likely acting as the initial seeds to continuously grow the 1-D nanostructures (Jian et al. 2006). Mainly, the stable nuclei of the originally formed metallic droplet has the size larger than that of a critical radius, which relates to the critical energy barrier height of the droplet as formerly given in equation (3.23).

3.4.1.2.2 Step 2: Metallic Droplet Oxidization

Once, the O_2 reactant gas is flowed into the reduced pressure system, it is exposed to the metallic zinc liquid droplet. The droplet then encounters O_2 reactant gas and is oxidized in a thermodynamically equilibrium condition. That metallic zinc droplet is converted to a preferred zinc sub-oxide, ZnO_x which is still the liquid phase (Jian et al. 2006) due to lack of oxygen in the system as shown in Figure 3.12 (step 2). At the initially reached stage, the oxygen molecules filled into the established system will gradually react to the metal surface in which x concentration of oxygen increases reasonably from a range of 0–0.5 concentrations. From the phase diagram,

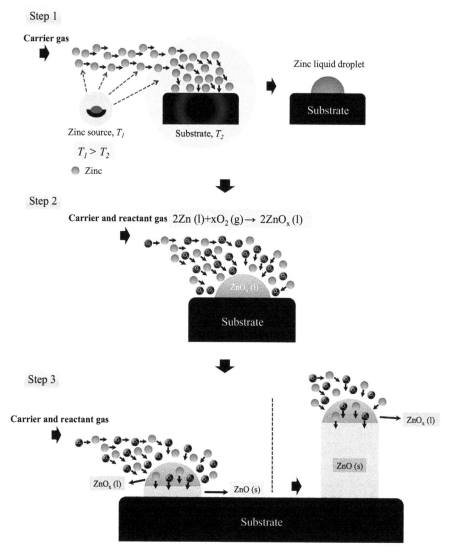

FIGURE 3.12 Schematic diagram of ZnO 1-D nanostructures formation synthesized via chemical vapor deposition from the metallic zinc with a 3-step approach.

this region clearly expresses that the ZnO_x is still to be a liquid phase because its melting point is equal to metallic zinc substance (Ellmer KaK 2008). The chemical reaction is generally written as:

$$2Zn(l) + xO_2(g) \rightarrow 2ZnOx(l) \qquad (3.30)$$

Interestingly, the ZnO_x liquid droplet will adjust the minimized surface energy to continuously grow ZnO solid phase in the next step.

3.4.1.2.3   Step 3: 1-D Nanostructures Formation

In a final step, the nanostructure formation comes from the chemical potential difference producing the driving force during a growth process (Schmidt et al. 2007, Koto 2015). From step 2, the precursor metallic nuclei of ZnO_x is still a liquid phase. Typically, the outside of an entire obtained system of the nuclei consists of vapor surrounding both zinc and oxygen vapor where it possesses the chemical potential μ_v. Whereas, the system of ZnO_x liquid droplet and solid substrate belong to the chemical potential μ_l and μ_0^s, respectively. Eventually, if the 1-D nanostructures are absolutely formed as solid phase, its chemical potential belongs to μ_s. Repeatedly, an explanation from step 2, the obtained system includes vapor surrounding, ZnO_x liquid droplet and substrate regions. The chemical potential difference leads to a thermodynamically driving force pushing the 1-D nanostructures out from the substrate. Essentially, the chemical potential parameter is significantly expected with a magnitude arrangement in the order of $\mu_v > \mu_l > \mu_0^s$. Therefore, when those systems are closely connected, the mass transfer phenomenon occurs by the diffusion mechanism to adjust the equivalent chemical potential. It can be seen that the chemical potential of liquid droplet μ_l is higher than that of solid substrate μ_0^s. This causes crystallization creation via the droplet supersaturation effect (Schmidt et al. 2007, Koto 2015). Subsequently, the solid ZnO with 1-D nanostructure seed is formed at the middle of a liquid droplet and the substrate as expressed in Figure 3.12 (left-hand of step 3). To continuously explain the nanostructures length growth, the chemical potential is still valid with an order of $\mu_v > \mu_l > \mu_s > \mu_0^s$, which the added μ_s is the chemical potential of an early established ZnO nanostructure with a solid phase. The continuous vapor of zinc and oxygen supply are constantly encountered by the zinc liquid droplet and they incorporate by cracking the surface of the liquid droplet (Schmidt et al. 2007). Then they diffuse through the liquid droplet and progressively pass to the liquid-solid boundary. Lastly, the solid phase as crystallization is completely augmented in height direction at the liquid-solid interfacial region (Kwon 2006, Schmidt et al. 2007, Li et al. 2014) as observed in Figure 3.12 (right-hand of step 3). The ZnO 1-D nanostructures are ultimately advanced in the height direction.

3.4.2   GROWTH KINETICS OF 3-D GROWTH ZnO NANOWIRE FOR CURRENT HEATING OXIDATION TECHNIQUE

The previous section presented growth mechanism explained that the growth of wire-like and belt-like nanostructures started with the equation of nucleation probability based on the case of supersaturation ratio parameter. The growth of nanostructures previously presented is on the substrates, suggesting two dimensions of growth kinetics. However, in this case, the growth of ZnO nanowire is not grown on substrate, implying a different growth kinetics. The growth kinetics of ZnO nanowire as three dimensions of growth kinetics with growth direction still preferred along the *c*-axis is suggested.

In general, the growth mechanism of the belt-like, wire-like nanostructures can be explained by the kinetics of anisotropic growth via a vapor–solid mechanism. The nucleation probability P on the surface of a nanostructure is given by modified equation (3.27) and is generally written by (Dai et al. 2003):

$$P = B \exp\left(\frac{-\pi\sigma^2}{k_B{}^2 T^2 \ln(\alpha)}\right) \qquad (3.31)$$

where B is a parameter constant, σ is the surface energy of the solid tetrapod, k_B is the Boltzmann's constant, T is the absolute temperature, and α is the supersaturation ratio between the actual vapor pressure and the equilibrium vapor pressure corresponding to temperature T (usually, $\alpha > 1$). For the same supersaturation ratio, the growth direction of a 1-D nanostructure is normally along a low crystal index, for example [0001], [10$\bar{1}$0], [11$\bar{2}$0] etc. The [0001] c-axis direction of ZnO has the lowest surface energy, σ_{min}, suggesting the highest nucleation probability P. Therefore, the c-axis of the ZnO structure exhibits the fastest growth rate and the highest possibility for the growth direction.

For 3-D growth ZnO nanowire, the case study here is the ZnO tetrapod structure from current heating and microwave-assisted thermal oxidation technique. The growth mechanism is started at the center core of a tetrapod and consisted of four grains of a hexagonal structure (Hongsith et al. 2009). These four grains are derived from the basic hexagonal structure of ZnO from a unit cell of ZnO hexagonal structure as shown in Figure 3.14a. It can be seen that the zinc ions are surrounded by a tetrahedral of oxygen ions and vice-versa. By considering a zinc atom only, the tetrahedral structure can be drawn as shown in Figure 3.14b having a theoretically angle between each axis of 109.5°. Therefore, this tetrahedral structure (Figure 3.14b) is the starting nuclei, acting as a seed of 3-D growth ZnO nanowire to form a tetrapod structure (Ronning et al. 2005). This tetrahedral structure is the preferred nucleation sites to form 3-D structure due to low surface energy in <0001>$_{ZnO}$ direction. The 3-D structure can form by accumulation and then condensation of ZnO vapor due to supersaturation conditions, growing along the c-axis as shown in Figure 3.13.

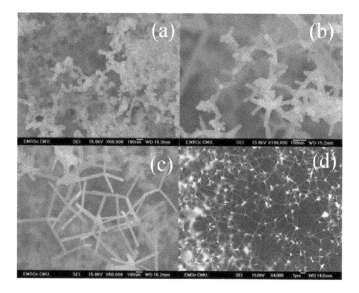

FIGURE 3.13   Tetrapod growth evolution; (a) the starting tetrahedral nuclei (b)–(c) growing along c-axis and (d) the connecting with neighbor tetrapod and forming network.

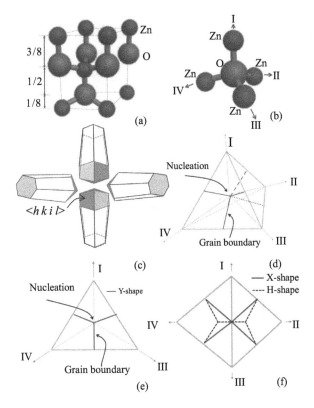

FIGURE 3.14 (a) the unit cell of ZnO structure; (b) the tetrahedral construction of Zn atom related with ZnO unit cell; (c) four legs of T-ZnO with wedge-like shape join together and form four grain boundaries; (d) the grain boundary image of tetrapod nucleation; (e) side view image on triangle plane; and (f) the edge of tetrapod nucleation.

The growth direction can be along I, II, III, and IV of zinc atoms, which has an angle between each axis of 109.5°, forming grain boundaries ideally on [hkil]$_{ZnO}$ as shown in Figure 3.14c and d, and the growth to c-axis nanowires becomes the four-legged structure of the tetrapod. In the case of 2-D growth on substrate, the structure can grow only along one c-axis direction, because the other three c-axis directions are obstructed by the substrate. Thus, the nanowire is formed as previously presented. At the center core, four legs of the tetrapod, having a wedge-like shape, join together and form four grain boundaries as shown in Figure 3.14c. These looks similarly to the TEM image in Figure 3.5d. It should be noted that the tip of the wedge-like structure is sharp (Hongsith et al. 2009). Figure 3.14e and f show the schematic view of a tetrahedral nucleus with four hexagonal ZnO grains, for Figure 3.14e from bottom view, and Figure 3.14f side view along an edge between triangle plans, respectively. The kinked Y-shape and X-shape of the grain boundaries can be observed in these views accordingly to TEM real images as shown in Figure 3.15a and b, respectively.

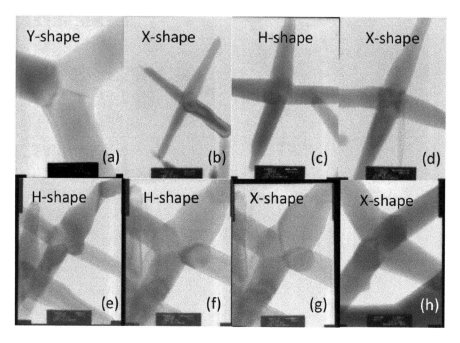

FIGURE 3.15 (a–h) TEM bright field image of ZnO tetrapods showed the grain boundary of each leg of tetrapod in the Y-shape, X-shape and H-shape.

According to the model described of tetrahedral nucleation, the angle between ideal grains boundaries is 120° and 90° for Y-shape and X-shape, respectively. But we have observed the slightly different angles of 120°, 130°, and 110° for Y-shape and 92°–98.5° for X-shape due to tilting in TEM (Figure 3.15). However, during 3-D formation the growth along *ab* plane still exists even though it is obstructed by the other adjacent grains. Some legs may therefore grow faster than the others due to non-uniform growth rate parameters (σ, α, and T) as discussed above during growth process. This can cause a formation of imperfect tetrahedral nucleation where grain boundaries do not meet at a point. Therefore, the kinked H-shape projection image would be observed similarly to the Ronning model (Ronning et al. 2005) as shown by dash lines in Figure 3.14f. In addition, all the angles for the ideal *T*-ZnO grown from tetrahedron nucleation in this model are 109.5°. However, *T*-ZnO particles with a set of angles of 109.5° have not been reported. Fujii and co-worker measured the angles between the *T*-ZnO legs using optical microscope and reported the principal angles of 102°, 110°, 116° and 129° (Fujiia et al. 1993).

In addition, we focus on the angles between grain boundary plane and *c*-axis direction of hexagonal structure space group of P6₃mc by using stereo graphic projection for ten orders of mirror index {*hkil*} and shown all possible planes of twin grain boundary in Figure 3.16.

From this analysis, the angle between *T*-ZnO legs varies from 3°–79° depending on the mirror index of the ZnO hexagonal structure as shown in Figure 3.16a. The circle lines 1–4 in Figure 3.16b were focused on plane from earlier reports

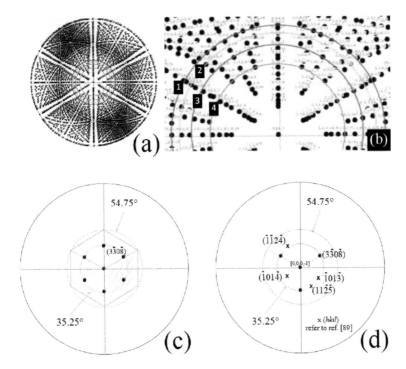

FIGURE 3.16 Stereographic projection for ten orders of mirror index (a) varies from 3°–79° focus the angles between grain boundary plane and *c*-axis direction (b) the circle line of focus plane from earlier reports (c) twin grain boundary plane index of < 30$\bar{3}$8 > in perfect ZnO tetrapod and (d) past discovery of twin grain boundary plane index.

(Nishio et al. 1997). For ideal T-ZnO, the angle should be about 54.75°, corresponding to a twin grain boundary plane index of < 30$\bar{3}$8 > in the ZnO hexagonal structure as shown in Figure 3.16c. However, Nishio and co-workers suggested < 11$\bar{2}$2 > and < 11$\bar{2}$4 > twin plane index and the angles distributions between *T*-ZnO legs around 102° to 130° have been discussed (Nishio et al. 1997). This can be explained by our kinetics of anisotropic growth. The lower index of twin plane index exhibit higher probability. However, another plane index cannot be neglected because they have possible probability, for example, < 10$\bar{1}$3 > twin plane exhibit legs angle about of 116° or < 20$\bar{2}$5 > twin plane about of 106°, etc. So, from this growth mechanism the legs also transform to the imperfect *T*-ZnO.

3.4.3 GROWTH KINETICS OF ITN-ZNO FOR MICROWAVE-ASSISTED THERMAL OXIDATION TECHNIQUE AT ULTRA-HIGH SUPERSATURATION RATIO

As previously presented, we have proposed the growth model of ZnO tetrapods by the kinetics of anisotropic growth via a vapor–solid mechanism with 3-D tetrahedral nuclei as a seed. The 3-D tetrahedral nuclei form by accumulation and then, condensation of ZnO vapor due to a supersaturation condition. These 3-D nuclei first grow along c-axis in four various tetrahedral axes due to symmetrical structure before finally forming the four-leg structure of the ZnO tetrapod. However, at ultra-high supersaturation ratio conditions, we realized a novel ZnO network in our technique. The arm of the ZnO tetrapod is perfectly connected with neighbor tetrapods, forming networks that we call "interlinked tetrapod network ZnO, ITN-ZnO" (Thepnurat et al. 2015).

To understand the growth kinetic of interlinked tetrapod network, TEM and crystal structures analysis is performed for the explanation. Figure 3.17a shows a brightfield (BF) image of the two interlinked tetrapod networks between tetrapod "A" and tetrapod "B." Figure 3.17b shows the high magnification BF image of a leg marked I of tetrapod A connecting with a leg marked II of tetrapod B together. From selected electron diffraction and trace analysis, it is found that the legs of tetrapod A and B grow along c-axis direction, but the growth direction is in reverse direction for each other. Figure 3.17c presents the HR-TEM image on the twin grain boundary connection area of tetrapod leg A and B, which grow oppositely along c-axis and minus c-axis direction. The spacing of the fringes is determined to be about 0.26 nm or

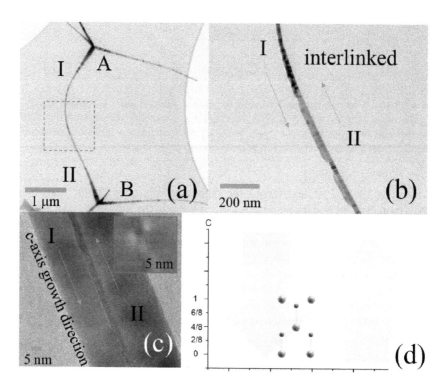

FIGURE 3.17 TEM results that focus on the interlinked tetrapod network ZnO, (a) bright field TEM image of two interlinked tetrapod network ZnO with label A and B, (b) high magnification image between legs I and II with the inset SADP which corresponding to both c-axis growth direction, (c) high resolution TEM image at the boundary between leg I and II with a line shift of atomic plane and (d) atomic structure model for interlinking of tetrapod legs.

c/2 where *c* is the lattice parameter in ZnO hexagonal structure equal to 5.20Å. It is found that the fringe lines of leg A and B are not perfectly connected together. There is an overlap mismatch about *c*/8.

To explain the mismatch spacing, it is best to consider the basic hexagonal wurtzite structure of a unit cell model of ZnO in Figure 3.17d that is composed of an alternating Zn and O atomic binding layers along the *c*-axis. There is the Zn atom layer to the bottom of the model with O atom at the next layer of *c*/8 and the distance from the second-floor layers to the base layers of Zn is *c*/2 and next with O atom layer of 5*c*/8, respectively. When a unit cell A is connected by unit cell B with opposite growth direction and the plane of the incoming is {01-10} plane of their surface as in ITN-ZnO, a binding must be at the position of ionic bonding between Zn and O atom due to coulomb interaction. Thus, a Zn atom layer of A is mismatched with a Zn atom layer of B and theoretically, it is equal to *c*/8 as shown in Figure 3.17d. It can be seen that the fringe lines of high-resolution TEM image of leg A and leg B had overlapping lattices of about *c*/8 corresponding to the growth model in Figure 3.17d.

3.5 CONCLUSIONS

In this chapter, we have successfully demonstrated that the thermal oxidation process is a simple and powerful technique for fabricating 1-D metal oxide nanostructures. Also, the fabrication of the 1-D metal oxide nanostructures by thermal oxidation process can be simply and completely explained via growth kinetics in term of thermodynamic parameters. ZnO is given as an example for this demonstration. However, it can be applied and explored to all metal oxide materials, which could be useful for nanofabrication and nanoengineering of novel nanodevices.

ACKNOWLEDGMENTS

Ekasiddh Wongrat would like to thank Thailand Research Fund and University of Phayao for financial support. We would like to thank M. Thepnurat for technical assistance and fruitful discussion.

REFERENCES

Benjwal, P., De, B., Kar, K.K. 2018. 1-D and 2-D morphology of metal cation co-doped (Zn, Mn) TiO₂ and investigation of their photocatalytic activity. *Appl. Surf. Sci.* 427:262–272.
Cabrera, N., and Mott, N.F. 1949. Theory of the oxidation of metals. *Rep. Prog. Phys.* 12:163.
Cai, X., Han, B., Deng, S. et al. 2014. Hydrothermal growth of ZnO nanorods on Zn substrates and their application in degradation of azo dyes under ambient conditions. *CrystEngComm* 16:7761–7770.
Carley, A.F., Davies, P.R., Kulkarni, G.U., Roberts, M.W. 1999. Oxygen chemisorption at Cu(110) at 120 K: Dimers, clusters and mono-atomic oxygen states. *Catal. Lett.* 58:93–97.
Chernavskii, P.A., Peskov, N.V., Mugtasimov, A.V., Lunin, V.V. 2007. Oxidation of metal nanoparticles: Experiment and model. *Russ. J. Phys. Chem. B* 1:394–411.
Choi, H.J. 2012. Vapor–liquid–solid growth of semiconductor nanowires. In: *Semiconductor Nanostructures for Optoelectronic Devices: Processing, Characterization and Applications*, G.C. Yi. (Ed.). Springer, Berlin, Germany.

Choopun, S., Hongsith, N., Wongrat, E. 2010. Metal-oxide nanowires by thermal oxidation reaction technique. In: *Nanowires*, P. Prete. (Ed.). InTech, Rijeka, Croatia.

Ciacchi, L.C., and Payne, M.C. 2004. "Hot-Atom" O_2 dissociation and oxide nucleation on Al(111). *Phys. Rev. Lett.* 92:176104.

Dai, Z.R., Pan, Z.W., Wang, Z.L. 2003. Novel nanostructures of functional oxides synthesized by thermal evaporation. *Adv. Funct. Mater.* 13:9–24.

Das, D., Datta, A.K., Kumbhakar, D.V., Ghosh, B., Pramanik, A., Gupta, S. 2017. Conditional optimisation of wet chemical synthesis for pioneered ZnO nanostructures. *Nano-Struct. Nano-Objects* 9:26–30.

Delalu, H., Vignalou, J.R., Elkhatib, M., Metz, R. 2000. Kinetics and modeling of diffusion phenomena occurring during the complete oxidation of zinc powder: Influence of granulometry, temperature and relative humidity of the oxidizing fluid. *Solid State Sci.* 2:229–235.

Devan, R.S., Patil, R.A., Lin, J.H., Ma, Y.R. 2012. One-dimensional metal-oxide nanostructures: Recent developments in synthesis, characterization, and applications. *Adv. Funct. Mater.* 22:3326–3370.

Ellmer KaK, A. 2008. *Transparent Conductive ZnO: Basics and Applications in Thin Film Solar Cells*, B. Rech (Ed.). Springer, Berlin, Germany.

Fan, H.J., Scholz, R., Kolb, F.M., Zacharias, M. 2004. Two-dimensional dendritic ZnO nanowires from oxidation of Zn microcrystals. *Appl. Phys. Lett.* 85:4142–4144.

Fehlner, F.P., and Mott, N.F. 1970. Low-temperature oxidation. *Oxid. Met.* 2:59–99.

Fletcher, N.H. 1958. Size effect in heterogeneous nucleation. *J. Chem. Phys.* 29:572–576.

Fu, Q., and Wagner, T. 2007. Interaction of nanostructured metal overlayers with oxide surfaces. *Surf. Sci. Rep.* 62:431–498.

Fujiia, M., Iwanagab, H., Ichiharac, M., Takeuchic, S. 1993. Structure of tetrapod-like ZnO crystals. *J. Cryst. Growth* 128:1095–1098.

García, J.F., Sánchez, S., Metz, R. 2008. Complete oxidation of zinc powder. Validation of kinetics models. *Oxid. Met.* 69:317–325.

Gaskell, D.V. 2003. *Introduction to the Thermodynamics of Materials*, 4 ed. Taylor & Francis Group, Great Britain.

Grela, M.A., and Colussi, A.J. 1996. Kinetics of stochastic charge transfer and recombination events in semiconductor colloids. Relevance to photocatalysis efficiency. *J. Phys. Chem.* 100:18214–18221.

Hongsith, N., Chairuangsri, T., Phaechamud, T., Choopun, S. 2009. Growth kinetic and characterization of tetrapod ZnO nanostructures. *Solid State Commun.* 149:1184–1187.

Hongsith, N., Wongrat, E., Kerdcharoen, T., Choopun, S. 2010. Sensor response formula for sensor based on ZnO nanostructures. *Sens. Actuators B: Chem.* 144:67–72.

Hsueh, T.J., and Hsu, C.L. 2008. Fabrication of gas sensing devices with ZnO nanostructure by the low-temperature oxidation of zinc particles. *Sens. Actuators B: Chem.* 131:572–576.

Hu, P., Han, N., Zhang, D., Ho, J.C., Chen, Y. 2012. Highly formaldehyde-sensitive, transition-metal doped ZnO nanorods prepared by plasma-enhanced chemical vapor deposition. *Sens. Actuators B: Chem.* 169:74–80.

Jian, J.K., Wang, C., Zhang, Z.H., Chen, X.L., Xu, L.H., Wang, T.M. 2006. Necktie-like ZnO nanobelts grown by a self-catalytic VLS process. *Mater. Lett.* 60:3809–3812.

Jiang, X., Jia, C.L., Szyszka, B. 2002. Manufacture of specific structure of aluminum-doped zinc oxide films by patterning the substrate surface. *Appl. Phys. Lett.* 80:3090–3092.

Johnson, D.L., and Stracher, G.B. 1995. *Thermodynamic Loop Applications in Materials Systems*. The Minerals, Metals and Materials Society, Warrendale, PA.

Kalyanaraman, R. 2008. Nucleation energetics during homogeneous solidification in elemental metallic liquids. *J. Appl. Phys.* 104:033506.

Kang, S., Chatterjee, U., Um, D.Y., Seo, I.S., Lee, C.R. 2017. Growth and characterization of n-AlGaN 1-D structures with varying Al composition using u-GaN seeds. *J. Cryst. Growth* 480:108–114.

Kim, T.W., Kawazoe, T., Yamazaki, S., Ohtsu, M., Sekiguchi, T. 2004. Low-temperature orientation-selective growth and ultraviolet emission of single-crystal ZnO nanowires. *Appl. Phys. Lett.* 84:3358–3360.

Kimura, T., and Maruyama, S. 2002. Molecular dynamics simulation of heterogeneous nucleation of a liquid droplet on a solid surface. *Microscale Thermophys. Eng.* 6:3–13.

Kofstad, P. 1988. *High Temperature Corrosion*. Elsevier Applied Science Publishers, London, UK.

Kong, J., Fan, D., Zhu, Y. 2012. Synthesis and Raman spectra of hammer-shaped ZnO nanostructures via thermal evaporation growth. *Mater. Sci. Semicond. Process.* 15:258–263.

Koto, M. 2015. Thermodynamics and kinetics of the growth mechanism of vapor–liquid–solid grown nanowires. *J. Cryst. Growth* 424:49–54.

Kwon, S.J. 2006. Theoretical analysis of non-catalytic growth of nanorods on a substrate. *J. Phys. Chem. B* 110:3876–3882.

Lad, R.J. 1995. Interactions at metal/oxide and oxide/oxide interfaces studied by ultrathin film growth on single-crystal oxide substrates. *Surf. Rev. Lett.* 02:109–126.

Li, X.L., Wang, C.X., Yang, G.W. 2014. Thermodynamic theory of growth of nanostructures. *Prog. Mater Sci.* 64:121–199.

Liang, H.Q., Pan, L.Z., Liu, Z.J. 2008. Synthesis and photoluminescence properties of ZnO nanowires and nanorods by thermal oxidation of Zn precursors. *Mater. Lett.* 62:1797–1800.

Liu, X.Y. 1999. A new kinetic model for three-dimensional heterogeneous nucleation. *J. Chem. Phys.* 111:1628–1635.

Liu, X.Y. 2000. Heterogeneous nucleation or homogeneous nucleation. *J. Chem. Phys.* 112:9949–9955.

Liu, X.Y. 2001. Interfacial process of nucleation and molecular nucleation templator. *Appl. Phys. Lett.* 79:39–41.

Liu, X.Y. 2002a. Effect of foreign particles: A comprehensive understanding of 3D heterogeneous nucleation. *J. Cryst. Growth* 237–239:1806–1812.

Liu, X.Y. 2002b. Heterogeneous 2D nucleation-induced surface instability. *J. Cryst. Growth* 237–239:101–105.

Machado, C., Aidel, S., Elkhatib, M., Delalu, H., Metz, R. 2002. Validation of a kinetic model of diffusion for complete oxidation of bismuth powder: Influence of granulometry and temperature. *Solid State Ionics* 149:147–152.

Martin, M., and Fromm, E. 1997. Low-temperature oxidation of metal surfaces. *J. Alloys Compd.* 258:7–16.

Mihailova, I., Gerbreders, V., Tamanis, E., Sledevskis, E., Viter, R., Sarajevs, P. 2013. Synthesis of ZnO nanoneedles by thermal oxidation of Zn thin films. *J. Non-Cryst. Solids* 377:212–216.

Mott, N.F. 1940. Oxidation of metals and the formation of protective films. *Nature* 145:996.

Nishio, K., Isshiki, T., Kitano, M., Shiojiri, M. 1997. Structure and growth mechanism of tetrapod-like ZnO particles. *Philos. Mag. A* 76:889–904.

Pilling, N.B., and Bedworth, R.E. 1923. The oxidation of metals at high temperatures. *J. Inst. Metals* 29:529.

Purushothaman, V., Ramakrishnan, V., Jeganathan, K. 2012. Interplay of VLS and VS growth mechanism for GaN nanowires by a self-catalytic approach. *RSC Adv.* 2:4802–4806.

Qurashi, A., El-Maghraby, E.M., Yamazaki, T., Kikuta, T. 2010. Catalyst supported growth of In_2O_3 nanostructures and their hydrogen gas sensing properties. *Sens. Actuators B: Chem.* 147:48–54.

Raksa, P., Gardchareon, A., Chairuangsri, T., Mangkorntong, P., Mangkorntong, N., Choopun, S. 2009. Ethanol sensing properties of CuO nanowires prepared by an oxidation reaction. *Ceram. Int.* 35:649–652.

Rapp, R.A. 1984. The high temperature oxidation of metals forming cation-diffusing scales. *Metall. Trans. B* 15:195–212.

Ren, S., Bai, Y.F., Chen, J. et al. 2007. Catalyst-free synthesis of ZnO nanowire arrays on zinc substrate by low temperature thermal oxidation. *Mater. Lett.* 61:666–670.

Revie, R.W., and Uhlig, H.H. 2008. *Corrosion and Corrosion Control: An Introduction to Corrosion Science and Engineering.* Wiley, New York.

Ronning, C., Shang, N.G., Gerhards, I., Hofsäss, H. 2005. Nucleation mechanism of the seed of tetrapod ZnO nanostructures. *J. Appl. Phys.* 98:034307.

Samal, S. 2016. High-temperature oxidation of metals. In: *High Temperature Corrosion,* Z. Ahmad. (Ed.). InTech, Rijeka, Croatia.

Samanta, P.K., and Saha, A. 2015. Wet chemical synthesis of ZnO nanoflakes and photoluminescence. photoluminescence. *Optik–Int. J. Light Electron Opt.* 126:3786–3788.

Schmid, M., Leonardelli, G., Tscheließnig, R., Biedermann, A., Varga, P. 2001. Oxygen adsorption on Al(111): Low transient mobility. *Surf. Sci.* 478:L355–L62.

Schmidt, V., Senz, S., Gösele, U. 2007. Diameter dependence of the growth velocity of silicon nanowires synthesized via the vapor-liquid-solid mechanism. *Phys. Rev. B* 75:045335.

Schroeder, P., Kast, M., Halwax, E., Edtmaier, C., Bethge, O., Brückl, H. 2009. Morphology alterations during postsynthesis oxidation of Zn nanowires. *J. Appl. Phys.* 105:104307.

Sekar, A., Kim, S.H., Umar, A., Hahn, Y.B. 2005. Catalyst-free synthesis of ZnO nanowires on Si by oxidation of Zn powders. *J. Cryst. Growth* 277:471–478.

Shen, J.J., Zhu, T.J., Zhao, X.B., Zhang, S.N., Yang, S.H., Yin, Z.Z. 2010. Recrystallization induced in situ nanostructures in bulk bismuth antimony tellurides: A simple top down route and improved thermoelectric properties. *Energy Environ. Sci.* 3:1519–1523.

Shewmon, P. 1989. *Diffusion in Solids.* A Publication of The Minerals, Metal & Materials Society, Warrendale, PA.

Stolen, S., Grande, T., Allan, N.L. 2004. *Chemical Thermodynamics of Materials Macroscopic and Microscopic Aspects.* John Wiley & Sons, Chichester, UK.

Tammann, G. 1920. Über Anlauffarben von Metallen. *Zeitschrift für anorganische und allgemeine Chemie* 111:78–89.

Thepnurat, M., Chairuangsri, T., Hongsith, N., Ruankham, P., Choopun, S. 2015. Realization of interlinked ZnO tetrapod networks for UV sensor and room-temperature gas sensor. *ACS Appl. Mater. Interfaces* 7:24177–24184.

Tu, N., Trung, D.Q., Kien, N.D.T., Huy, P.T., Nguyen, D.H. 2017. Effect of substrate temperature on structural and optical properties of ZnO nanostructures grown by thermal evaporation method. *Physica E* 85:174–179.

Wagner, C. 1935. Beitrag zur Theorie des Anlaufvorgangs. *Zeitschrift für Physikalische Chemie* 21B:25.

Wagner, C. 1936. Beitrag zur Theorie des Anlaufvorganges. II. *Zeitschrift für Physikalische Chemie* 32B:447.

Wagner, C., and Grünewald, K. 1938. Beitrag zur Theorie des Anlauf Vorganges. III. *Zeitschrift für Physikalische Chemie* 40B:455.

Wintterlin, J., Schuster, R., Ertl, G. 1996. Existence of a "Hot" atom mechanism for the dissociation O_2 on Pt(111). *Phys. Rev. Lett.* 77:123–126.

Wongrat, E., Chanlek, N., Chueaiarrom, C., Samransuksamer, B., Hongsith, N., Choopun, S. 2016. Low temperature ethanol response enhancement of ZnO nanostructures sensor decorated with gold nanoparticles exposed to UV illumination. *Sens. Actuators A: Phys.* 251:188–197.

Wongrat, E., and Choopun, S. 2011. Sensitivity improvement of ethanol sensor based on ZnO nanostructure by metal impregnation. *Sens. Lett.* 9:936–939.

Wongrat, E., Hongsith, N., Wongratanaphisan, D., Gardchareon, A., Choopun, S. 2012. Control of depletion layer width via amount of AuNPs for sensor response enhancement in ZnO nanostructure sensor. *Sens. Actuators B: Chem.* 171–172:230–237.

Wongrat, E., Pimpang, P., Choopun, S. 2009. Comparative study of ethanol sensor based on gold nanoparticles: ZnO nanostructure and gold: ZnO nanostructure. *Appl. Surf. Sci.* 256:968–971.

Wongrat, E., Umma, K., Gardchareon, A., Wongratanaphisan, D., Choopun, S. 2011. Growth kinetic and characterization of $Mg_xZn_{1-x}O$ nanoneedles synthesized by thermal oxidation. *J. Nanosci. Nanotechnol.* 11:8498–8503.

Wu, Z.W., Tyan, S.L., Chen, H.H. et al. 2017. Temperature-dependent photoluminescence and XPS study of ZnO nanowires grown on flexible Zn foil via thermal oxidation. *Superlattices Microstruct.* 107:38–43.

Xiang, B., Wang, P., Zhang, X. et al. 2007. Rational synthesis of p-type zinc oxide nanowire arrays using simple chemical vapor deposition. *Nano Lett.* 7:323–328.

Xu, C.H., Lui, H.F., Surya, C. 2011. Synthetics of ZnO nanostructures by thermal oxidation in water vapor containing environments. *Mater. Lett.* 65:27–30.

Xu, Q., Hong, R., Chen, X., Wei, J., Wu, Z. 2017. Synthesis of ZnO nanoporous structure materials by two-step thermal oxidation of Zn film. *Ceram. Int.* 43:16391–16394.

Yawong, O., Choopun, S., Mangkorntong, P., Mangkorntong, N. 2005. Zinc oxide nanostructure by oxidization of zinc thin films. *CMU Journal Special Issue on Nanotechnology* 4:7–10.

Young, D.J. 2008. Oxidation of pure metals. In: *Corrosion Series. 1*, D.J. Young. (Ed.). Elsevier: Amsterdam, the Netherlands.

Yu, W., and Pan, C. 2009. Low temperature thermal oxidation synthesis of ZnO nanoneedles and the growth mechanism. *Mater. Chem. Phys.* 115:74–79.

Zhang, Y., Yang, Y., Gu, Y. et al. 2015. Performance and service behavior in 1-D nanostructured energy conversion devices. *Nano Energy* 14:30–48.

Zhou, G. 2009. Nucleation thermodynamics of oxide during metal oxidation. *Appl. Phys. Lett.* 94:201905.

Zhou, G., Dai, W., Yang, J.C. 2008. Crater formation via homoepitaxy of adatoms dislodged from reducing oxide islands on metal surfaces. *Phys. Rev. B* 77:245427.

Zhou, G., and Yang, J.C. 2004. Reduction of Cu_2O islands grown on a Cu(100) surface through vacuum annealing. *Phys. Rev. Lett.* 93:226101.

4 Progress, Perspectives, and Applications of 1-D ZnO Fabrication by Chemical Methods

Tan Wai Kian, Hiroyuki Muto,
Go Kawamura, and Atsunori Matsuda

CONTENTS

4.1 INTRODUCTION

Since 1990, the study on one-dimensional (1-D) nanostructures boomed since Hitachi scientists picked up vapor-liquid-solid (VLS) whisker growth technique for the formation of III-V first p-n junction heterostructured nanowhiskers, with demonstration of good orientation control by adopting R. S. Wagner's equation of VLS growth (Yang et al. 2010). Since then, the development and work on nanowires and nanorods had been increasing significantly which led to various applications. Physical methods such as physical vapor deposition, metalorganic vapor phase epitaxial growth and thermal evaporation usually involve the usage of high temperature and expensive sophisticated equipment, but nevertheless the control and precision at the current stage are better compared to chemical fabrication methods (Li et al. 2012). Thermal oxidation of a metallic Zn layer is deemed as the simplest way to generate complex ZnO nanostructures such as nanorods, nanosheets and nanowires in a relatively short period, but the inability to obtain controlled uniform formation remains its biggest disadvantage (Tan et al. 2011, 2014b). On the other hand, the advantages of chemical fabrication methods are lower fabrication cost, simpler apparatus set-up for large scale production and also the feasibility of low temperature fabrication. As the focus of this chapter is on chemical methods, the fabrication involving physical methods will be only touched on briefly to give the readers an idea of the pro and cons of both methods. This chapter will cover the methodologies via chemical methods for one dimensional growth of ZnO nanostructures such as nanorods, nanowires and nanotubes. The growth mechanisms, morphologies, functional properties and applications will be discussed. As there are several detailed reviews focusing on specific topics on ZnO, this book chapter will provide an insight on the current outlook of 1-D ZnO nanostructures formation via chemical methods.

As ZnO is an environmentally friendly material with a wide band gap of 60 meV and possesses exceptional properties in term of semiconductor and electrochemical phenomena, it is utilized for many applications such as bio/chemical sensors (Bhat et al. 2017), light emitting devices (Pearton 2005, Djurisic and Leung 2006), ferromagnetism (Heo et al. 2004), photocatalyst (Sun et al. 2009, Anandan et al. 2010, Hung et al. 2011), sensors (Ahsanulhaq et al. 2010, Al-Hardan et al. 2010, Chang et al. 2010, Geng et al. 2010, Kim et al. 2011), nanopiezotronics (Wang et al. 2006, Wang 2007) and dye-sensitized solar cells (DSSCs) (Zhang et al. 2009, Ameen et al. 2012, Tan et al. 2013a). ZnO is also a better light-emitting material compared to the widely used GaN due to its easy fabrication of high-quality single crystals as opposed to GaN (Djurišić et al. 2010). The general properties of ZnO are shown in Table 4.1.

ZnO is an amphoteric oxide which tends to crystallize and form its common wurtzite structure that portrays a hexagonal unit cell with the space group C6mc, which exhibits lattice parameters of a = 0.3249 nm and c = 0.5207 nm (Tan et al. 2016). The wurtzite ZnO consists of atoms forming hexagonal-close-pack sub-lattices which stack alternatively along the c-axis. Each Zn^{2+} sub-lattice contains four Zn^{2+} ions and is surrounded by four O^{2-} ions and vice versa, coordinated at the edges of a tetrahedral as

TABLE 4.1
General Properties of ZnO

Molecular weight	Zn: 65.38; O: 16.00; ZnO: 81.38
Lattice	Hexagonal wurtzite
Lattice constants	a = 0.324 nm, c = 0.519 nm, c/a = 1.60
Density	5.78 g/cm^3, or 4.21×10^{22} ZnO molecules/cm^3
Dielectric constant	8.54
Refractive index	2.008
Energy bandgap	3.2 ~ 3.37 eV (Direct band-gap)
Enthalpy of formation	Zn (s) + 1/2O$_2$(g) → ZnO(s) = −83.17 kcal/mol
Solubility of H$_2$O	1.6×10^{-6} g per gram of H$_2$O at 25°C

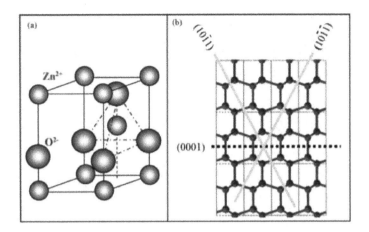

FIGURE 4.1 (a) Common ZnO wurtzite structure and (b) facets of ZnO nanostructures. (Reproduced from Wang, Z.L., *Mater. Sci. Eng. R: Rep.*, 64, 33–71, 2009. With permission.)

shown in Figure 4.1. This tetrahedral coordination will form polar symmetry along the hexagonal axis which induces the effect of piezoelectricity and spontaneous polarization in the ZnO wurtzite crystal. The polarization effect is one of the major factors influencing the crystal growth during the synthesis of ZnO nanostructures especially in the formation of 1-D nanostructures such as ZnO nanowires and nanorods.

4.1.1 ELECTRONIC AND OPTICAL PROPERTIES OF ZNO

In nanoscale form, ZnO exhibits electrical and optical properties differently than its bulk counterpart. ZnO possesses a wide band gap of 3.37 eV and is an intrinsic n-type semiconductor as electrons are excited from ionized zinc interstitials existing in the crystal lattice. Therefore, the design and control of ZnO semiconducting properties could be made by introduction of Fermi levels within the band gap intrinsically or extrinsically. Intrinsic generation of Fermi level generally lies between the conduction band and valence band of the band gap at room temperature and ZnO possesses a low formation energy, large amount of point defects and native donors, which could contribute to its intrinsic Fermi level (Djurišić et al. 2010). As for extrinsic doping of either n-type or p-type materials into ZnO, the Fermi level could be altered to be nearer to the conduction or valence band, respectively. ZnO exhibits conductivity of approximately 10^{-17} to 10^3 $\Omega^{-1}cm^{-1}$ which is dependent on its sample preparations as its semiconducting properties are largely dependent on the defects that are present in the oxide's lattice. Two distinguished defects in its crystal structure are interstitial and substituted Zn atoms. In interstitial defects, the partial reduction by reactive agents causes the Zn atoms to enter the void space by becoming interstitial atoms in the form of Zn, Zn^+ or Zn^{2+}. As for substitutional defect, the Zn atoms in ZnO crystal lattice are replaced by other metals or salts, and these substituted Zn atoms would diffuse to the surface and then vaporize. The conductivity of ZnO can be either increased or decreased depending on the metallic substitution atoms that are induced. Table 4.2 shows general electrical properties of ZnO.

TABLE 4.2

Electrical Properties of Single-Crystal Wurtzite ZnO

Properties	Values
Effective electron mass (M*)	0.24–0.30 m_e
Effective hole mass (m_h*)	0.45–0.60 m_e
Electron Hall mobility at 300 K for n-type (μ_e)	200 $cm^2V^{-1}s^{-1}$
Electron Hall mobility at 300 K for p-type	5–50 $cm^2V^{-1}s^{-1}$
Intrinsic carrier concentration (n)	$<10^6$ cm^{-3}
Background carrier doping	n-type: $\sim10^{20}$ electron cm^{-3}
	p-type: 10^{19} holes cm^{-3}
Optical transmission, T (1/α)	80%–95%

Source: Pearton, S., *Prog. Mater. Sci.*, 50, 293–340, 2005.

ZnO naturally appears in white and its color could be changed by doping, heat-treatment and impurities impregnation. ZnO also absorbs light at ultraviolet (UV) and near UV regions. The properties of ZnO can be characterized using photoluminescence (PL) and Raman spectrometry. Typical PL spectra exhibited by ZnO nanostructures consist of two regions which are the UV and visible region. UV emission exhibited is also termed as deep-level emission which is attributed to the recombination of excitons (electron-hole pair recombination or band-to-band recombination). Highly crystalline ZnO would exhibit strong UV emission. The origin of the green band in the visible region of ZnO is attributed to various impurities and defects (Pearton et al. 2007). The visible emissions are mostly related to the recombination of electrons with oxygen vacancies or with photo-excited holes in the valence band; hence, higher defect concentrations may lead to higher emission intensity in this region (Kurbanov et al. 2011). Feasibility of ZnO nanostructures to exhibit PL emissions at low or room-temperature, as well as the ease of the optical properties alteration by doping especially with rare-earth elements have been further explored for optoelectronic devices application (Tan et al. 2013b).

Raman spectroscopy is commonly used to detect the single-crystal ZnO wurtzite lattice dynamics. In a perfect ZnO wurtzite crystal, the 4 atoms per unit cell will correspond to 12 phonon modes. The modes consist of one longitudinal-acoustic (LA), two transverse-acoustic (TA), three longitudinal-optical (LO) and six transverse-optical (TO) branches. Table 4.3 shows the Raman peaks exhibited by ZnO wurtzite crystal. The A_1 and E_1 branches are Raman and infra-red active, while the two E_2 branches (non-polar) are only Raman active. The E_2-low mode is associated with the vibrations of the Zn sub-lattice whilst the E_2-high mode is associated with the oxygen atoms only. The B_1 branches are always inactive. Both A_1 and E_1 modes are polar and split into TO and LO phonons. Non-polar modes with symmetry E_2 have two frequencies: high (H) and low (L) which is associated with oxygen atoms and with Zn sub-lattices respectively. For a ZnO single crystal, among the eight sets of optical modes, A_1, E_1, and E_2 are Raman active. It is typical to have a

TABLE 4.3

Phonon Modes of Wurtize ZnO at Room Temperature

Phonon Mode	Value (cm⁻¹): Single Crystal
E_2^{low}	101
E_2^{high}	437
$TO(A_1)$	380
$LO(A_1)$	574
$TO(E_1)$	591

Source: Zhang, Q. et al., *Adv. Mater.*, 21, 4087–4108, 2009.

band at around 570 cm⁻¹ i.e. the LO phonon mode which is often associated with the presence of surface defects.

The feasibility of 1-D ZnO nanostructures formation has been the backbone for the novel development of complex nano-architectured ZnO generation such as 3-D framework structures by chemical methods. Nano-patterning and precise controlled growth of ZnO nanostructures on a substrate will also be discussed later in this chapter. Finally, the utilization of these 1-D ZnO nanostructures in innovative technologies breakthrough will also be described.

4.2 FORMATION OF ONE-DIMENSIONAL ZnO NANOSTRUCTURES

Chemical process methods for ZnO growth in aqueous solution involve minimization of free energy of the whole system intrinsically and are considered to be in a reversible equilibrium (Xu and Wang 2011). As for the formation of ZnO wurtzite structure, the growth along the c-axis, which possesses higher energy polar surfaces, is preferred, specifically at the ±(0001) surfaces with the alternating stacking of Zn^{2+} and O^{2-} terminated surfaces (Sugunan et al. 2006, Tan et al. 2013a). The fast growth along the c-axis is the essential growth mechanism leading to the formation of 1-D nanostructure. It is well known that 1-D nanostructures such as nanorods, nanowires and nanotubes exhibit large surface areas and nanotubes would be the most ideal structure to provide the highest surface-to-volume ratio. Works were reported on the nanotubes formation via a two-step formation with etching of the nanowires'/ nanorods' core in order to generate the tubular structure, or the use of a template such as anodic aluminum oxide membranes that would require its removal after the ZnO growth step. One step fabrication is also possible by altering growth parameters such as solvent composition, seed layer thickness, post pH adjustment and ultra-sonification of the reaction solution (Xu and Wang 2011). By using a combination of electrochemical and chemical method in neutral pH solution, Elias et al. had demonstrated the feasibility of ZnO nanotubes generation in a three-step formation with control of the nanotubes' wall thickness by adjusting the electrodeposition at 80°C within a couple of hours timeframe (Elias et al. 2008). The initial electrodeposition step of ZnO nanowire arrays was carried out with O_2 reduction in an aqueous solution of zinc chloride ($ZnCl_2$) and potassium chloride (KCl). The subsequent step was the etching of the ZnO nanowires' core by KCl solution to obtain the tubular structure and therefore KCl concentration, temperature and immersion time were crucial parameters investigated in the ZnO nanotube formation. Final precise achievement of the nanotube wall thickness was fine-tuned by the electrodeposition step.

With the robust progress in technology as well as research interest in the development of ZnO nanostructures, various methods were reported for ZnO fabrication as well as modification techniques. Organic and inorganic compounds, as well as polymer matrixes were reportedly used for ZnO modification in order to generate the desired morphologies and properties required for specific application. By using inorganic compounds such as SiO_2, Al_2O_3 and metal ions, the particle size and surface area can be changed and the dispersion degree of the ZnO nanoparticles can also be improved. Organic compound such as carboxylic acid and silanes could be used to alter its physicochemical properties and increase ZnO compatibility with organic matrix. The long-term stability in an organic matrix is also reported to improve with reduction of particles aggregation. Meanwhile, introduction of polymer matrices such as poly(ethylene glycol), polystyrene, poly(methyl methacrylate) and chitin are reported to improve properties such as electrical, thermal and optical of the fabricated ZnO/polymer composite (Kolodziejczak-Radzimska and Jesionowski 2014).

Interesting morphology such as rotor-like ZnO nanorods formation was also reported by Rai et al. (2010) without any alloying element, template or surfactant by sono-chemical route. The rotor-like nanostructures demonstrated stronger photoluminescence emission in the green region due to more oxygen vacancies in the polar planes. By using the hydrothermal method and utilizing naturally abundant bio-template "alginic acid," Chetia et al. had demonstrated the feasibility of controlling the morphological structure in the fabrication of 1-D ZnO nanorods for DSSCs application (Chetia et al. 2016).

4.3 PROGRESS AND PERSPECTIVES

As ZnO nanostructures can be easily designed especially using chemical processes, vast morphologies such as nanosheets, nanobelts, nanotubes and nanowires can be fabricated. The anisotropic growth behavior of ZnO wurtzite helps to promote oriented growth that led to these nanostructures. It would be ideal if ZnO nanostructures could be generated with the minimum use of a chemical precursor or catalyst, as this could avoid contamination and lower the fabrication cost. Hot-water treatment was used in the study of natural oxidation of metallic Zn foils and sol-gel coatings by Tan et al. (2013a, 2015, 2016). Often, the subsequent Ostwald ripening process that occurs would cause the adhesion of adjacent nanostructures. This phenomenon usually occurs at prolonged processing times in the chemical process due to the repetitive dissolution and re-deposition cycles. In pure hot-water treatment of etched Zn foils reported by Tan et al., prolonged hot-water treatment time without any chemical addition for 24 h led to the formation of rough and larger ZnO nanorods due to Ostwald ripening (Tan et al. 2011). Although ZnO nanostructures were formed, the formation controllability was poor. In an attempt to control the ZnO nanorods formation and its branching degree, field assisted hot-water treatment was reported by Matsuda et al. where ZnO nanoflowers were obtained by changing the substrate and applied voltage (Matsuda et al. 2013). Due to the OH⁻ ions generation and accumulation at the vicinity of the electrode with applied voltage, branching and formation of ZnO nanoflowers were promoted.

As ZnO wurtzite is a stable oxide, formation of ZnO nanotubes in one step-process remains a challenge and not many works are reported (Elias et al. 2008). A more commonly used method is a two-step process that involves the formation of a 1-D nanorod followed by selective etching of its polar surface. As the point defects of the basal planes are non-uniform, they are etched at a higher rate than the peripheral region leading to tubular structure formation (She et al. 2008). Figure 4.2 shows the progressive change from nanorod to nanotube with different etching time up to 120 min.

One dimensional ZnO nanorods and nanowires are useful as a platform for secondary growth in order to achieve more complex hierarchical structure for various application. Zhang et al. had demonstrated the formation of three-dimensional (3-D) ZnO nanowires in a series of multiple step growth. First, ZnO nanowires were grown on a seeded substrate followed by shorter secondary ZnO nanowire branches formation on the primary ones. This complex structure which is shown in Figure 4.3 had enabled higher amounts of Au nanoparticles deposition, enabling improved solar-to-hydrogen conversion efficiency (Zhang et al. 2014).

Meanwhile, in a work reported by Xu et al. (2012), one step hydrothermal formation of comb-like ZnO nano-superstructures (NSSs) was demonstrated with the addition of Au nanowire as a catalyst which is shown in Figure 4.4. Due to the difference in surface kinetic energy of the ZnO NSSs based on density functional theory, the

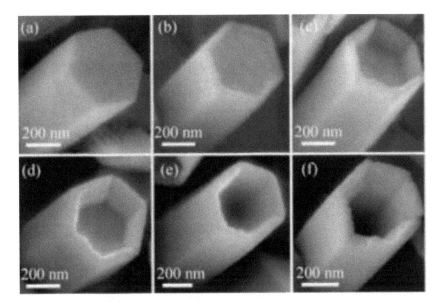

FIGURE 4.2 SEM images of ZnO nanotubes obtained at a different etching stages of (a) 0 min, (b) 5 mins, (c) 10 mins, (d) 15 mins, (e) 60 mins and (f) 120 mins. (Reproduced from She, G.-W. et al., *Appl. Phys. Lett.*, 92, 053111, 2008. With permission.)

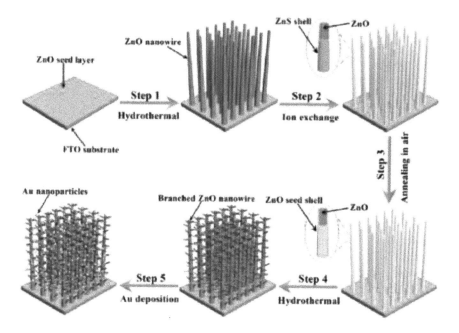

FIGURE 4.3 Schematic diagram of processes involved in the 3-D Au nanoparticles modified ZnO nanowires. (Reproduced from Zhang, X. et al., *ACS Appl. Mater. Interfaces*, 6, 4480–4489, 2014. With permission.)

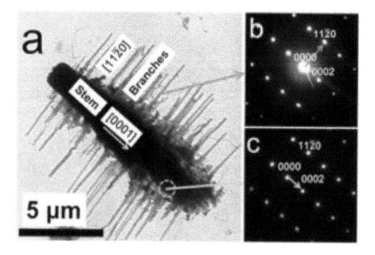

FIGURE 4.4 (a) Bright field TEM image showing ZnO NSS with SAED indicating the (b) branch formation in [11$\bar{2}$0] direction and (c) stem formation in [0001] direction. (Reproduced from Xu, X. et al., *Cryst. Growth Des.*, 12, 4829–4833, 2012. With permission.)

growth mechanism is revealed that $E\{0001\} > E\{11\bar{2}0\} > E\{\bar{1}100\}$. Growth rate was faster at [0001] direction in the initial reaction stage when there was sufficient supply of Zn^{2+}. The formation of $(\bar{1}100)$ plane would promote stems formation of NSSs and as the concentration of Zn^{2+} becomes insufficient for further stems growth, the existing smaller crystallite unit would promote the preferable nano-branch growth on both sides of the stems (second fastest growth direction). The understanding of the surface kinetic energy would be useful for nano-architecture design with catalyst.

Lee et al. (2011) had further demonstrated the feasibility of synthesizing 3-D ZnO nano-architectures using hydrothermal growth of a crystallographic oriented seed layer obtained via sputtering and pulsed laser deposition of ZnO on (001) Si and Al_2O_3 (0001) by growth mask patterning. Although their method involved the physical method, the principle of having an oriented seed layer provides an important insight towards ZnO nanostructure design by chemical process methods. In their finding, polycrystalline seed layer led to the formation of flower-like ZnO nano-architectures while c-axis oriented seed layers led to the formation of polygonal ZnO pillars. It is noteworthy that multi-domain columnar-joint structures were demonstrated in their work emphasizing the controlled morphology formation that could be used to alter light propagation paths for 1-D ZnO nanostructures. Figure 4.5 summarizes the morphological ZnO nanostructures obtained by Lee et al.

FIGURE 4.5 Morphological evolution SEM images of ZnO nano-architectures using (a) sputtered ZnO, (b) PLD deposited ZnO/Si (001) and (c) PLD deposited ZnO/Al_2O_3 (0001) with varying diameter of the circular holes in the growth masks. The diameters of the holes from the left to right are 100 nm, 0.5 μm, 1 μm, and 2 μm. (Reproduced from Lee, W.W. et al., *Cryst. Growth Des.*, 11, 4927–4932, 2011. With permission.)

4.4 APPLICATION OF ONE-DIMENSIONAL (1-D) NANOSTRUCTURES

4.4.1 DYE-SENSITIZED SOLAR CELLS (DSSCs) AND SURFACE PLASMON RESONANCE ENHANCED DSSCs

Since ZnO nanowires (NWs) were first used by Law et al. as replacement for nanoparticles film with the aim of improving the electron diffusion efficiency, many subsequent developments had followed until an efficiency plateau was obtained. Superiority of the ZnO NW as a more direct electron pathway was demonstrated by the near linear increase in short-circuit density indicating excellent electron transport in nanowires with a reduced recombination rate compared to nanoparticles films (Zhang et al. 2009). In the working principle of DSSCs, dye molecules (commonly ruthenium complexes) are used to absorb the incident photons in order to excite the electron from valence band to conduction band. Then, the oxidized dye molecules are reduced by the electrolyte (I^-/I_3^-) in a quick manner to avoid recombination with the excited electrons in the conduction band. The collected electrons are then transferred into an external circuit through ZnO which acts as an electron pathway before returning to the counter electrode for the reduction of electrolyte. It is mentioned in a report that ZnO nanowires obtained using hydrothermal methods demonstrated a higher power conversion efficiency than those obtained via vapor phase methods due to higher dye adsorption behavior. Studies by Law et al. also reported a smaller time constant for electron injection from photo-excited ruthenium dye into the nanowires especially on non-polar side facets compared to nanoparticles based on femtosecond transient adsorption spectroscopy investigation (Law et al. 2005). Higher strength and resistance of nanowires are also useful for flexible devices fabrication. Despite the advantages of good electron conduction pathway, low cost fabrication and vast fabrication methods of 1-D ZnO nanostructures by chemical processes, the highest efficiency achieved still remains low compared to TiO_2. This is due to several factors such as formation of stable Zn^{2+}/dye complexes, high carrier recombination from the surface trap states of ZnO nanowires, low electron injection efficiency due to the energy level mismatch between the lowest unoccupied molecular orbital (LUMO) of the dye and photoanode's conduction band, low dye regeneration efficiency due to the energy level mismatch between the highest occupied molecular orbital (HOMO) of the dye and redox mediator potential in the electrolyte (Xu and Wang 2011).

Further increments of the length or film thickness of the 1-D ZnO nanostructures is no longer an effective way to enhance the photo-conversion efficiency of DSSCs. Despite the efforts, the highest efficiency reported for ZnO DSSCs is 6.58% which is only half of TiO_2 (Li et al. 2012). The 1-D ZnO nanostructures are also used as a scaffold for metal nanoparticles deposition such as Au or Ag nanoparticles to induce surface plasmon resonance effect, which would enable the harvesting of visible light. High aspect ratio ZnO NWs were used for deposition of Au nanoparticles and an improved efficiency of 75% was reported by Tan et al. (2017). With the formation of high aspect ratio tapered ZnO nanorods, Au nanoparticles agglomeration at the surface of the nanowires could be avoided and the Au nanoparticles could penetrate deeper into the mid-section of ZnO NWs.

4.4.2 Ultraviolet and Gas Sensors

ZnO has excellent potential to be used as nano-sensors such as UV sensor and gas sensors. The presence of oxygen-related hole-trapping states at ZnO nanowire surface could promote high sensitivity for UV detection. In a work reported by Wang et al., they managed to achieve enhanced UV detection up to 5 magnitude orders by functionalizing the ZnO surface with high UV absorbing polymer to counter the point defects that were present in a confined dimensionality of ZnO nanowires and nanobelts that would limit its UV sensitivity. This photo-conductance enhancement is due to the new energy levels introduced by the polymer that lied in the corresponding band gap and in the conduction band of ZnO which serve as "hopping" platforms and increased the excitation probability of an electron to the conduction band (Wang 2009). As for gas sensing applications, adsorption of a molecule such as ethanol gas (C_2H_5OH) could alter the dielectric property of the surface and this would change the surface conductance. The large surface to volume ratio of 1-D ZnO nanostructures' surface conductance has high sensitivity towards surface chemistry changes (Wang 2009). Kim et al. (2011) demonstrated 7–9 times improvement of 10–1000 ppm ethanol sensing using ZnO hierarchical nanostructures compared to agglomerated nanoparticles, which were prepared by hydrolysis of zinc salt with oleic acid dissolved in n-hexane solution.

4.4.3 Nanopiezotronics

Since the introduction of nanopiezotronics concept in 2007, coupled piezoelectric and semiconducting properties of 1-D ZnO nanowires and nanobelts for devices and component fabrications, such as field effect transistors and diodes, have been expanding. When an external force is applied on a ZnO NW, piezoelectric potential is generated and with a strong enough potential, it could act as a gate voltage for field effect transistor. In a work by Wang et al. (2006), an electric field by piezoelectricity across two electrodes obtained from bent nanowires that serve as the gate for electric current flow control was demonstrated, as shown in Figure 4.6. The deformation led to generation of external force and free electrons in the n-type ZnO nanowire that may be trapped at the positive surface, which are immobile, leading to lower effective carrier density. Higher degrees of nanowire bending would result in narrower conducting channels of ZnO and a wider depletion region that would lead to reduced conductivity in this piezoelectric field effect transistor application. As ZnO nanowires fabrication is fairly achievable through chemical processes, it is predicted that applications in piezoelectric devices will further increase.

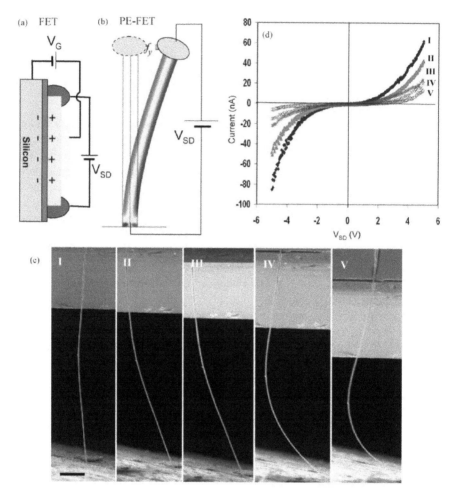

FIGURE 4.6 The physical principle of the piezoelectric-field effect transistor. (a) Schematics of a conventional field effect transistors using a single nanowire/nanobelt, with gate, source and drain. (b) The principle of the piezoelectric-field effect transistor, in which the piezoelectric potential across the nanowire created by the bending force replacing the gate as in conventional FET. Ohmic contacts were used at both ends. (c) SEM images with the same magnification showing the five typical consecutive bending cases of a ZnO wire; the scale bar represents 10 mm. (d) Corresponding I–V characteristics of the ZnO nanowire for the five different bending cases. (Reproduced from Wang, Z.L., *Mater. Sci. Eng. R: Rep.*, 64, 33–71, 2009. With permission.)

4.4.4 PHOTOCATALYST

There are many works reported on the usage of ZnO as a photocatalyst for deg-radation of organic compounds such as methyl orange and methylene blue (Yang et al. 2010, Ma et al. 2011, Stanković et al. 2012, Tan et al. 2014a). Hung's group

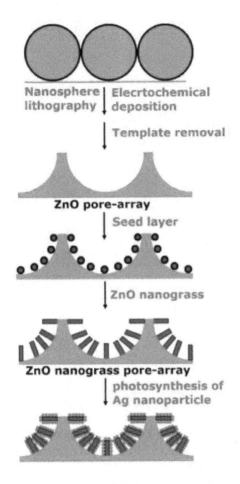

FIGURE 4.7 Schematic illustrations of ZnO nanograss decorated pore-array films. (Reproduced from Hung, S.T. et al., *J. Hazard Mater.*, 198, 307–316, 2011. With permission.)

had demonstrated an interesting formation of ZnO nanograss-like structures on pore-array films and then further deposited Ag nanoparticles onto the ZnO surfaces as shown in Figure 4.7. Their report had demonstrated the feasibility of nano-architecture design to maximize the potential of 1-D ZnO nanorods. The as-obtained nano-architectures marked an improved photocatalytic activity for degradation of methyl orange under UV light irradiation as light scattering within the unique structure would promote more light absorption (Hung et al. 2011). With the carefully designed nanograss architecture as shown in Figure 4.8, further deposition of Ag nanoparticles was plausible and the effect of sizes as well as densities of Ag nanoparticles were also reported.

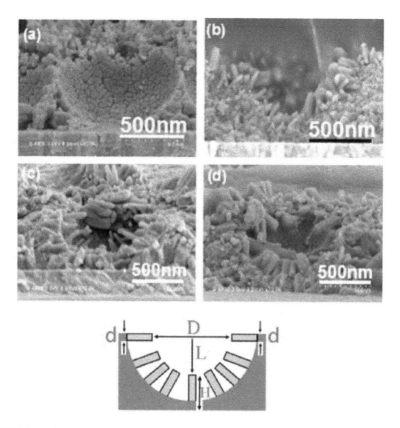

FIGURE 4.8 Cross sectional SEM images of the (a) ZnO pore array film and (b), (c), (d) nanograss-decorated pore array films for 75, 90 and 105 mins, respectively.

4.4.5 BIOSENSORS

As healthcare becomes a main concern of the society nowadays, technology has also been evolving towards the monitoring and promotion of medical awareness. In versatile ZnO applications, ZnO is also widely used for biological applications such as anti-bacterial applications, drug delivery, bio-sensors, bio-imaging probes, etc. (Bhat et al. 2017). In bio-sensing applications, the key is to focus on the mechanism for the immobilization of particular protein or biomolecules on ZnO surfaces. Recently, Bhat et al. had reported a comprehensive review on this topic, and the following are a few important highlights to emphasize the potential of ZnO for biosensing technology. Firstly, ZnO nanorod arrays are usually functionalized as shown in Figure 4.9. The multistep functionalization of ZnO nanorod arrays exhibit uniform and homogenous conjugation of ATES/GA surface (step II) prior to angiotensin II and bovine serum albumin surface attachment occurs (Ahsanulhaq et al. 2010). Intermediates that were obtained could be confirmed using X-ray photoelectron spectroscopy and Coomassie assay.

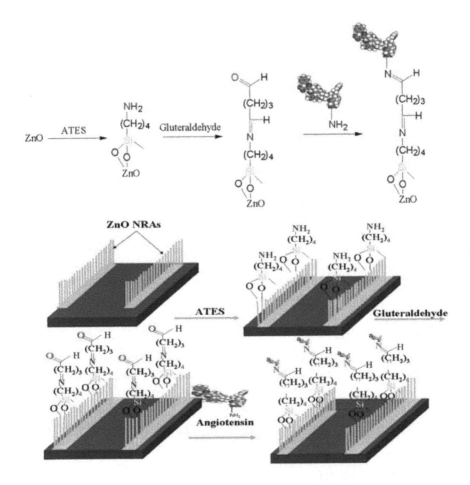

FIGURE 4.9 An illustration example of immobilization process with angiotensin II on ZnO nanorod arrays. (Reproduced from Bhat, S.S. et al., *Trends Analyt. Chem.*, 86, 1–13, 2017. With permission.)

In a separate work reported by Liu et al., ZnO nanorods obtained were functionalized on a transparent thermoplastic surface by coupling of dimercapto-succinic acid (DMSA) and 1-ethyl-3-(3dimethylaminopropyl) carbodiimide (EDC). After that, by covalently attaching human serum albumin and bovine serum albumin with DMSA and EDC onto the ZnO nanorods, the targeted biomolecules could be detected by exploring the functionalized ZnO nanorods using PL and advanced optical techniques (Liu et al. 2006). Recent findings have shown that organic-inorganic heterostructures would be able to present some unique function-alities compared to only pure organic or inorganic materials. It is envisioned that more biosensors based on protein-functionalized semiconductor could be achieved. With the rapid progress related to functionalization of carbon nanotubes, graphene and other nanomaterials with proteins through bio-conjugation, it is also believed that 1-D ZnO nanostructures are also one of the potential materials to be considered due to its feasibility of precise nano-architecture design, properties alteration and organic functionalization.

4.5 FUTURE OUTLOOK

Looking forward, the future of 1-D nanostructures development will be more focused on the balance between cost, performance and stability of the nanostructures for its vast applications. As lightweight and wearable devices becoming more popular, precise fabrication of ZnO nanostructures in order to obtain the best-desired properties have never been more crucial. Fortunately, with the advancement of inkjet printing technology, precise seed crystals can be obtained using sol-gel derived zinc precursor from micrometer to nanometer scale. Development of pattern transfer using selective adhesion polymer also enable the transfer of one dimensional ZnO nanostructures for subsequent secondary or more complex formation of ZnO nanostructures which has a huge potential for application in flexible and foldable devices.

4.6 CONCLUSION

Fundamental breakthroughs have enabled discoveries and fabrication of 1-D ZnO nanostructures and its development has been a unique nanoscale building block. With ZnO's unique electronic and optical properties, feasibility of low-temperature chemical process fabrication will only further propel its research and study towards applications in optoelectronics, biomedical and environmental sectors focusing on lightweight and flexible wearable devices.

ACKNOWLEDGMENT

The authors would like to acknowledge the support of the laboratory members in Matsuda-Muto-Kawamura-Tan at Toyohashi University of Technology, Japan and the effort of Assoc. Prof. Zainovia Lockman in the research collaboration between both institutions.

REFERENCES

Ahsanulhaq, Q., J. H. Kim, and Y. B. Hahn. 2010. "Immobilization of angiotensin II and bovine serum albumin on strip-patterned ZnO nanorod arrays." *Journal of Nanoscience and Nanotechnology 10* (7):4159–4165. doi:10.1166/jnn.2010.2404.
Al-Hardan, N. H., M. J. Abdullah, and A. Abdul Aziz. 2010. "Electron transport mechanism of thermally oxidized ZnO gas sensors." *Physica B: Condensed Matter 405* (21):4509–4512. doi:10.1016/j.physb.2010.08.027.
Ameen, S., M. S. Akhtar, M. Song, and H. S. Shin. 2012. "Vertically aligned ZnO nanorods on hot filament chemical vapor deposition grown graphene oxide thin film substrate: Solar energy conversion." *ACS Applied Materials & Interfaces 4* (8):4405–4412. doi:10.1021/am301064j.

Anandan, S., N. Ohashi, and M. Miyauchi. 2010. "ZnO-based visible-light photocatalyst: Band-gap engineering and multi-electron reduction by co-catalyst." *Applied Catalysis B: Environmental 100* (3–4):502–509. doi:10.1016/j.apcatb.2010.08.029.

Bhat, S. S., A. Qurashi, and F. A. Khanday. 2017. "ZnO nanostructures based biosensors for cancer and infectious disease applications: Perspectives, prospects and promises." *TrAC Trends in Analytical Chemistry 86*:1–13. doi:10.1016/j.trac.2016.10.001.

Chang, C. M., M. H. Hon, and I. C. Leu. 2010. "Preparation of ZnO nanorod arrays with tailored defect-related characterisitcs and their effect on the ethanol gas sensing performance." *Sensors and Actuators B: Chemical 151* (1):15–20. doi:10.1016/j.snb.2010.09.072.

Chetia, T. R., M. S. Ansari, and M. Qureshi. 2016. "Rational design of hierarchical ZnO superstructures for efficient charge transfer: Mechanistic and photovoltaic studies of hollow, mesoporous, cage-like nanostructures with compacted 1D building blocks." *Physical Chemistry Chemical Physics 18* (7):5344–5357. doi:10.1039/c5cp07687k.

Djurišić, A. B., and Y. H. Leung. 2006. "Optical properties of ZnO nanostructures." *Small 2* (8–9):944–961. doi:10.1002/smll.200600134.

Djurišić, A. B., A. M. C. Ng, and X. Y. Chen. 2010. "ZnO nanostructures for optoelectronics: Material properties and device applications." *Progress in Quantum Electronics 34* (4):191–259. doi:10.1016/j.pquantelec.2010.04.001.

Elias, J., R. Tena-Zaera, G.-Y. Wang, and C. Lévy-Clément. 2008. "Conversion of ZnO nanowires into nanotubes with tailored dimensions." *Chemistry of Materials 20* (21):6633–6637. doi:10.1021/cm801131t.

Geng, B., J. Liu, and C. Wang. 2010. "Multi-layer ZnO architectures: Polymer induced synthesis and their application as gas sensors." *Sensors and Actuators B: Chemical 150* (2):742–748. doi:10.1016/j.snb.2010.08.008.

Heo, Y. W., D. P. Norton, L. C. Tien, Y. Kwon, B. S. Kang, F. Ren, S. J. Pearton, and J. R. LaRoche. 2004. "ZnO nanowire growth and devices." *Materials Science and Engineering: R: Reports 47* (1–2):1–47. doi:10.1016/j.mser.2004.09.001.

Hung, S. T., C. J. Chang, and M. H. Hsu. 2011. "Improved photocatalytic performance of ZnO nanograss decorated pore-array films by surface texture modification and silver nanoparticle deposition." *Journal of Hazardous Materials 198*:307–316. doi:10.1016/j.jhazmat.2011.10.043.

Kim, K.-M., H.-R. Kim, K.-I. Choi, H.-J. Kim, and J.-H. Lee. 2011. "ZnO hierarchical nanostructures grown at room temperature and their C_2H_2OH sensor applications." *Sensors and Actuators B: Chemical 155* (2):745–751. doi:10.1016/j.snb.2011.01.040.

Kolodziejczak-Radzimska, A., and T. Jesionowski. 2014. "Zinc oxide-from synthesis to application: A Review." *Materials (Basel) 7* (4):2833–2881. doi:10.3390/ma7042833.

Kurbanov, S. S., H. D. Cho, and T. W. Kang. 2011. "Effect of excitation and detection angles on photoluminescence spectrum from ZnO nanorod array." *Optics Communications 284* (1):240–244. doi:10.1016/j.optcom.2010.09.011.

Law, M., L. E. Greene, J. C. Johnson, R. Saykally, and P. Yang. 2005. "Nanowire dye-sensitized solar cells." *Nature Materials 4*:455. doi:10.1038/nmat1387.

Lee, W. W., J. Yi, S. B. Kim, Y.-H. Kim, H.-G. Park, and W. I. Park. 2011. "Morphology-Controlled three-dimensional nanoarchitectures produced by exploiting vertical and in-plane crystallographic orientations in hydrothermal ZnO crystals." *Crystal Growth & Design 11* (11):4927–4932. doi:10.1021/cg200806a.

Li, L., T. Zhai, Y. Bando, and D. Golberg. 2012. "Recent progress of one-dimensional ZnO nanostructured solar cells." *Nano Energy 1* (1):91–106. doi:10.1016/j.nanoen.2011.10.005.

Liu, T.-Y., H.-C. Liao, C.-C. Lin, S.-H. Hu, and S.-Y. Chen. 2006. "Biofunctional ZnO nanorod arrays grown on flexible substrates." *Langmuir 22* (13):5804–5809. doi:10.1021/la052363o.

Ma, S., R. Li, C. Lv, W. Xu, and X. Gou. 2011. "Facile synthesis of ZnO nanorod arrays and hierarchical nanostructures for photocatalysis and gas sensor applications." *Journal of Hazardous Materials 192* (2):730–740. doi:10.1016/j.jhazmat.2011.05.082.

Matsuda, A., W. K. Tan, S. Furukawa, and H. Muto. 2013. "Morphology-control of crystallites precipitated from ZnO gel films by applying electric field during hot-water treatment." *Materials Science in Semiconductor Processing 16* (5):1232–1239. doi:10.1016/j.mssp.2012.12.018.

Pearton, S. 2005. "Recent progress in processing and properties of ZnO." *Progress in Materials Science 50* (3):293–340. doi:10.1016/j.pmatsci.2004.04.001.

Pearton, S. J., D. P. Norton, M. P. Ivill, A. F. Hebard, J. M. Zavada, W. M. Chen, and I. A. Buyanova. 2007. "ZnO doped with transition metal ions." *IEEE Transactions on Electron Devices 54* (5):1040–1048. doi:10.1109/ted.2007.894371.

Rai, P., J.-N. Jo, I.-H. Lee, and Y.-T. Yu. 2010. "Fabrication of 3D rotor-like ZnO nanostructure from 1D ZnO nanorods and their morphology dependent photoluminescence property." *Solid State Sciences 12* (10):1703–1710. doi:10.1016/j.solidstatesciences.2010.07.009.

She, G.-W., X.-H. Zhang, W.-S. Shi, X. Fan, J. C. Chang, C.-S. Lee, S.-T. Lee, and C.-H. Liu. 2008. "Controlled synthesis of oriented single-crystal ZnO nanotube arrays on transparent conductive substrates." *Applied Physics Letters 92* (5):053111. doi:10.1063/1.2842386.

Stanković, A., Z. Stojanović, L. Veselinović, S. Davor Škapin, I. Bračko, S. Marković, and D. Uskoković. 2012. "ZnO micro and nanocrystals with enhanced visible light absorption." *Materials Science and Engineering: B 177* (13):1038–1045. doi:10.1016/j.mseb.2012.05.013.

Sugunan, A., H. C. Warad, M. Boman, and J. Dutta. 2006. "Zinc oxide nanowires in chemical bath on seeded substrates: Role of hexamine." *Journal of Sol-Gel Science and Technology 39* (1):49–56. doi:10.1007/s10971-006-6969-y.

Sun, J. H., S. Y. Dong, Y. K. Wang, and S. P. Sun. 2009. "Preparation and photocatalytic property of a novel dumbbell-shaped ZnO microcrystal photocatalyst." *Journal of Hazardous Materials 172* (2–3):1520–1526. doi:10.1016/j.jhazmat.2009.08.022.

Tan, W. K., K. A. Razak, K. Ibrahim, and Z. Lockman. 2011. "Oxidation of etched Zn foil for the formation of ZnO nanostructure." *Journal of Alloys and Compounds 509* (24):6806–6811. doi:10.1016/j.jallcom.2011.03.055.

Tan, W. K., K. A. Razak, Z. Lockman, G. Kawamura, H. Muto, and A. Matsuda. 2013a. "Formation of highly crystallized ZnO nanostructures by hot-water treatment of etched Zn foils." *Materials Letters 91*:111–114. doi:10.1016/j.matlet.2012.08.103.

Tan, W. K., K. A. Razak, Z. Lockman, G. Kawamura, H. Muto, and A. Matsuda. 2013b. "Photoluminescence properties of rod-like Ce-doped ZnO nanostructured films formed by hot-water treatment of sol–gel derived coating." *Optical Materials 35* (11):1902–1907. doi:10.1016/j.optmat.2013.01.011.

Tan, W. K., K. A. Razak, Z. Lockman, G. Kawamura, H. Muto, and A. Matsuda. 2014a. "Synthesis of ZnO nanorod–nanosheet composite via facile hydrothermal method and their photocatalytic activities under visible-light irradiation." *Journal of Solid State Chemistry 211*:146–153. doi:10.1016/j.jssc.2013.12.026.

Tan, W. K., T. Ito, G. Kawamura, H. Muto, Z. Lockman, and A. Matsuda. 2017. "Controlled facile fabrication of plasmonic enhanced Au-decorated ZnO nanowire arrays dye-sensitized solar cells." *Materials Today Communications 13*:354–358. doi:10.1016/j.mtcomm.2017.11.004.

Tan, W. K., G. Kawamura, and A. Matsuda. 2016. Design of ZnO nano-architectures and its applications. *Two-Dimensional Nanostructures for Energy-Related Applications*. Boca Raton, FL: CRC Press.

Tan, W. K., G. Kawamura, H. Muto, K. A. Razak, Z. Lockman, and A. Matsuda. 2015. "Blue-emitting photoluminescence of rod-like and needle-like ZnO nanostructures formed by hot-water treatment of sol–gel derived coatings." *Journal of Luminescence 158*:44–49. doi: 10.1016/j.jlumin.2014.09.028.

Tan, W. K., L. C. Li, K. A. Razak, G. Kawamura, H. Muto, A. Matsuda, and Z. Lockman. 2014b. "Formation of two-dimensional ZnO nanosheets by rapid thermal oxidation in oxygenated environment." *Journal of Nanoscience and Nanotechnology 14* (4):2960–2967. doi:10.1166/jnn.2014.8559.

Tan, W. K., Z. Lockman, K. A. Razak, G. Kawamura, H. Muto, and A. Matsuda. 2013a. "Enhanced dye-sensitized solar cells performance of ZnO nanorod arrays grown by low-temperature hydrothermal reaction." *International Journal of Energy Research:* 1992–2000. doi:10.1002/er.3026.

Tan, W. K., K. A. Razak, Z. Lockman, G. Kawamura, H. Muto, and A. Matsuda. 2013b. "Optical properties of two-dimensional ZnO nanosheets formed by hot-water treatment of Zn foils." *Solid State Communications 162*:43–47. doi:10.1016/j.ssc.2013.02.018.

Wang, X., J. Zhou, J. Song, J. Liu, N. Xu, and Z. L. Wang. 2006. "Piezoelectric field effect transistor and nanoforce sensor based on a single ZnO nanowire." *Nano Letters 6* (12):2768–2772. doi:10.1021/nl061802g.

Wang, Z. L. 2007. "The new field of nanopiezotronics." *Materials Today 10* (5):20–28. doi:10.1016/s1369-7021(07)70076-7.

Wang, Z. L. 2009. "ZnO nanowire and nanobelt platform for nanotechnology." *Materials Science and Engineering: R: Reports 64* (3–4):33–71. doi:10.1016/j.mser.2009.02.001.

Xu, S., and Z. L. Wang. 2011. "One-dimensional ZnO nanostructures: Solution growth and functional properties." *Nano Research 4* (11):1013–1098. doi:10.1007/s12274-011-0160-7.

Xu, X., M. Wu, M. Asoro, P. J. Ferreira, and D. L. Fan. 2012. "One-step hydrothermal synthesis of comb-like ZnO nanostructures." *Crystal Growth & Design 12* (10):4829–4833. doi:10.1021/cg3005773.

Yang, L. Y., S. Y. Dong, J. H. Sun, J. L. Feng, Q. H. Wu, and S. P. Sun. 2010. "Microwave-assisted preparation, characterization and photocatalytic properties of a dumbbell-shaped ZnO photocatalyst." *Journal of Hazardous Materials 179* (1–3):438–443. doi:10.1016/j.jhazmat.2010.03.023.

Yang, P., R. Yan, and M. Fardy. 2010. "Semiconductor nanowire: What's next?" *Nano Letters 10* (5):1529–1536. doi:10.1021/nl100665r.

Zhang, Q., C. S. Dandeneau, X. Zhou, and G. Cao. 2009. "ZnO nanostructures for dye-sensitized solar cells." *Advanced Materials 21* (41):4087–4108. doi:10.1002/adma.200803827.

Zhang, X., Y. Liu, and Z. Kang. 2014. "3D branched ZnO nanowire arrays decorated with plasmonic Au nanoparticles for high-performance photoelectrochemical water splitting." *ACS Appl Mater Interfaces 6* (6):4480–4489. doi:10.1021/am500234v.

5 One-Dimensional α-Fe_2O_3 Nanowires Formation by High Temperature Oxidation of Iron and Their Potential Use to Remove Cr(VI) Ions

Subagja Toto Rahmat, Monna Rozana, Tan Wai Kian, Go Kawamura, Atsunori Matsuda, and Zainovia Lockman

CONTENTS

5.1 INTRODUCTION

Protection of ground and surface water from heavy metal ions is an issue; despite being rather complex, it must be urgently addressed. Once contaminated, ground and surface water are difficult to clean and hence, strict control of industrial effluent contents is vital in ensuring minimum contamination (Barakat, 2011, Looi et al., 2013). The textile industry has a high water consumption and subsequently produces a high discharge rate of wastewater (Correia et al., 1994, Robinson et al., 2001, Sponza, 2002,

Bisschops and Spanjers, 2003). The wastewater inevitably will contain various chemical reagents ranging from inorganic compounds like dissolved metal ions as an example, to organic compounds like wax and dyes (Correia et al., 1994, Robinson et al., 2001, Bisschops and Spanjers, 2003). Some of these pollutants, for instance, heavy-metal loaded pigments, cannot be removed efficiently by typical municipal sewerage systems (Barakat, 2011). Considering their toxicity, a more precise method is therefore required. Reducing heavy metal ions to their more benign compounds is proposed to be one effective treatment method. There are various reductants that can be used but only recently, reduction via photocatalysis technology has gained more significant attention (Chen and Ray, 2001, 2015, Barakat, 2011, Chowdhury et al., 2015).

The term photocatalysis is a combination of the words *photo*chemistry and *catalysis* (Fujishima, 1972, Chen and Ray, 2001, Chen et al., 2008, Fujishima et al., 2000). This implies that both light and catalyst are required to bring about or to accelerate a chemical reaction (Fujishima, 1972, Chen and Ray, 2001, Chen et al., 2008). Referring to the International Union of Pure and Applied Chemistry (IUPAC) Recommendation 2011, the definition of photocatalysis is the change in the rate of a chemical reaction or its initiation under the action of ultraviolet, visible or infrared radiation in the presence of a substance—the photocatalyst—that absorbs light and is involved in the chemical transformation of the reaction partners. The IUPAC definition for photocatalyst is a substance able to produce, by absorption of ultraviolet, visible or infrared radiation, chemical transformations of the reaction partners, repeatedly joining them in intermediate chemical interactions and regenerating its chemical composition after each cycle of such interactions (Braslavsky et al., 2011).

An obvious photocatalyst is semiconductors as they have a specific electronic structure characterized by a filled valence band (VB) and an empty conduction band (CB) enabling them to absorb light (Fujishima, 1972, Hoffmann et al., 1995, Chen et al., 2008). The bands are separated by an energy gap, called band gap (E_g). Once irradiated, an electron from the filled valence band (VB) will be promoted (excited) into the conduction band (CB) forming free photogenerated electrons (e^-_{CB}). This process will leave a free hole (h^+_{VB}) in the VB. For the use in wastewater treatment, the semiconductor ought to be immersed in wastewater and must be continuously irradiated by photons of energy larger than its band gap. Therefore, at the semiconductor|liquid junction a charge transfer process will occur until equilibrium is achieved. The excess charge is distributed within a so-called space charge (depletion) region and built-in potential is created within this region. The potential is high enough to separate the electron-hole pairs (EHPs). Once separated, the e^-_{CB} and h^+_{VB} will migrate towards the particle surface as a result of the potential gradient that exists between the semiconductor and the liquid it is immersed in. A charge transfer processes will then occur (Chen et al., 2008, Kabra et al., 2008, Barakat, 2011).

Figure 5.1 is a schematic illustration of the EHP generation (G) process in a spherical semiconductor particle. As seen from the figure, EHP can recombine (R) followed by dissipation of the adsorbed energy with the liberation of heat (or light). They can also be separated and transported to the surface of the semiconductor. If heavy metal ions are adsorbed on the surface of a semiconductor, reduction (A → A$^+$) can happen provided the potential of e^-_{CB} is negative enough to initiate reduction. The process of reduction must be accompanied with the oxidation process (D → D$^-$) whereby a suitable hole scavenger

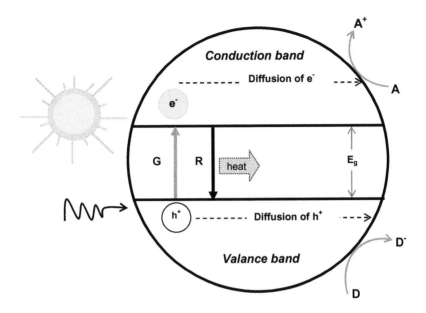

FIGURE 5.1 Schematic illustration of the generation of electron-hole pairs in a semiconductor particle under light illumination.

is needed as to make reduction reaction the principle reaction. Knowledge on the positions of CB and VB of a semiconductor of choice at different pH is therefore essential, as this can determine if the heavy metal ions can be reduced or not by the semiconductor (Chen and Ray, 2001, Cheng et al., 2015). Choice of the right scavenger is also a prerequisite in forming a catalysis system that can effectively reduce heavy metal ions.

The number of EHPs generated in a semiconductor particle is dependent on the intensity of the incident light that impinges on the semiconductor as well as on the its optoelectronic characteristics (Hoffmann et al., 1995). Since generation, recombination and catalysis processes are competing, it is important to design a photocatalysis system which can generate EHP with minimum loss hence catalytic reaction can be much enhanced. Recombination can be defined as disappearance of free electrons and holes due to the transition of electrons from the conduction band to the valence band. This is a so-called band-to-band transition. Recombination can also occur by electron transition to a defect centre then to valance band hence is influenced by impurities or defects in the crystalline structure (Braslavsky et al., 2011). As recombination result in the disappearance of free carriers, this process determines the rates and quantum yields of the photocatalytic reactions.

There are arrays of commercially available semiconducting materials that can be used as photocatalysts (Chen et al., 2008) but not all can reduce heavy metal ions. The most commonly studied wide band gap oxide-based semiconductors like TiO_2 have conduction band electrons positive enough for $Cr(VI)$ reduction to occur (Chowdhury et al., 2015). However, due to its wide band gap, TiO_2 needs ultraviolet (UV) radiation for it to become photoactive. Utilization of sunlight for semiconductor activation can reduce much of the complexity of a photocatalytic reactor to perform reduction of heavy metal ions in wastewater. Here, α-Fe$_2$O$_3$ is explored as it has a smaller energy band gap as TiO$_2$. In this chapter we first give a brief overview of textile as the main industry that would require photocatalytic technology for wastewater treatment, then on the properties of α-Fe$_2$O$_3$ as a potential photocatalyst to perform reduction of heavy metal ions with a main focus on Cr(VI) ions reduction to Cr(III). The formation of 1-D α-Fe$_2$O$_3$ nanowires by thermal oxidation process is presented in the subsequent heading.

5.2   HEAVY METAL IN TEXTILE WASTEWATER

Heavy metal pollution due to anthropogenic activities in aquatic ecosystems has been acknowledged as a worldwide environmental problem and has been receiving increasing attention over the last few decades. The contamination of aquatic systems by heavy metals has become one of the most challenging pollution issues as some of the heavy metals are classified as human carcinogens, according to the International Agency for Research on Cancer (IARC). Among various industries that contribute to heavy metal discharged, the textile industry has been identified as the main contributor. Water quality deterioration due to textile industries have been documented in literatures (Correia et al., 1994, Robinson et al., 2001, Sponza, 2002, Bisschops and Spanjers, 2003, Barakat, 2011, Ghaly et al., 2014) especially from countries where the industries are on the rise (Noor Syuhadah et al., 2015, Dey and Islam, 2015, Noreen et al., 2017). Textile effluent is known to be colored and has a high pH, temperature, biological oxygen demand, chemical oxygen demand, total suspended solids and total dissolved solids (Sponza, 2002). There are several reports on characterizations of textile wastewater that pointed on the existence of chromium (Cr), cadmium (Cd), lead (Pb), zinc (Zn), and copper (Cu) compounds from colored pigments used in the industry (Correia et al., 1994, Sponza, 2002, Bisschops and Spanjers, 2003). These heavy metal compounds can also arise from the metallic part of dye molecules and from the dyeing process (Menezes et al., 2010, Sungur and Gülmez, 2015). Cr(VI), Cd and Pb compounds, for example, have been widely used as pigments for textile dyes (IARC Monographs 100C (2012), Correia et al., 1994, Bisschops and Spanjers, 2003). Untreated textile effluents will obviously contain some of these compounds.

Heavy metals are elements with atomic weights between 63.5 and 200.6, and a specific gravity greater than 5.0 g/cm^3. They are persistent pollutants since they cannot be degraded; hence, they can unceasingly exist in water, or they can be absorbed by suspended solids, then accumulate in sediments and biomagnified along aquatic food chains. Among various aquatic organisms that may be affected are sediment-dwelling organisms like clams, oysters and mussels, which are prone to accumulate large quantities of heavy metals due to their habitat and feeding nature (Senthil Kumar et al., 2008, Rajeshkumar et al., 2018). Assessment of trace metal contamination levels, bioavailability and toxicity in sediments has been done by various authors in recent years indicating the rising concern of heavy metal pollution in aquatic organisms (Hashim et al., 2014, Tang et al., 2014, Hossen et al., 2015, Khan et al., 2016, Bhuyan et al., 2017, Rajeshkumar et al., 2018). Sediment quality is a good indicator of heavy metal pollution in water because it tends to concentrate the heavy metal compounds. Highly contaminated sediments can also be correlated to the increase of heavy metal ions in water (Venkatramanan et al., 2015, Wang et al., 2017). Consumption of heavy metal polluted water as drinking water can be very harmful to humans. Moreover, rivers are important sources of irrigation; contaminated rivers have led to heavy metal pollution in crops and vegetables (Cheng, 2003, Huang et al., 2013). Rice for example is a crop that is grown in flooded plains thus irrigating paddy fields with heavy metal–loaded water may result in As, Pb and/or Cd contaminated rice, as reported by many authors in recent years (Meharg et al., 2013, Norton et al., 2014, Praveena and Omar, 2017, Davis et al., 2017, Islam et al., 2017, Mukherjee et al., 2017, Rasheed et al., 2018). Cd, Cr(VI) and As have been classified as group 1 carcinogens (IRAC Monographs-100C (2012); Arsenic and Arsenic Compounds, Chromium and Chromium Compounds and Cadmium and Cadmium Compounds). Consuming them, either from eating contaminated seafood or from crops irrigated by contaminated water may lead to adverse health. Therefore, preventing them from entering the food chain is crucial by providing techniques for wastewater treatments at the source point.

5.3 CONCERN OF HEAVY METAL IN TEXTILE WASTEWATER IN MALAYSIA

According to Freedonia Group, an international business research company (Freedonia Group, 2009), global demand for dyes and pigments is expected to grow 6.0 percent per year to USD19.5 billion in 2019. There is also an increased demand for organic pigments since inorganic, heavy metal–based pigments are not environmentally friendly. Textile market accounted for over half of the global dye and organic pigment demand in 2014. Dye and pigment consumption will remain concentrated in the Asia/Pacific region where the majority of global textile product manufacturing can be found. Countries like China, India, Turkey, Pakistan and Indonesia are among major exporters in textile. Malaysia's textile industry had exports valued at RM6.6 billion (USD2.2 billion) in 2011 in accordance to Malaysian Investment Development Authority (MIDA Report 2012). After five years, textiles and textile product industries were recorded as the tenth largest export earner with RM13.9 billion, contributing approximately 1.8 percent to Malaysia's total export of manufactured goods (MIDA Report, 2016). This translates to more textile production in the country and inevitably more wastewater containing dyes and heavy metal compounds. A stringent requirement must therefore be put in place to avoid environmental degradation due to untreated textile wastewater (Pang and Abdullah, 2013). In Malaysia, the Department of Environment, Ministry of Natural Resources & Environment is responsible in imposing regulations regarding the removal of dyes and heavy metals from industrial effluent (Muyibi et al., 2008, Afroz et al., 2014). Industries violating the Environmental Quality Act (1974) will be penalized accordingly. Enforcement of this law is to ensure textile industries treat their dye and heavy metal–containing effluents to the required standard. Often, big companies have incorporated an environmental management system (complied with ISO14000) in their practices, but small to medium enterprises (SMEs) may have neglected to impose strict control of their wastewater (Visvanathan and Kumar, 1999). On-site wastewater treatment is a real challenge for the SMEs, perhaps due to limited resources for acquiring such facilities. A cheaper, simpler and smaller-scaled apparatus must therefore be invented that can be fixed to such industries without incurring too much cost. Currently the main methods of textile wastewater treatments are by physical and chemical means. Oxidative, Fenton's reagent, ozonation, photochemical and electrochemical destruction are examples of chemical methods. On the other hand, physical treatments include adsorption on activated carbon (Birgani et al., 2016), fly ash and coal, membrane separation, ion exchange and electrokinetic coagulation. Each of these processes have their own advantages and disadvantages and not one process can remove all pollutants effectively (Vandevivere et al., 1999, Barakat, 2011, Carolin et al., 2017). In this chapter the use of 1-D α-Fe$_2$O$_3$ for Cr(VI) removal will be discussed as an example of a simple technique for heavy metal removal.

5.4 α-Fe$_2$O$_3$ FOR Cr(VI) REMOVAL

Among all of the heavy metal ions mentioned previously, concerns regarding the presence of Cr in the environment have been focused on the adverse health effects caused by Cr(VI) contaminated water and soil. Cr(VI) has been documented since the 1920s as carcinogenic in nature after multiple cases denoted the increase of nasal and lung cancer among industrial workers in direct contact with Cr(VI) compounds. The most affected would be those working on chromate production, chromate pigments production and chromate plating in metal industries; inhaling chromate containing fumes and dust constantly. A number of studies were collated and by the 1980s considerable evidence had accumulated on the cancer risk of Cr(VI) leading to the identification of Cr(VI) as a human carcinogen by inhalation. There has been concern on the hazards related to ingestion of Cr(VI) as well, with the conclusion of Cr(VI) to be potentially carcinogenic by ingestion (Costa, 2003). In addition, chromium compounds have also been documented to induce genotoxicity and DNA damage (Wani et al., 2018), effecting the renal system (kidney and liver damage) and immune system (ATSDR, 2011, Wilbur et al., 2012). Due to its obvious dangers, exposure of humans to Cr(VI) must be avoided. Nonetheless, Cr(VI) compounds are important in modern life as they have been widely utilized as inorganic pigments for textile; for instance, lead chromate is a well known yellow pigment for coloring fibers. Moreover, chromate compounds including lead chromates are well known corrosion preventative compounds in essentially all industries. Cr(VI) is a diprotic acid with pK_a values of 0.8 and 6.5. In textile effluent (with pH of 6–8) Cr(VI) can exist as oxyanions of chromate (CrO$_4^{2-}$) and bichromate (HCrO$_4^{1-}$). Both Cr(VI) oxyanions are highly mobile and strong oxidants. As chromates are highly mobile in aqueous environments, especially at higher pH conditions, sorption is rather difficult, making removal via adsorption to be less efficient. On the other hand, Cr(III) forms sparingly soluble solids like Cr(OH)$_3$ in the environment and hence is less mobile. Therefore, redox transformation of Cr(VI) to Cr(III) have been seen as an effective process for controlling the transport and fate of chromium in the environment. Cr(VI) can be reduced by several well-known various reductants. Cr(VI) reduction by Fe(II) and other inorganic reductants has been long acknowledged (Espenson, 1970). Reduction can also occur by Cr(VI) by organic compounds such as alkanes, alcohols, aldehydes, ketones, and aliphatic and aromatic acids. Reduction by bacteria has also been reported whereby microbial remediation has been seen successful for the removal of Cr(VI) by reduction. Photocatalytic reduction, as introduced is an alternative route in achieving redox transformation of Cr(VI) to Cr(III). To date TiO_2 is the most experimented oxide used as a photocatalyst (Cheng et al., 2015). Nevertheless, TiO_2 has drawbacks in visible light activity whereby it can only be activated under UV range (280–390 nm) which accounts for only 4% of solar radiation. This is due to the wide bandgap the oxide has (Roy et al., 2010). By selecting a narrow band gap semiconductor, energy from the sun can be harvested to create a system which can convert sun energy into useful electrons with the right potential energy for reduction of heavy metal ions. Reduced heavy metal ions can then be precipitated out. α-Fe$_2$O$_3$ (hematite) is a promising candidate as it has a small energy gap of ~2 eV (Turner et al., 1984, Sivula et al., 2012). Moreover, amongst the series of iron oxides, α-Fe$_2$O$_3$ is thermodynamically the most stable under ambient conditions and hence, synthesizing them is not that difficult.

Mishra and Chun (2015) have summarized in their review paper on the possibilities of the use of α-Fe$_2$O$_3$ as photocatalysts, but are more focused on dye degradation, whereas Dave and Chopda (2014) reviewed on the application of magnetic properties of iron oxide nanoparticles as nanoadsorbent for heavy metal removal from water/wastewater. A more comprehensive review is by Hua et al. (2012) on the use of metal oxide including iron oxide for heavy metal removal via both reduction and absorption. According to Hua et al., the most reported iron oxide for heavy metal removal included goethite (α-FeOOH), hematite (α-Fe$_2$O$_3$), amorphous hydrous Fe oxides, maghemite (γ-Fe$_2$O$_3$), magnetite (Fe$_3$O$_4$) and iron/iron oxide (Fe@Fe$_x$O$_y$). Iron oxides are concluded to have an effective and specific adsorption toward heavy metal including Cu(II), Pb, Cr(VI) and Ni. As for reduction of Cr(VI) via photocatalytic iron, only several documents are available as summarized in Table 5.1.

5.4.1 Properties of α-Fe$_2$O$_3$

In α-Fe$_2$O$_3$, the anions, O^{2-} are arranged in a hexagonal closed-packed lattice along [001] direction whereas the cations, Fe^{3+} occupy two-thirds of the octahedral interstices in the (001) basal planes. The cations can also be thought as pairs of FeO$_6$ octahedra that share edges with three neighboring octahedrals. Such arrangements influence the magnetic, electronic and optical properties of the oxide. The electronic properties of α-Fe$_2$O$_3$ is of great interest in order to understand the performance of this material as photocatalyst. Morin (1951) described the electrical properties of α-Fe$_2$O$_3$ and suggested two conduction models: conduction occurring in the d-level of the Fe ions and in the sp bonds of oxygen. The exact conduction mechanism remains unclear, but it is well accepted that α-Fe$_2$O$_3$ is photoactive and can produce a reasonable photocurrent, which means that electrons can be produced on the surface of α-Fe$_2$O$_3$. The n-type property of α-Fe$_2$O$_3$, originates from oxygen deficient in α-Fe$_2$O$_3$ crystals thus, synthesis process of the oxide will influence the number of electrons produced. The same electrons can also be expected to reduce certain heavy metal ions. It is interesting that the n-type conduction occurs due to thermally activated hopping movements of electrons from 2+ to 3+ ions within the oxide. Such phonon assisted mechanisms will result in the increase of mobility of carriers with temperature and hence, α-Fe$_2$O$_3$ is expected to produce an increased number of free electrons when heated up. This will indeed create a

TABLE 5.1
Cr(VI) Photoreduction on Fe$_2$O$_3$-Based Photocatalysts

Photocatalyst	Fabrication Method	Light Source	C$_o$ (mg/L)	pH	Catalyst Loading (g/L)	Hole Scavenger	Removal Efficiency (%)	Time (min)
Nanosized Fe$_2$O$_3$ supported on natural Algerian clay (Mekatel et al., 2012)	Impregnation	Visible light (200 W W lamp)	100	2	1	Oxalic acid	100	40
Nanosized Fe$_2$O$_3$ supported on natural Algerian clay (Mekatel et al., 2012)	Impregnation	Sunlight	100	2	1	Oxalic acid	80	300
CdS/α-Fe$_2$O$_3$ nanocomposites (Zhang et al., 2013)	Chemical bath method	Visible light (500 W Xe lamp)	50	3	0.25	Formic acid	100	60
α-Fe$_2$O$_3$/g-C$_3$N$_4$ composite (Xiao et al., 2015)	Hydrothermal	Visible light (300 W Xe lamp)	10	2	2	—	98	150
α-Fe$_2$O$_3$/g-C$_3$N$_4$ composite (Xiao et al., 2015)	Hydrothermal	Visible light (300 W Xe lamp)	10	2	2	Citric acid	99	15
α-Fe$_2$O$_3$/g-C$_3$N$_4$ composite (Xiao et al., 2015)	Hydrothermal	Visible light (300 W Xe lamp)	10	2	2	Oxalic acid	97	20
α-Fe$_2$O$_3$/activated graphene composite (Du et al., 2015)	Microwave-assisted	Visible light	10	2	1	—	95.28	160
Pure α-Fe$_2$O$_3$ (Du et al., 2015)	Microwave-assisted	Visible light	10	2	1	—	25.26	160
Fe$_3$O$_4$/rGO nanocomposite (Boruah et al., 2016)	Chemical approach	Sunlight	25	3	0.5	—	96	25

(Continued)

TABLE 5.1 (Continued)
Cr(VI) Photoreduction on Fe$_2$O$_3$-Based Photocatalysts

Photocatalyst	Fabrication Method	Light Source	C_o (mg/L)	pH	Catalyst Loading (g/L)	Hole Scavenger	Removal Efficiency (%)	Time (min)
ZrO$_2$/Fe$_3$O$_4$ (Kumar et al., 2016)	Co-precipitation	Sunlight	70	2	0.5	—	71.2	120
ZrO$_2$/Fe$_3$O$_4$/chitosan (Kumar et al., 2016)	Co-precipitation	Sunlight	70	2	0.5	—	84.7	120
ZrO$_2$/Fe$_3$O$_4$/chitosan (Kumar et al., 2016)	Co-precipitation	Sunlight	70	2	0.5	Ethanol	98.5	120
TiO$_2$/Fe$_3$O$_4$ (Challagulla et al., 2016)	Sol-gel	UV (125 W Hg lamp)	10	3	0.3	Oxalic acid	100	30
Nano-Fe$_3$O$_4$@Fe$_2$O$_3$/Al$_2$O$_3$ (Nagarjuna et al., 2017)	Co-precipitation	Visible light	50	3	0.3	Oxalic acid	100	25

rather effective photocatalyst under sunlight illumination, which can induce both electron generation and faster electron movement.

The highest occupied energy states in α-Fe_2O_3 are primarily O_p in character and the lowest unoccupied states are from an empty Fe_d band. The energy gap measurements have been reported by various authors with the majority of authors reporting on the indirect band gap of ~2 eV. The absorption of photons by α-Fe_2O_3 begins near the infra-red spectral region with rather weak absorption coefficient, a of ~10^3 cm^{-1}. α increases significantly to 105 cm^{-1} in between 1.9–2.2 eV. Yeh and Hackerman in 1977 reported on the generation of photocurrent from iron oxide grown by thermal oxidation. The photocurrent for the iron oxide electrode was recorded to begin at 400 nm and the value was increased with reduction of wavelength until around 850 nm whereby the photocurrent was reduced (Yeh and Hackerman 1977). This indicates electrons generation by the oxide with coverage of the whole visible region in the sun spectra.

In recent years, progress has been made in increasing the photocurrent (and hence hydrogen evolution) on α-Fe_2O_3 by nanostructuring and doping (Beermann et al., 2000, Lindgren et al., 2002, Chernomordik et al., 2012, Sivula, 2012, Rozana et al., 2014, 2015, 2016 and 2017). This is done to reduce recombination, as it is known that the carrier mobility in α-Fe_2O_3 is rather small and hence, the oxide suffers from a high rate of recombination. Moreover, short diffusion lengths of minority carriers in the oxide makes the oxide suffer from lack of free carriers at the surface (Rozana et al., 2017). This is done to reduce recombination as it is known that the carrier mobility in α-Fe_2O_3 is rather small and hence the oxide suffers from high rate of recombination. Similarly, to improve on the photocatalytic reduction process that may occur on α-Fe_2O_3, nanostructuring the oxide can be done. Among various nanostructures that can be synthesized, 1-D nanowires are thought to have many advantages (Li et al., 2012) especially when used as a photocatalyst. A photocatalytic process to reduce heavy metal ions can be described as a solid-liquid system, whereby α-Fe_2O_3 is the solid heterogeneous photocatalyst that will be immersed in liquid (solution containing heavy metal ions). The main objective of the photoreduction process is to reduce the concentration of heavy metal ions from the solution and by utilizing nanowires; total removal can be achieved considering all the advantages nanowire structures possess. The first advantage of using α-Fe_2O_3 nanowires is that the specific surface area will increase. The concomitant increase in the number of surface sites greatly enhances the efficiency of heavy metal ion absorption and hence, removal as the overall charge transfer kinetics at the semiconductor|solution interface is thought to have increased significantly. 1-D nanowires have high surface energy which makes them more reactive (Liao et al., 2008, Li et al., 2012). Moreover, sufficient light absorption can also be achieved using such fine nanostructures translating into effective electron-hole pairs generation process. Another advantage is the shorter diffusion path lengths for the photogenerated charge carriers; thus, they will not be scattered or recombined before successfully reaching the liquid. With a very small diameter, carriers will only have to travel an extremely short distance in order to reach the semiconductor|solution interface. This can further improve on electron transfer process for efficient reduction to occur.

5.4.2 MEASURING Cr(VI) REDUCTION IN AQUEOUS SOLUTION

When α-Fe$_2$O$_3$ suspension or supported α-Fe$_2$O$_3$ immersed in Cr(VI) solution is irradiated with light of 500 nm wavelength, electron-hole pairs are formed within the oxide (eq. 5.1). The photogenerated carriers responsible for the redox process occurring on the surface of the α-Fe$_2$O$_3$ are holes for oxidation of water (eq. 5.2) and electrons for reduction of Cr(VI) (eq. 5.3). The reaction scheme can be summarised as:

Step 1: Carrier generation

$$Fe_2O_3 \rightarrow Fe_2O_3(e^- + h^-) \tag{5.1}$$

Step 2: Oxidation of water

$$H_2O + 4h^+ \rightarrow O_2 + 4H^+ \tag{5.2}$$

Step 3: Oxidation of water

$$Cr^{6+} + e^- \rightarrow Cr^{3+} \tag{5.3}$$

Reduction of Cr(VI) compounds on a photocatalyst surface is pH dependent and in aqueous solutions with pH 6–8, the most probable Cr(VI) species are chromate, CrO_4^{2-}; and bichromate, $HCrO_4^-$; both of which are very mobile with reduction potentials ($E°$) greater than 1 V as shown in equations (5.4) and (5.5).

$$CrO_4^{2-} + 8H^+ + 3e^- \rightarrow Cr^{3+} + 4H_2O \quad E° = 1.07 \text{ V} \tag{5.4}$$

$$HCrO_4^{1-} + 7H^+ + 3e^- \rightarrow Cr^{3+} + 4H_2O \quad E° = 1.20 \text{ V} \tag{5.5}$$

In a more acidic environment, $Cr_2O_7^{2-}$ ions are the dominating species and the reduction and concurrent water oxidation are shown in equation (5.6). Figure 5.2a shows a schematic of band edges of α-Fe$_2$O$_3$ whereas Figure 5.2b depicts the process of reduction occurring on a single oxide particle.

$$Cr_2O_7^{2-} + 6e^- + 14H^+ \rightarrow 2Cr^{3+}(aq) + 7H_2O \quad E° = 1.33 \text{ V} \tag{5.6}$$

From equation (5.4) through equation (5.6), it is evident that Cr(VI) reduction is a multiproton reduction process; thus, the reaction must be done in an acidic environment as they are less favorable in alkaline environments. As also seen, reduction ought to occur in the presence of electrons of adequate potential. Moreover, as the point zero charge (PZC) of oxide semiconductors are rather low, adsorption can occur more at a lower pH. However, even with a good absorbing surface and suitable band edge energy, due to the slow kinetics of water oxidation, photoreduction of Cr(VI) can be a sluggish process. Therefore, a more effective hole scavenger that can capture holes from the VB is required. Apart from pH of the solution to-be-treated and hole scavengers, there are various other parameters that ought to be studied in ensuring rapid reduction on heterogeneous photocatalysis: nature of the solution,

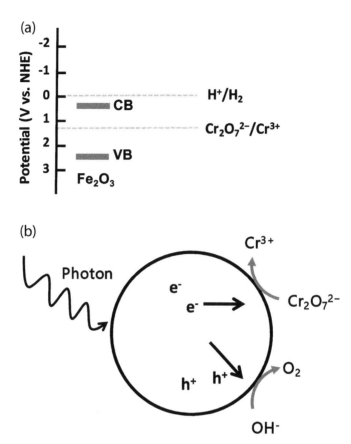

FIGURE 5.2   (a) Band edges of Fe_2O_3 at standard conditions (pH 0) with respect to the redox potentials for H_2 generation and $Cr(VI)/Cr^{3+}$. (b) Schematic depicting photocatalytic reduction of hexavalent chromium.

initial concentration of the solute, photocatalyst loading, light intensity, temperature and reactor design.

Water matrices in real wastewater treatment plants are rather complex and hence, the exact nature of the chromate species and their concentration are not known. This could interferes with the reduction process inducing possibility of Cr(III) reduction to Cr(VI).

Moreover, the complexity of the water to-be treated will obviously influence on the kinetic of the reduction process. The kinetic is also determined by the light intensity as well as the source of light. Increasing photon-flux (number of incident photons per unit time per unit area) can increase electron-hole pairs generation rate, subsequently, the reduction rate. Nevertheless, if the generation process exceeds the reduction rate, recombination will occur reducing the effectiveness of the catalytic system. In addition, intensity of light from artificial sources can be well controlled, but intensity of sunlight varies from one location to another and from one day to another, depending on weather and thus affecting reduction process. In an open system, heat from the sun can also evaporate water away; altering the concentration of the heavy metal ions or promoting other thermally activated chemical reactions. Catalyst loading is also an important parameter that determines reduction kinetics; higher loading of catalyst can lead to faster reduction process. Nevertheless, with too much catalyst a so-called shielding effect can be induced whereby light cannot reach some of the catalysts reducing generation.

Often a reactor is used for the reduction process to take place. An example of a photocatalyst reactor utilizing artificial light and TiO_2 as catalysts for Cr(VI) reduction is shown in Figure 5.3 (a – open) and (b – closed and lights on). The lamps are arranged in a circle and a vial (shown as a bottle) containing wastewater and photocatalysts are placed right in the middle of the reactor. The lamps surround the reaction vial and the distance between the vial and the lamps can be controlled. The system is covered on their periphery by thick aluminum foil as to further improve on the light absorb efficiency of the photocatalysts.

After reduction occurs, the concentration of Cr(VI) in aqueous solution is measured to determine the success in the removal by reduction process. A typical analytical technique to measure total Cr in solution is inductively coupled plasma (ICP), paired with mass spectrometry or atomic emission spectrometry. A graphite furnace and flame atomic absorption spectrophotometer can also be used (Marqués et al., 2000, Rakhunde et al., 2011). Nonetheless, to obtain Cr(VI) and Cr(III) concentration, it is best to separate the two oxidation states using high-performance liquid chromatography. They then will be detected and quantified with ICP and mass spectrometry. Such techniques can be used to deduce the percentage of Cr(VI) reduction indicating the success of the reduction process. A more common approach as well as one of the most used is via a colorimetric reaction of Cr(VI) with diphenylcarbazide (DPC) in solution. This method is referred to 7196 A by the United Stated Environmental Protection Agency, US EPA (EPA, 1992). In this method, extract to be tested will be added in with DCP solution and acid (as to reduce pH to 2). Once the color of the solution has developed, the solution will be transferred to an absorption cell and absorbance at 540 nm will be measured by using a spectrometer. Absorbance reading will be corrected and from the corrected absorbance, the concentration of Cr(VI) in

(a)   (b)

FIGURE 5.3   An example of a lab-scale Cr(VI) reduction photocatalysis system (a) irradiation source and catalysts-carrying bottle places in the middle and (b) when the system is closed with lights on. (Reprinted from Athanasekou et al. 2017. Copyright 2018, with permission from Elsevier.)

mg/L can be deduced by reference to a calibration curve. A plot of C_t (concentration at a given time) over C_o (initial concentration) is plotted based on the DCP colorimetric method, in order to assess on the feasibility of $\alpha\text{-Fe}_2O_3$ as a reducing photocatalysts. A UV-visible spectrometer has been used to measure the absorbance for this case.

A series of experiments of Cr(VI) reduction of $\alpha\text{-Fe}_2O_3$ nanowires formed by thermal oxidation have been conducted in our laboratories utilising the above mentioned DCP method. The first indication to denote success in reduction is the color changes from pink to colorless. A typical C_t over C_o plot against time is shown in Figure 5.4a. As can be seen, reduction on nanowires is rather rapid indicating the feasibility of $\alpha\text{-Fe}_2O_3$ nanowires to successfully reduce Cr(VI) under sunlight. Figure 5.4b is the scanning electron microscope (SEM) image of oxidized iron comprising of aligned nanowires utilised as photocatalyst for this experiment. The nanowires are relatively well aligned with a length of ~1 μm and diameter varies from 10 to 100 nm. Majority of the nanowires are tapered with larger base and extremely small diameter at the top.

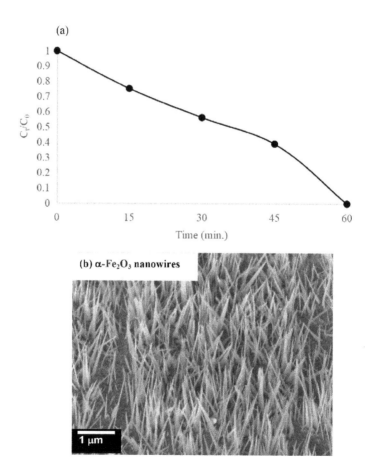

FIGURE 5.4 (a) Photocatalysis reduction of Cr(VI) under sunlight on $\alpha\text{-Fe}_2O_3$ nanowires formed by thermal oxidation of iron (b) SEM micrograph of the $\alpha\text{-Fe}_2O_3$ nanowires.

5.5   OXIDATION OF IRON FOR α-Fe₂O₃ NANOWIRES FORMATION

Thermal oxidation can be regarded as a type of corrosion but, unlike electrochemical corrosion (or anodization), the process occurs in a dry state. The result of thermal oxidation is an oxide film comprising metal cations and oxygen (Hauffe, 1967, Kofstad and Stiedel, 1967, Sedriks, 1996, Khanna, 2002, McCafferty, 2010). It is known that iron at ambient temperature is prone to develop a native oxide (and hydroxide) on its surface due to oxidation. The oxidation process is active, often leading to the degradation of iron. At elevated temperatures, the kinetic energy of iron atoms is higher and the oxidation rate is said to be rectilinear whereby the interfacial reaction is rate determining. The oxide film will continue to thicken and an oxygen gradient across the film is established which may affect and alter the composition of the oxide scale (Hauffe, 1967, Kofstad and Stiedel, 1967, Khanna, 2002, McCafferty, 2010).

Unlike zinc, for example whereby only ZnO is formed, (Tan et al., 2011), the growth of the oxide scale is determined by diffusion of anions inwards and metal ions outwards. Figure 5.5 illustrates a typical oxide film (MO) growth process by ionic transport (M^{2+} and O^{2-}) along with electrons movement going in the same direction as O^{2-}. The ionic movement is determined by diffusion of ions between vacant lattice sites (vacancies) and the process can only occur at high temperature (Wagner, 1938). At intermediate temperatures, ionic diffusion occurs via grain boundaries. This process is termed as short-circuit diffusion process and will be dependent on the grain size of the oxide and the feature of the grain boundaries.

Unlike zinc, for example whereby only ZnO is formed, (Tan et al., 2011), the scale formed during oxidation of iron is rather complex as the scale formed is typically consisting of discrete layers with very different phases and microstructures (Wood, 1962, Cornell and Schwertman, 2003, Chen and Yuen, 2003, Dunnington et al., 1952). For iron oxidation above 570°C, the oxide scale consists of three phases: α-Fe₂O₃ (hematite), Fe₃O₄ (magnetite) and FeO (wustite) whereas, below 570°C the scale only consists of hematite and magnetite (Cornell and Schwertman, 2003, Enache et al., 2011). The hematite is the surface oxide whereas magnetite and wustite form the inner layer. The gradation of phase arises due to the difference on the oxygen partial pressure requirement. Wustite is rich of metal and requires the lowest oxygen partial pressure

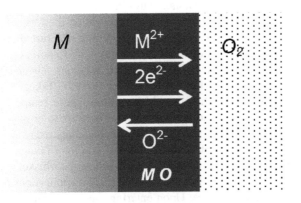

FIGURE 5.5   Transport processes operating during growth of oxide.

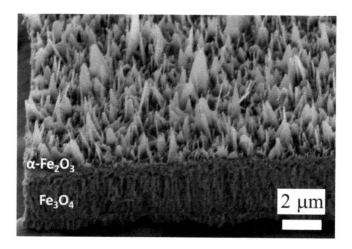

FIGURE 5.6 Cross sectional SEM image of oxidised iron at 500°C for 2 hours.

whereas hematite (Fe$_2$O$_3$) is oxygen rich thus requiring a higher partial pressure of oxygen to form (Chen and Yuen, 2003). The relative thickness of these layers is sensitive to the oxidation conditions: temperature and oxygen content as well as the surface finish (pre-treatment procedure) and the orientation (and texture) of the surface grains of the metal to be oxidized (Juricic et al., 2006). The composition of the parent metal also influences the thickness of the multilayer oxide formation. The intriguing thing about oxidized iron is, despite the multi-layered scale, the outer hematite layer is actually comprised of two different structures; a thin inner hematite layer which is compact and arrays of the 1-D nanowires (or whiskers as reported by Raynaud and Rapp, 1984) at the surface along with 2-D nanosheets structure as observed recently by Budiman et al., 2016a. Figure 5.6 shows one example of cross sectional image of oxidized iron at 500°C for 2 hours grown in our laboratory. The thin inner hematite layer has an equiaxed structure whereas the inner Fe$_3$O$_4$ layer is comprised of large columnar grains. The inner layer is porous.

To produce only α-Fe$_2$O$_3$ nanowires, oxidation temperatures can be varied. 400°C–800°C is a range in which nanowires have been observed. We have observed however that, for 2 hours oxidation in dry air condition the formation of nanowires can be seen at 400°C to 600°C (Figure 5.7a–c) but the structure disappears at 700°C (Figure 5.7d) and emerges again at 800°C (Figure 5.7e). Reports on the formation of iron oxide "blades" or "whiskers" by thermal oxidation are few and perhaps the first comprehensive study was performed by Takagi in 1957. Takagi suggested several factors which affect the growth of the whiskers: atmosphere, temperature and surface preparation. The optimum condition for whisker structure to grow is at temperatures ranging from 400°C to 850°C in oxygen atmosphere (compared to air) and the surface preparation must be done by electropolishing the iron foil (Takagi, 1957). Following this work, several other reports emerged to explain how the whiskers formed and grow on the surface of the iron foil as shown in Table 5.2. No applications on such whiskers were reported in early literature. Upon entering the era of nanotechnology (around late 1990s), the term whiskers somehow slowly disappeared and the microstructure

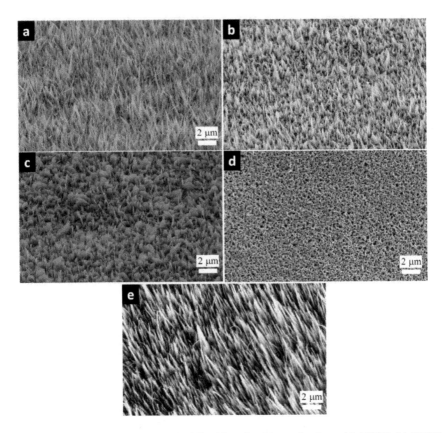

FIGURE 5.7   SEM micrographs of oxidised iron for 2 hours in air at: (a) 400°C, (b) 500°C, (c) 600°C, (d) 700°C and (e) 800°C.

formed has been termed "nanowires." Nanowires are, by definition, conducting fibers with diameter in nanoscale (as described in Chapter 1).

Recent works on the formation of nanowires have focused on the formation of nanowires with high areal density, i.e. a greater number of wires per given area at as short an oxidation time as possible. Fu et al. successfully synthesized large arrays of aligned α-Fe$_2$O$_3$ nanowires by oxidizing an iron at 600°C under CO_2 + SO_2 + NO_2 + H_2O atmosphere for 10 hours (Fu et al., 2003). Hiralal et al. (2011) studied the effect of oxygen pressure, temperature and annealing time (Hiralal et al., 2008). They reported that the increase of time and temperature lead to the increase of density and size of nanowires whereas the oxidation pressure did not affect the nanowires formation. Srivastava et al. reported on the effect of iron texture on the formation of nanowires and reported that the nanowires tend to grow on [1 1 0] oriented grain (Srivastava et al., 2011). The length and width of the nanowires depends on the oxidation conditions. Table 5.2 depicts several important works on this subject. As can be seen, the width of the nanowires can be as large as 150 nm while the length varies from 0.1 to 5 µm. Nonetheless, the measurement of the diameter and the length can be a bit difficult as the nanowires tend to bend when they are very long and the bottom diameter at the

TABLE 5.2
Literature Review on Oxidation of Iron for α-Fe_2O_3 Formation

	Thermal Oxidation Conditions			
Author (Year)	Atmosphere	Temperature (°C)	Time (hour)	Morphology
Takagi (1957)	Air	400–850	—	α-Fe_2O_3 whiskers
Tallman and Gulbransen (1968)	Dry O_2	100–500		α-Fe_2O_3 whiskers
Voss et al. (1982)	20 Torr O_2	600	1	Micron-size α-Fe_2O_3 blades
Fu et al. (2001)	19.4% CO_2, 80.43% N_2, 0.164% SO_2, H_2O vapor	550–650	10–120	α-Fe_2O_3 nanowires
Fu et al. (2003)	19.3% CO_2, 80.56% N_2, 0.14% SO_2, H_2O (steam)	540–600	10–30	α-Fe_2O_3 nanowires
Wang et al., 2005	(CO_2/N_2/SO_2, 19.3:80.56:0.14) and H_2O vapor	550–600	Several days	α-Fe_2O_3 bicrystalline nanowires
Wen et al. (2005)	N_2 (20 sccm) O_2 (2–5 sccm)	800	10	α-Fe_2O_3 nanowires
Han et al. (2006)	19.4% CO_2, 80.43% N_2, 0.164% SO_2, H_2O vapor	550–650	10–120	α-Fe_2O_3 nanowires
Srivastava et al. 2007	Ozone-rich	700	2–4	α-Fe_2O_3 nanowires
Zheng et al. 2007	Air	260–400 (hot plate)	10	α-Fe_2O_3 nanoflake
Hiralal et al. (2008)	Ar & O_2 50 sccm	400–620	1 to 10	α-Fe_2O_3 nanowires
Hsu et al. (2008) Bertrand et al. (2010)	Air	350	10	α-Fe_2O_3 nanowires
Nasibulin et al. (2009)	Air	700	A few seconds	α-Fe_2O_3 nanowires
Hiralal et al. (2011)	Air	255	24	
Srivastava et al. (2011)	Moist O_2	700	16	α-Fe_2O_3 nanowires
Zhong et al. (2011)	Air	280–480	—	Quasi 1-D α-Fe_2O_3
Yuan et al. (2012)	200 Torr O_2	(i) 400 (ii) 600	1	α-Fe_2O_3 nanowires
Vincent et al. (2012)	1:1 O_2:Ar gases flow of 100 sccm	600–800	8–10	α-Fe_2O_3 nanorods and nanocorals

(Continued)

TABLE 5.2 (*Continued*)
Literature Review on Oxidation of Iron for α-Fe$_2$O$_3$ Formation

Author (Year)	Thermal Oxidation Conditions			Morphology
	Atmosphere	Temperature (°C)	Time (hour)	
Grigorescu et al. (2012)	Air	600	1	α-Fe$_2$O$_3$ platelet, corals and nanowires
Vesel and Balat-Pichelin (2014)	Non-equilibrium O$_2$ plasma	927	—	Iron oxide nanowires
Budiman et al. (2016a)	Water vapor	500 and 800	2	α-Fe$_2$O$_3$ nanorods and nanosheets
Budiman et al. (2016b)	Dry air versus water vapor	300–500	2	α-Fe$_2$O$_3$ nanorods and nanosheets

substrate|oxide interface and at the oxide|air interface are different. α-Fe$_2$O$_3$ nanowires are often tapered to an extremely fine diameter at the oxide|air interface.

A typical mechanism describing whiskers formation was published by Wagner (1967) via a process called Vapour-Liquid-Solid (VLS). 1-D nanostructures grown due to VLS process have then been constantly reviewed and investigated by many authors including Wang et al. (2015) whereby mechanism of 1-D with variations in morphologies formation was proposed. Nevertheless, in oxidation process, source of vapour really depends on the nature of the metal and as of the case of iron which has rather high vapour pressure, the process is unlikely.

Two main mechanisms for nanowires formation by oxidation are (i) internal diffusion along a centre tunnel of a nanowires and (ii) surface diffusion at the sides of a nanowires. These two mechanisms are widely accepted despite the primary process is still not very clear. Suffice to say however, both mechanisms are probable, or occur simultaneously or in succession, but based on electron micrographs of the nanowires a central tunnel was observed by Voss et al. (1982) supporting an "easy diffusion" path mechanism for iron ions to move outward to form the nanowire as illustrated in Figure 5.8a. Here, the main process occurring is surface diffusion along this central tunnel of a nanowire. M represents metal ions which in our case are Fe ions. These ions are moving through the base layer outward reacting with adsorbed oxygen from the environment to form a new layer of oxide at the tip of the nanowire. As the surface diffusion rate is magnitudes faster than grain boundary diffusion, the occurrence may have promoted the formation of the elongated structure, developing further into nanowires as a function of temperature and time. On a contrary, in recent years Hiralal et al. (2008) proposed on the formation via an alternative model whereby the mechanism for the iron to reach the tip was proposed to be similar to the quasi-liquid layer on the surface of ice, where at temperatures well below 0°C a thin surface layer of water is always present. For the case of α-Fe$_2$O$_3$ nanowires, "bulbs" are often seen to exist at the tip of nanowires as illustrated in Figure 5.8b and such curved surfaces will have a much lower energy of dissociation than in the bulk, allowing for a rapid diffusion. The surface mobility is said to be a function of

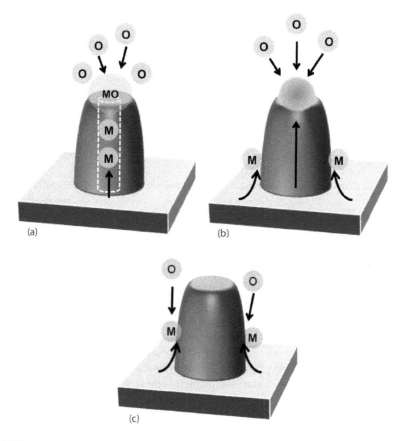

FIGURE 5.8 Illustrations on mechanisms proposed for the formation of oxide NW: (a) internal diffusion along a center tunnel of a nanowire, (b) surface diffusion at the sides of nanowires and oxygen dissociation at the tip, and (c) surface diffusion and oxygen dissociation at the sides.

time, with faster mobility at higher temperatures. Ion transport may have occurred by surface diffusion and growth occurs on steps on the side of the growing nanowire as shown in Figure 5.8c. Based on this, there is also the possibility that as the tip of the nanowire is in nanoscale, the tip may have possessed high curvature and surface energy that adsorption and dissociation of oxygen gasses are much faster than on bulk grains. Therefore, growth is accelerated at the tip forming the c-axis (long axis) growth compared to the a–b planes of the oxide.

Yuan et al. (2012) proposed an alternative mechanism of α-Fe$_2$O$_3$ nanowires formation. The mode of transportation of ions is thought to be evoked by the compressive stress generated in the oxide layer. They coined the term "stress-driven mechanism." Based on transmission electron microscopy (TEM) images from their work, Yuan et al. argued on how solid-state transformation of α-Fe$_2$O$_3$ to Fe$_3$O$_4$ results in generation of stresses at the interfacial regions. Stress generated can be released by several mechanisms like crack, spalling of the oxide scales and plastic deformation (Mitchell et al.,

1982). The nucleation of the nanowire occurs on the existing equiaxed α-Fe$_2$O$_3$ surface grains. And underneath this layer, larger grains of Fe$_3$O$_4$ and/or FeO exist dependent on the oxidation temperature. They concluded that screw dislocation possesses the steps necessary for nucleation of nanowires, then growth occurs by flux driven due to a stress gradient across the scale. Figure 5.8c shows surface diffusion along the sides of a nanowire. For this mechanism, Yuan et al. proposed the transfer of iron atoms onto the side wall of the nanowire via an adatom-nanowire exchange by surface diffusion.

All mechanisms above demonstrate the importance of the diffusion process in the formation of nanowires by oxidation process. The activation energy for surface diffusion is lower than the bulk (lattice) or grain boundary diffusion; thus, surface diffusion is regarded as the main process for nanowires formation. All these will however depend on the temperature at which the iron substrate is oxidized. Low and intermediate temperature mechanisms above demonstrate the importance. On the other hand, stress generated across the oxide scale can also induce flux of iron ions outward. The magnitude of the stress will therefore determine the concentration of reactants transported across the scale for nanowires growth. Nasibulin et al. (2009), on the other hand, proposed the electric field–assisted diffusion whereby, the outward diffusion of iron ions occur due to an electric field developed across the oxide layer.

Crystals often have imperfections such as screw dislocation line defects, which upon intersection with the crystal surface would make steps that propagate as spirals and thus become an endless source of crystal steps (Jin et al., 2010). This linear defect is usually produced as the response to tensile/compressive stress when applied and a "slip" in a single crystal commences when the shear stress achieves some critical value. Assuming that the surface of the oxide consists of such crystal steps, the formation of nanowires can be thought to have "nucleated" from such steps via surface diffusion (Figure 5.9a). The slipping crystal assists the nanowires to grow curl up and form a spiral-like structure. The TEM image in Figure 5.9b shows a single α-Fe$_2$O$_3$ nanowire. It can be observed that the nanowires are rather tapered with a larger width at the bottom. Once the nanowire structure has nucleated on steps at

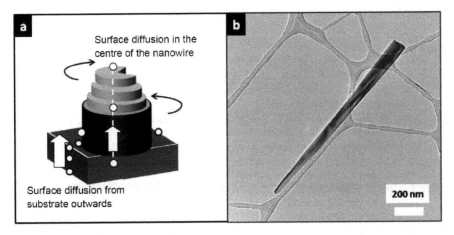

FIGURE 5.9 (a) Nanowire growth via screw dislocation and surface diffusion and (b) TEM image of a single α-Fe$_2$O$_3$ nanowire.

the surface grains, an internal diffusion mechanism occurs in which iron ions are transported along the screw dislocation core outwards to form new oxide at the tip. The elongated nanostructure always grows in [101] growth direction. There has also been some argument on lateral growth of nanowires forming blades or so-called nanosheets structure (Budiman et al., 2016a, 2016b).

5.6 CONCLUSION

The existence of pollutants in surface water, particularly from hazardous heavy metals and metalloids such as lead (Pb), mercury (Hg), chromium (Cr), cadmium (Cd) and arsenic (As) discharged from industrial establishments must be minimized as some of these heavy metals are carcinogens and some can induce social changes in human behaviour. These metals can enter humans through the food chain either by direct consumption of heavy metal-laden aquatic organisms or by consuming heavy metal-contaminated crops irrigated by polluted water. A treatment is therefore required to ensure a reduction of heavy metal ions polluting the environment. One treatment method is by a reduction process and so far reduction of Cr(VI) has been shown successful via photocatalysis technology utilizing nanostructured semiconductor oxide. In this chapter, we presented a brief overview on heavy metal pollutions and the possibilities of treatment with α-Fe_2O_3 nanowires. The nanowires can be formed by thermal oxidation of iron and the topic is reviewed in this chapter. The growth of the nanowires can be explained by the surface diffusion process, quasi-liquid process and stress-driven mechanism, and the formation can be as rapid as a few minutes and can be done at low temperature (<400°C).

ACKNOWLEDGMENTS

Heavy Metal Mitigation project is supported by USM-Research University Grant for Toyohashi University of Technology, Japan-USM collaboration; 1001/PBAHAN/870048.

REFERENCES

Afroz, R., M. M. Masud, R. Akhtar, and J. B. Duasa. 2014. Water pollution: Challenges and future direction for water resource management policies in Malaysia. *Environment and Urbanization ASIA* 5 (1):63–81.
Athanasekou, C., G. E. Romanos, S. K. Papageorgiou, G. K. Manolis, F. Katsaros, and P. Falaras. 2017. Photocatalytic degradation of hexavalent chromium emerging contaminant via advanced titanium dioxide nanostructures. *Chemical Engineering Journal* 318:171–180.
Barakat, M. A. 2011. New trends in removing heavy metals from industrial wastewater. *Arabian Journal of Chemistry* 4 (4):361–377.
Beermann, N., L. Vayssieres, S.-E. Lindquist, and A. Hagfeldt. 2000. Photoelectrochemical studies of oriented nanorod thin films of hematite. *Journal of the Electrochemical Society* 147 (7):2456–2461.
Bertrand, N., C. Desgranges, D. Poquillon, M. C. Lafont, and D. Monceau. 2010. Iron oxidation at low temperature (260–500°C) in air and the effect of water vapor. *Oxidation of Metals* 73 (1):139–162.

Bhuyan, M. S., M. A. Bakar, A. Akhtar, M. B. Hossain, M. M. Ali, and M. S. Islam. 2017. Heavy metal contamination in surface water and sediment of the Meghna River, Bangladesh. *Environmental Nanotechnology, Monitoring & Management* 8:273–279.

Birgani, P. M., N. Ranjbar, R. C. Abdullah et al. 2016. An efficient and economical treatment for batik textile wastewater containing high levels of silicate and organic pollutants using a sequential process of acidification, magnesium oxide, and palm shell-based activated carbon application. *Journal of Environmental Management* 184:229–239.

Bisschops, I., and H. Spanjers. 2003. Literature review on textile wastewater characterisation. *Environmental Technology* 24 (11):1399–1411.

Boruah, P. K., P. Borthakur, G. Darabdhara et al. 2016. Sunlight assisted degradation of dye molecules and reduction of toxic Cr(VI) in aqueous medium using magnetically recoverable Fe$_3$O$_4$/reduced graphene oxide nanocomposite. *RSC Advances* 6 (13):11049–11063.

Budiman, F., N. Bashirom, W. K. Tan, K. Abdul Razak, A. Matsuda, and Z. Lockman. 2016a. Rapid nanosheets and nanowires formation by thermal oxidation of iron in water vapour and their applications as Cr(VI) adsorbent. *Applied Surface Science* 380:172–177.

Budiman, F., N. Bashirom, W. K. Tan, K. Abdul Razak, A. Matsuda, and Z. Lockman. 2016b. *Procedia Chemistry* 19:586–593

Braslavsky, S. E., A. M. Braun, A. E. Cassano et al. 2011. Glossary of terms used in photocatalysis and radiation catalysis (IUPAC Recommendations 2011). *Pure and Applied Chemistry* 83 (4):931–1014.

Carolin, C. F., P. S. Kumar, A. Saravanan, G. J. Joshiba, and M. Naushad. 2017. Efficient techniques for the removal of toxic heavy metals from aquatic environment: A review. *Journal of Environmental Chemical Engineering* 5 (3):2782–2799.

Challagulla, S., R. Nagarjuna, R. Ganesan, and S. Roy. 2016. Acrylate-based polymerizable sol–gel synthesis of magnetically recoverable TiO$_2$ supported Fe$_3$O$_4$ for Cr(VI) photoreduction in aerobic atmosphere. *ACS Sustainable Chemistry & Engineering* 4 (3):974–982.

Chen, D., and A. K. Ray. 2001. Removal of toxic metal ions from wastewater by semiconductor photocatalysis. *Chemical Engineering Science* 56 (4):1561–1570.

Chen, D., M. Sivakumar, and A. K. Ray. 2008. Heterogeneous photocatalysis in environmental remediation. *Developments in Chemical Engineering and Mineral Processing* 8 (5–6):505–550.

Chen, R. Y., and W. Y. D. Yuen. 2003. Review of the high-temperature oxidation of iron and carbon steels in air or oxygen. *Oxidation of Metals* 59 (5):433–468.

Cheng, Q., C. Wang, K. Doudrick, and C. K. Chan. 2015. Hexavalent chromium removal using metal oxide photocatalysts. *Applied Catalysis B: Environmental* 176–177:740–748.

Cheng, S. 2003. Heavy metal pollution in China: Origin, pattern and control. *Environmental Science and Pollution Research* 10 (3):192–198.

Chernomordik, B. D., H. B. Russell, U. Cvelbar et al. 2012. Photoelectrochemical activity of as-grown, α-Fe$_2$O$_3$ nanowire array electrodes for water splitting. *Nanotechnology* 23 (19):194009.

Chowdhury, P., A. Elkamel, and A. K. Ray. 2015. Chapter 2 photocatalytic processes for the removal of toxic metal ions. In *Heavy Metals In Water: Presence, Removal and Safety*, S. Sharma (Ed.). Cambridge, UK: The Royal Society of Chemistry.

Cornell, R. M., and U. Schwertmann. 2003. *The Iron Oxides: Structure, Properties, Reactions, Occurences, and Uses*. Weinheim, Germany: Wiley-VCH.

Correia, V. M., T. Stephenson, and S. J. Judd. 1994. Characterisation of textile wastewaters–A review. *Environmental Technology* 15 (10):917–929.

Costa, M. 2003. Potential hazards of hexavalent chromate in our drinking water. *Toxicology and Applied Pharmacology* 188 (1):1–5.

Dave, P. N., and L. V. Chopda. 2014. Application of iron oxide nanomaterials for the removal of heavy metals. *Journal of Nanotechnology* 2014, Article ID 398569: 14 pages.

Davis, M. A., A. J. Signes-Pastor, M. Argos et al. 2017. Assessment of human dietary expo-
sure to arsenic through rice. *Science of the Total Environment* 586:1237–1244.
Dey, S., and A. Islam. 2015. A review on textile wastewater characterization in Bangladesh.
Resources and Environment 5 (1):15–44.
Du, Y., Z. Tao, J. Guan et al. 2015. Microwave-assisted synthesis of hematite/activated gra-
phene composites with superior performance for photocatalytic reduction of Cr(VI).
RSC Advances 5 (99):81438–81444.
Dunnington, B. W., F. H. Beck, and M. G. Fontana. 1952. The mechanism of scale formation
on iron at high temperature. *Corrosion* 8 (1):2–12.
Enache, C. S., Y. Q. Liang, and R. van de Krol. 2011. Characterization of structured α-Fe$_2$O$_3$
photoanodes prepared via electrodeposition and thermal oxidation of iron. *Thin Solid
Films* 520 (3):1034–1040.
Espenson, J. H. 1970. Oxidation of transition metal complexes by chromium (VI). *Accounts
of Chemical Research* 3 (10):347–353.
EPA. 1992: Environmental Proptection Agency (EPA), US. 2012. Method 7196 Chromium
Hexavalent (Colorimetric). Retrieved from: https://www.epa.gov/sites/production/
files/2015-12/documents/7196a.pdf
The Freedonia Group, 2009. World Dyes & Organic Pigments. Excess from: https://www.
freedoniagroup.com/
Fu, Y. Y., R. M. Wang, J. Xu et al. 2003. Synthesis of large arrays of aligned α-Fe$_2$O$_3$ nanow-
ires. *Chemical Physics Letters* 379 (3):373–379.
Fu, Y., J. Chen, and H. Zhang. 2001. Synthesis of Fe$_2$O$_3$ nanowires by oxidation of iron.
Chemical Physics Letters 350 (5):491–494.
Fujishima, A., and K. Honda. 1972. Electrochemical photolysis of water at a semiconductor
electrode. *Nature* 238 (5358):37–38.
Fujishima, A., T. N. Rao, and D. A. Tryk. 2000. Titanium dioxide photocatalysis. *Journal of
Photochemistry and Photobiology C: Photochemistry Reviews* 1 (1):1–21.
Ghaly, A. E., R. Ananthashankar, M. Alhattab, and V. V. Ramakrishnan. 2014. Production,
characterization and treatment of textile effluents: A critical review. *Journal of
Chemical Engineering & Process Technology* 5 (182): 1–18.
Grigorescu, S., C.-Y. Lee, K. Lee et al. 2012. Thermal air oxidation of Fe: rapid hematite nanow-
ire growth and photoelectrochemical water splitting performance. *Electrochemistry
Communications* 23:59–62.
Han, Q., Y. Y. Xu, Y. Y. Fu et al. 2006. Defects and growing mechanisms of α-Fe$_2$O$_3$ nanowires.
Chemical Physics Letters 431 (1):100–103.
Hashim, R., T. H. Song, N. Z. M. Muslim, and T. P. Yen. 2014. Determination of heavy metal
levels in fishes from the lower reach of the kelantan river, Kelantan, Malaysia. *Tropical
Life Sciences Research* 25 (2):21–39.
Hauffe. 1967. High temperature oxidation of metals. P. Kofstad John Wiley & Son, New York
1966, 340 S. *Materials and Corrosion* 18 (10):956–957.
Hiralal, P., H. E. Unalan, K. G. U. Wijayantha et al. 2008. Growth and process conditions of
aligned and patternable films of iron(III) oxide nanowires by thermal oxidation of iron.
Nanotechnology 19 (45):455608.
Hiralal, P., S. Saremi-Yarahmadi, B. C. Bayer et al. 2011. Nanostructured hematite photo-
electrochemical electrodes prepared by the low temperature thermal oxidation of iron.
Solar Energy Materials and Solar Cells 95 (7):1819–1825.
Hoffmann, M. R., S. T. Martin, W. Choi, and D. W. Bahnemann. 1995. Environmental appli-
cations of semiconductor photocatalysis. *Chemical Reviews* 95 (1):69–96.
Hossen, M. F., S. Hamdan, and M. R. Rahman. 2015. Review on the risk assessment of heavy
metals in Malaysian clams. *The Scientific World Journal* 2015, Article ID 905497:7 pages.

Hsu, L.-C., Y.-Y. Li, C.-G. Lo, C.-W. Huan and G. Chern. 2008. Thermal growth and magnetic characterization of α-Fe$_2$O$_3$ nanowires. *Journal of Physics D: Applied Physics* 41 (185003):5pp.

Hua, M., S. Zhang, B. Pan, W. Zhang, L. Lv, Q. Zhang. 2012. Heavy metal removal from water/wastewater by nanosized metal oxides: *A review. Journal of Hazardous Materials* 211–212:317–331.

Huang, Z., X.-D. Pan, P.-G. Wu, J.-L. Han, and Q. Chen. 2013. Health risk assessment of heavy metals in rice to the population in Zhejiang, China. *PLoS One* 8 (9):e75007.

International Agency for Research on Cancer. 2012. IARC Monographs-100C on Chromium Compounds.

International Agency for Research on Cancer. 2012. IARC Monographs-100C on Arsenic and Arsenic Compounds.

Islam, S., M. M. Rahman, M. R. Islam, and R. Naidu. 2017. Geographical variation and age-related dietary exposure to arsenic in rice from Bangladesh. *Science of the Total Environment* 601–602:122–131.

Jin, S., M. J. Bierman, and S. A. Morin. 2010. A new twist on nanowire formation: Screw-dislocation-driven growth of nanowires and nanotubes. *The Journal of Physical Chemistry Letters* 1 (9):1472–1480.

Juricic, C., H. Pinto, T. Wroblewski, and A. Pyzalla. 2006. Dependence of oxidation behavior and residual stresses in oxide layers on armco iron substrate surface condition. *Materials Science Forum* 524–525:963–968.

Kabra, K., R. Chaudhary, and R. L. Sawhney. 2008. Solar photocatalytic removal of metal ions from industrial wastewater. *Environmental Progress* 27 (4):487–495.

Khan, B., H. Ullah, S. Khan, M. Aamir, A. Khan, and W. Khan. 2016. Sources and contamination of heavy metals in sediments of Kabul River: The role of organic matter in metals retention and accumulation. *Soil and Sediment Contamination: An International Journal* 25 (8):891–904.

Khanna, A. S. 2002. *Introduction to High Temperature Oxidation and Corrosion.* Materials Park, OH: ASM International.

Kofstad, P., and C. A. Steidel. 1967. High temperature oxidation of metals. *Journal of The Electrochemical Society* 114 (7):167C–172C.

Kumar, A., C. Guo, G. Sharma et al. 2016. Magnetically recoverable ZrO$_2$/Fe$_3$O$_4$/chitosan nanomaterials for enhanced sunlight driven photoreduction of carcinogenic Cr(VI) and dechlorination & mineralization of 4-chlorophenol from simulated waste water. *RSC Advances* 6 (16):13251–13263.

Li, Y., X.-Y. Yang, Y. Feng, Z.-Y. Yuan, and B.-L. Su. 2012. One-dimensional metal oxide nanotubes, nanowires, nanoribbons, and nanorods: Synthesis, characterizations, properties and applications. *Critical Reviews in Solid State and Materials Sciences* 37 (1):1–74.

Liao, L., Z. Zheng, B. Yan et al. 2008. Morphology controllable synthesis of α-Fe$_2$O$_3$ 1D nanostructures: Growth mechanism and nanodevice based on single nanowire. *The Journal of Physical Chemistry C* 112 (29):10784–10788.

Lindgren, T., H. Wang, N. Beermann, L. Vayssieres, A. Hagfeldt, and S.-E. Lindquist. 2002. Aqueous photoelectrochemistry of hematite nanorod array. *Solar Energy Materials and Solar Cells* 71 (2):231–243.

Looi, L. J., A. Z. Aris, W. L. W. Johari, F. M. Yusoff, and Z. Hashim. 2013. Baseline metals pollution profile of tropical estuaries and coastal waters of the Straits of Malacca. *Marine Pollution Bulletin* 74 (1):471–476.

Marqués, M. J., A. Salvador, A. Morales-Rubio, and M. de la Guardia. 2000. Chromium speciation in liquid matrices: A survey of the literature. *Fresenius Journal of Analytical Chemistry* 367:601–613.

McCafferty, E. 2010. *Introduction to Corrosion Science*. New York: Springer-Verlag.

Meharg, A. A., G. Norton, C. Deacon et al. 2013. Variation in rice cadmium related to human exposure. *Environmental Science & Technology* 47 (11):5613–5618.

Mekatel, H., S. Amokrane, B. Bellal, M. Trari, and D. Nibou. 2012. Photocatalytic reduction of Cr(VI) on nanosized Fe_2O_3 supported on natural Algerian clay: Characteristics, kinetic and thermodynamic study. *Chemical Engineering Journal* 200–202:611–618.

Menezes, E. A., R. Carapelli, S. R. Bianchi et al. 2010. Evaluation of the mineral profile of textile materials using inductively coupled plasma optical emission spectrometry and chemometrics. *Journal of Hazardous Materials* 182 (1):325–330.

MIDA. 2012. Malaysia Investment Performance 2012. edited by Malaysian Investment Development Authority (MIDA).

MIDA. 2016. Malaysia Investment Performance 2016. edited by M. I. D. A. (MIDA): Malaysian Investment Development Authority (MIDA).

Mishra, M., and D.-M. Chun. 2015. α-Fe_2O_3 as a photocatalytic material: A review. *Applied Catalysis A: General* 498:126–141.

Mitchell, T. E., D. A. Voss, and E. P. Butler. 1982. The observation of stress effects during the high temperature oxidation of iron. *Journal of Materials Science* 17 (6):1825–1833.

Morin, F. J. 1951. Electrical properties of α-Fe_2O_3 and α-Fe_2O_3 containing titanium. *Physical Review* 83 (5):1005.

Mukherjee, A., M. Kundu, B. Basu et al. 2017. Arsenic load in rice ecosystem and its mitigation through deficit irrigation. *Journal of Environmental Management* 197:89–95.

Muyibi, S. A., A. R. Ambali, and G. S. Eissa. 2008. The impact of economic development on water pollution: Trends and policy actions in Malaysia. *Water Resources Management* 22 (4):485–508.

Nagarjuna, R., S. Challagulla, R. Ganesan, and S. Roy. 2017. High rates of Cr(VI) photoreduction with magnetically recoverable nano-Fe_3O_4@Fe_2O_3/Al_2O_3 catalyst under visible light. *Chemical Engineering Journal* 308:59–66.

Nasibulin, A. G., S. Rackauskas, H. Jiang, Y. Tian, P. Reddy Mudimela, S. D. Shandakov, L. I. Nasibulina, J. Sainio, and E. I. Kauppinen. 2009. Simple and rapid synthesis of α-Fe_2O_3 nanowires under ambient conditions. *Nano Research* 2:373–379.

Noreen, M., M. Shahid, M. Iqbal, and J. Nisar. 2017. Measurement of cytotoxicity and heavy metal load in drains water receiving textile effluents and drinking water in vicinity of drains. *Measurement* 109:88–99.

Norton, G. J., P. N. Williams, E. E. Adomako et al. 2014. Lead in rice: Analysis of baseline lead levels in market and field collected rice grains. *Science of the Total Environment* 485–486:428–434.

Pang, Y. L., and A. Z. Abdullah. 2013. Current status of textile industry wastewater management and research progress in Malaysia: A review. *Clean–Soil, Air, Water* 41 (8):751–764.

Praveena, S. M., and N. A. Omar. 2017. Heavy metal exposure from cooked rice grain ingestion and its potential health risks to humans from total and bioavailable forms analysis. *Food Chemistry* 235:203–211.

Rajeshkumar, S., Y. Liu, X. Zhang, B. Ravikumar, G. Bai, and X. Li. 2018. Studies on seasonal pollution of heavy metals in water, sediment, fish and oyster from the Meiliang Bay of Taihu Lake in China. *Chemosphere* 191:626–638.

Rakhunde, R., L. Deshpande, and H. D. Juneja. 2012. Chemical speciation of chromium in water: A review. *Critical Reviews in Environmental Science and Technology* 42 (7):776–810.

Rasheed, H., P. Kay, R. Slack, and Y. Y. Gong. 2018. Arsenic species in wheat, raw and cooked rice: Exposure and associated health implications. *Science of the Total Environment* 634:366–373.

Raynaud, G. M., and R. A. Rapp. 1984. In situ observation of whiskers, pyramids and pits during the high-temperature oxidation of metals. *Oxidation of Metals* 21 (1):89–102.

Robinson, T., G. McMullan, R. Marchant, and P. Nigam. 2001. Remediation of dyes in textile effluent: A critical review on current treatment technologies with a proposed alternative. *Bioresource Technology* 77 (3):247–255.

Roy, P., D. Kim, K. Lee, E. Spiecker, and P. Schmuki. 2010. TiO_2 nanotubes and their application in dye-sensitized solar cells. *Nanoscale* 2 (1):45–59.

Rozana, M., K. Abdul Razak, G. Kawamura, A. Matsuda, and Z. Lockman. 2015. Formation of aligned iron oxide nanopores as Cr adsorbent material. *Advanced Materials Research* 1087:460–464.

Rozana, M., K. A. Razak, C. K. Yew, Z. Lockman, G. Kawamura, and A. Matsuda. 2016. Annealing temperature-dependent crystallinity and photocurrent response of anodic nanoporous iron oxide film. *Journal of Materials Research* 31 (12):1681–1690.

Rozana, M., M. A. Azhar, D. M. Anwar et al. 2014. Effect of applied voltage on the formation of self-organized iron oxide nanoporous film in organic electrolyte via anodic oxidation process and their photocurrent performance. *Advanced Materials Research* 1024:99–103.

Rozana, M., A. Matsuda, G. Kawamura, W. K. Tan, and Z. Lockman. 2017. Formation of nanoporous-Fe₂O₃ thin film as photoanode by anodic oxidation on iron. In *Two-Dimensional Nanostructures for Energy-Related Applications*, C. K. Yew (Ed.). CRC Press: Boca Raton, FL.

Noor Syuhadah, S., N. Z. M. Muslim, and H. Rohasliney. 2015. Determination of heavy metal contamination from batik factory effluents to the surrounding area. *International Journal of Chemical, Environmental & Biological Sciences* 3 (1):7–9.

Sedriks, A. J. 1996. *Corrosion of Stainless Steels.* Wiley: New York.

Senthil Kumar, K., K. S. Sajwan, J. P. Richardson, and K. Kannan. 2008. Contamination profiles of heavy metals, organochlorine pesticides, polycyclic aromatic hydrocarbons and alkylphenols in sediment and oyster collected from marsh/estuarine Savannah GA, USA. *Marine Pollution Bulletin* 56 (1):136–149.

Sivula, K. 2012. Nanostructured α-Fe₂O₃ photoanodes. In *Photoelectrochemical Hydrogen Production*, R. V. D. Krol, and M. Grätzel (Eds.). New York: Springer.

Sponza, D. T. 2002. Necessity of toxicity assessment in Turkish industrial discharges (examples from metal and textile industry effluents). *Environmental Monitoring and Assessment* 73 (1):41–66.

Srivastava, H., P. Tiwari, A. K. Srivastava, and R. V. Nandedkar. 2007. Growth and characterization of α-Fe₂O₃ nanowires. *Journal of Applied Physics* 102 (5):054303.

Srivastava, H., P. Tiwari, A. K. Srivastava, S. Rai, T. Ganguli, and S. K. Deb. 2011. Effect of substrate texture on the growth of hematite nanowires. *Applied Surface Science* 258 (1):494–500.

Sungur, S., and F. Gülmez. 2015. Determination of metal contents of various fibers used in textile industry by MP-AES. *Journal of Spectroscopy* 2015:5.

Takagi, R. 1957. Growth of oxide whiskers on metals at high temperature. *Journal of the Physical Society of Japan* 12 (11):1212–1218.

Tan, W. K., K. A. Razak, K. Ibrahim, and Z. Lockman. 2011. Oxidation of etched Zn foil for the formation of ZnO nanostructure. *Journal of Alloys and Compounds* 509 (24):6806–6811.

Tallman, R. L., and E. A. Gulbransen. 1968. Dislocation and grain boundary diffusion in the growth of α-Fe₂O₃ Whiskers and twinned platelets peculiar to gaseous oxidation. *Nature* 218:1046–1047.

Tang, W., B. Shan, W. Zhang, H. Zhang, L. Wang, Y. Ding. 2014. Heavy metal pollution characteristics of surface sediments in different aquatic ecosystems in Eastern China: A comprehensive understanding. *PLoS One* 9 (9):e108996. doi:10.1371/journal.pone.0108996.

Turner, J. E., M. Hendewerk, J. Parmeter, D. Neiman, and G. A. Somorjai. 1984. The characterization of doped iron oxide electrodes for the photodissociation of water: Stability, optical, and electronic properties. *Journal of The Electrochemical Society* 131 (8):1777–1783.

Vandevivere, P. C., R. Bianchi, and W. Verstraete. 1999. Review: Treatment and reuse of wastewater from the textile wet-processing industry: Review of emerging technologies. *Journal of Chemical Technology & Biotechnology* 72 (4):289–302.

Venkatramanan, S., S. Y. Chung, T. Ramkumar, and S. Selvam. 2015. Environmental monitoring and assessment of heavy metals in surface sediments at Coleroon River Estuary in Tamil Nadu, India. *Environmental Monitoring and Assessment* 187 (8):505.

Vesela, A., and M. Balat-Pichelin. 2014. Synthesis of iron-oxide nanowires using industrial-grade iron substrates. *Vacuum* 100:71–73.

Vincent, T., M. Gross, H. Dotan, and A. Rothschild. 2012. Thermally oxidized iron oxide nanoarchitectures for hydrogen production by solar-induced water splitting. *International Journal of Hydrogen Energy* 37 (9):8102–8109.

Visvanathan, C., and S. Kumar. 1999. Issues for better implementation of cleaner production in Asian small and medium industries. *Journal of Cleaner Production* 7 (2):127–134.

Voss, D. A., E. P. Butler, and T. E. Mitchell. 1982. The growth of hematite blades during the high temperature oxidation of iron. *Metallurgical Transactions A* 13 (5):929–935.

Wagner, C. 1938. The mechanism of the movement of ions and electrons in solids and the interpretation of reactions between solids. *Transactions of the Faraday Society* 34:851–859.

Wagner, R. S. 1967. Defects in silicon crystals grown by the VLS technique. *Journal of Applied Physics* 38 (4):1554–1560.

Wang, H., J.-T. Wang, Z.-X. Cao et al. 2015. A surface curvature oscillation model for vapour–liquid–solid growth of periodic one-dimensional nanostructures. *Nature Communications* 6:6412.

Wang, R., Y. Chen, Y. Fu, H. Zhang, and C. Kisielowski. 2005. Bicrystalline hematite nanowires. *The Journal of Physical Chemistry B* 109 (25):12245–12249.

Wang, X., L. Zhang, Z. Zhao, and Y. Cai. 2017. Heavy metal contamination in surface sediments of representative reservoirs in the hilly area of southern China. *Environmental Science and Pollution Research* 24 (34):26574–26585.

Wani, P. A., J. A. Wani, and S. Wahid. 2018. Recent advances in the mechanism of detoxification of genotoxic and cytotoxic Cr (VI) by microbes. *Journal of Environmental Chemical Engineering* 6 (4):3798–3807.

Wen, X., S. Wang, Y. Ding, Z. L. Wang, and S. Yang. 2005. Controlled growth of large-area, uniform, vertically aligned arrays of α-Fe$_2$O$_3$ nanobelts and nanowires. *The Journal of Physical Chemistry B* 109 (1):215–220.

Wilbur, S., H. Abadin, M. Fay et al. 2012. Toxicological profile for chromium. Agency for Toxic Substances and Disease Registry: Atlanta, GA.

Wood, G. C. 1962. The oxidation of iron-chromium alloys and stainless steels at high temperatures. *Corrosion Science* 2 (3):173–196.

Xiao, D., K. Dai, Y. Qu, Y. Yin, and H. Chen. 2015. Hydrothermal synthesis of α-Fe$_2$O$_3$/ g-C$_3$N$_4$ composite and its efficient photocatalytic reduction of Cr(VI) under visible light. *Applied Surface Science* 358:181–187.

Yeh, L.-S. R., and N. Hackerman. 1977. Iron oxide semiconductor electrodes in photoassisted electrolysis of water. *Journal of The Electrochemical Society* 124 (6):833–836.

Yuan, L., Y. Wang, R. Cai et al. 2012. The origin of hematite nanowire growth during the thermal oxidation of iron. *Materials Science and Engineering: B* 177 (3):327–336.

Zhang, S., W. Xu, M. Zeng, J. Li, J. Xu, and X. Wang. 2013. Hierarchically grown CdS/Fe$_2$O$_3$ heterojunction nanocomposites with enhanced visible-light-driven photocatalytic performance. *Dalton Transactions* 42 (37):13417–13424.

Zheng, Z., Y. Chen, Z. Shen et al. 2007. Ultra-sharp α-Fe$_2$O$_3$ nanoflakes: growth mechanism and field-emission. *Applied Physics A* 89 (1):115–119.

Zhong, M., Z. Liu, X. Zhong, H. Yu, and D. Zeng. 2011. Thermal growth and nanomagnetism of the quasi-one dimensional iron oxide. *Journal of Materials Science & Technology* 27 (11):985–990.

6 Anodic ZrO$_2$ Nanotubes for Heavy Metal Ions Removal

*Nurulhuda Bashirom, Monna Rozana, Nurul Izza
Soaid, Khairunisak Abdul Razak, Andrey Berenov,
Syahriza Ismail, Tan Wai Kian, Go Kawamura,
Atsunori Matsuda, and Zainovia Lockman*

CONTENTS

6.1 INTRODUCTION

Heavy metals are metals in periodic tables with density over 5 g/cm^3 (Järup, 2003). Examples of heavy metals are arsenic (As), cadmium (Cd), chromium (Cr), copper (Cu), lead (Pb), mercury (Hg), nickel (Ni), and zinc (Zn). Cr has three oxidation states; +2, +3, and +6. Trivalent chromium, Cr(III) can be consumed by humans, while hexavalent chromium, Cr(VI) is highly toxic in nature and must be avoided. Pollution of Cr(VI) in surface and groundwater has long been known to be a serious global problem and must be removed. Excess amount of Cr(VI) causes severe health risks not only to humans, but also to aquatic life through bioaccumulation along food chains. Therefore, steps should be taken to identify the source of Cr(VI) discharge and to reduce its amount in water to an acceptable levels. One step that

143

can be taken is via utilization of nanotechnology whereby nanostructured materials or nanomaterials can provide the solution for Cr(VI) removal from contaminated wastewater prior to discharge. A number of studies on the utilization of nano-materials in industrial wastewater treatment have been reviewed by several authors (Wu et al., 2015, Mohamed et al., 2016, Nagarjuna et al., 2017). Nanomaterials are preferred as they have a high surface area, extreme reactivity, and unique electron confinement effect. These make them ideal as an adsorbent and as a catalyst. As for the former, adsorption on nanomaterials has been acknowledged as an effective and economic method for heavy metal ions removal whereas the latter is a technique to convert, for example from Cr(VI) to less toxic Cr(III) by the reduction process. Reduction can be done by electrons transferred from the catalyst surface to an adsorbed heavy metal ions. The catalyst used can be a semiconductor; a material which is characterized by a filled valence band (VB) and an empty conduction band (CB). The gap between the VB and CB is usually rather small and hence, when irradiated with light at appropriate energy, electrons (e^-) from the VB can be promoted to the CB leaving holes (h^+) behind. Electrons in the CB are free and they can diffuse to the surface of photocatalyst and transferred to the adsorbed heavy metal ions. Upon receiving the electrons (e^-), the adsorbed heavy metal ions are reduced, preferably to a more benign compound. For a reduction process to happen, the potential of the electrons in the CB of the photocatalyst must be more negative than the reduction potential of the adsorbed heavy metal ions. Not all heavy metal ions can be reduced and not all the photocatalysts can do the reduction process. Therefore, it is necessary to select the right material as the photocatalyst. There are arrays of semiconductor oxides that can be used as photocatalyst to reduce certain heavy metals ions.

The ultimate goal in the photoreduction process is indeed a complete removal of heavy metal ions, therefore the main requirement would be on the potential of the CB electrons of the semiconductor. Photogenerated electrons in ZrO_2 can reduce Cr(VI) to Cr(III) since the reduction potential of Cr(VI) ($E^0_{Cr(VI)/Cr(III)}$) = 1.33 V in acidic medium is more positive than the CB potential of ZrO_2. Oxidation of water takes place simultaneously due to the more negative oxidative potential of H_2O ($E^0_{O_2/H_2O}$) = 1.23 V than the VB potential of ZrO_2. Table 6.1 shows several other oxide semiconductors that can be used for Cr(VI) reduction. As obvious, among all those listed, ZrO_2 has the most negative CB edge; thus, it is expected that photogenerated electrons are rather efficient in reducing Cr(VI) to Cr(III).

Nonetheless, ZrO_2 is a wide band gap of semiconductor with an absorption edge at 248 nm and hence, sunlight activation is not possible. Creating highly defective ZrO_2 may however introduce defect states for sunlight activation. Defective ZrO_2 can be achieved by fabricating ZrO_2 by anodization of zirconium (Zr). Anodization is a process to produce surface oxide on a piece of metal elec-trochemically. Anodization can results in highly defective anodic film that dis-plays light absorption and emission in the visible range. In addition, the anodic film can be nanostructured to be in a form of self-organized 1-D nanotubes (NTs) (Bashirom et al., 2017, Mohajernia et al., 2017, Rozana et al., 2017, Soaid et al., 2017, Taib et al., 2017) when the process is done in fluoride electrolyte.

TABLE 6.1

Oxide Semiconductor Photocatalysts for Photoreduction of Cr(VI) with Calculated Band Edge Positions,[a] and the Corresponding Absorption Edge[b]

Semiconductor	Bandgap (eV)	Absorption Edge (nm)	E_{CB} (eV)	E_{VB} (eV)
TiO$_2$ (anatase)	3.2	413	−0.29	2.91
ZrO$_2$	5.0	248	−1.09	3.91
ZnO	3.2	388	−0.31	2.89
WO$_3$	2.7	443	0.74	3.44
Fe$_2$O$_3$	2.2	539	0.28	2.48

Source: [a]Xu and Schoonen, 2000; [b]Hernández-Ramírez and Medina-Ramírez, 2015.

Nanotubular materials are preferred as photocatalyst due to their inherent size-dependent properties including increased surface area, high surface energy and high reactivity.

6.2 HEAVY METAL REMOVAL

Table 6.2 shows the anthropogenic sources of heavy metal and their corresponding health hazards to humans. The maximum concentration limit (MCL) of heavy metals in drinking water as regulated by the World Health Organisation (WHO), United States Environmental Protection Agency (US-EPA), and Malaysian Drinking Water Quality Standard are also included in the table. Table 6.3 displays concentration of heavy metal ions detected in waters, soils, sediments, and dusts located in several states in Malaysia in recent years. The presence of the heavy metals can be ascribed to industrial activities, plantations, agricultural, mining, and shipping. As seen in Table 6.3, contamination from chromium is the highest near agricultural sites and electronic and automotive industries. However, contamination from agricultural sites may possibly come from fertilizer and pesticide and could be in a form of Cr(III) and hence, not too alarming despite the large concentration shown in the table. Contamination from industries on the other hand can be related to plating compounds and pigments released that can be one of the derivations of Cr(VI) compounds.

There are several methods to remove Cr(VI) from wastewater including adsorption, membrane filtration, ion exchange, electrochemical treatment and, as mentioned, photocatalytic reduction (Owlad et al., 2009). Each of these methods has their own advantages and disadvantages as listed in Table 6.4. Among all of these methods, photocatalytic reduction appears to be the most promising as it can directly eliminate Cr(VI) by reducing it to Cr(III). Cr(III) can then be precipitated into chromium hydroxide, Cr(OH)$_3$ in alkaline media and safely discharged into the environment or it can be recycled.

TABLE 6.2

Exposure and Health Effects of Heavy Metals and the Guidelines for Drinking Water Quality Standard by WHO, US-EPA, and Malaysian Standard

Heavy Metals	Use/Exposure	Health Effects	MCL (mg/L) WHO	MCL (mg/L) US-EPA	MCL (mg/L) Malaysian Drinking Water Quality Standard
Cr	Chrome electroplating, chrome alloy production, glassmaking, leather tanning, paints/pigments, photoengraving, porcelain, and ceramics manufacturing, production of high-fidelity magnetic audio tapes, textile, welding of alloys or steel, wood preservatives, anti-algae agents, antifreeze, and cement (ATSDR, 2013b)	Respiratory problems, upset stomachs and ulcers, lung cancer, organ damage, skin rashes, and death (Lenntech, 2018a)	0.05	0.1	0.05
Zn	Smelter slags and wastes, mine tailings, coal and bottom fly ash, fertilizers and wood preservatives (ATSDR)	Loss of appetite, disturbances in sensations, skin problems, stomach cramps, vomiting, nausea and anaemia, and organ damage (Lenntech, 2018b)	–	5.0	3.0
Cd	Metal plating, pigments, NiCd batteries, and plastic stabilizers (ATSDR, 2013a)	Lung damage, kidney disease, and lung cancer (ATSDR, 2015)	0.003	0.005	0.003
Hg	Artisanal and small scale gold mining, coal combustion, non-ferrous metals, cement production, consumer products, iron and steel, chlor-alkali, and oil refining (Agency, 2017)	Emotional changes, insomnia, tremors, neuromuscular changes, headaches, disturbances in sensations, changes in nerve responses, low mental function, and death (US-EPA, 2017)	0.006	0.002	0.001
Ni	Stainless steel, alloys, electroplating, rechargeable batteries, coins, steel and jewellery (Harasim and Filipek, 2015)	Allergic reaction, lung cancer, and respiratory problems (ATSDR, 2005)	0.07	–	0.02

(Continued)

TABLE 6.2 (Continued)
Exposure and Health Effects of Heavy Metals and the Guidelines for Drinking Water Quality Standard by WHO, US-EPA, and Malaysian Standard

Heavy Metals	Use/Exposure	Health Effects	MCL (mg/L) WHO	MCL (mg/L) US-EPA	Malaysian Drinking Water Quality Standard
Cu	Coal combustion, oil combustion, pyrometallurgical, secondary nonferrous metal production, steel and iron manufacturing, refuse incineration, phosphate fertilizers, and wood combustion (Registry, N/A)	Nose, mouth, and eyes irritations, headaches, dizziness, nausea, diarrhea, vomiting, stomach cramps, organ damage and death (ATSDR, 2004)	2.0	1.3	1.0
As	Construction/contracting, ore smelting, semiconductor manufacture, insecticides, algaecides, desiccants used in mechanical cotton harvesting, glass manufacturing, herbicides, and nonferrous alloys (Gehle, 2009)	Neurologic, respiratory, hematologic, cardiovascular, gastrointestinal, and other systems problems (Gehle, 2009)	0.01	0.0	0.01

Source: Standards for drinking water quality WHO 2011, USEPA 2018, Malaysian Division, 2012.

Note: ATSDR: Agency for Toxic Substances and Disease Registry; USEPA: US Environmental Protection Agency; WH: World Health Organization.

TABLE 6.3

Concentrations of Heavy Metals in Several States in Malaysia (mg/kg) from Year 2013–2017

Polluted Area	As	Cd	Cu	Pb	Ni	Hg	Cr	Zn	Fe	Co	Sn	Mn	Ref.	Possible sources
Kuala Perlis Coast, Perlis	0.01	0.1	41.9	–	–	–	–	121.0	–	–	–	–	Lias et al. (2013)	Shipping activities
Port Klang, Selangor	60.4	0.8	17.4	59.5	11.4	0.2	46.4	51.1	–	–	–	–	Sany et al. (2013)	Industrial wastewater and port activities
Langat River, Selangor	14.5	0.1	2.7	15.5	–	–	–	29.7	–	–	–	–	Shafie et al. (2013)	Oil palm plantations, shipping activities, and steelmaking industries
Bandar Baru Bangi and Kajang, Selangor	–	3.5	–	5.2	–	–	2.0	0.8	1.0	–	–	–	Latif et al. (2013)	Electronic and chemical industries
Kuala Lumpur	–	0.7	137.3	144.3	28.6	–	51.8	292.6	9888	–	–	243.4	Han et al. (2013)	Electronic and chemical industries
Langat River, Selangor	16.2	–	5.7	30.4	4.5	–	15.9	35.9	–	–	–	–	Lim et al. (2013)	Oil palm plantations, shipping activities, and steelmaking industries
Gebeng industrial city, Pahang	46.9	0.4	25.2	32.8	3.4	2.6	10.7	44.8	–	285.9	–	–	Hossain et al. (2014)	Petrochemical plant
Ranau, Sabah	–	–	–	–	2802	–	15145	–	–	101.0	–	–	Tashakor et al. (2014)	Cocoa plantation
Bukit Rokan and Petasih, Negeri Sembilan	–	–	–	–	938.0	–	6614	–	–	279.5	–	–	Tashakor et al. (2014)	Palm oil and rubber plantation

(Continued)

TABLE 6.3 (Continued)
Concentrations of Heavy Metals in Several States in Malaysia (mg/kg) from Year 2013–2017

Polluted Area	As	Cd	Cu	Pb	Ni	Hg	Cr	Zn	Fe	Co	Sn	Mn	Ref.	Possible sources
Klang, Selangor	–	45.0	150.0	1300	–	–	14.0	335.0	15900	8.5	–	–	Yuswir et al. (2015)	Electronic and chemical industries
Langat River, Selangor	–	0.6	–	–	7.8	–	21.0	–	–	–	114.3	–	Kadhum et al. (2015)	Oil palm plantations, shipping activities, and steelmaking industries
Bayan Lepas Free Industrial Zone, Penang	–	1.7	194.2	35.5	51.5	–	38.0	56.0	15.2	6.2	–	–	Khodami et al. (2016)	Electronic and chemical industries
Kelantan River, Kelantan	–	0.1	25.2	65.5	26.9	–	63.1	62.2	–	–	–	–	Wang et al. (2017)	Logging and mining activities
Rawang, Selangor	18.4	71.7	425.1	593.3	113.3	–	501.3	526.8	19337	5.6	–	–	Praveena and Aris (2017)	Automotive industry

TABLE 6.4
Advantages and Disadvantages of Heavy Metal Removal Methods from Wastewater

Method	Advantages	Disadvantages
Adsorption	Low-cost, easy operating conditions, wide pH range, and high metal binding capacities	Low selectivity and production of waste products
Membrane filtration	Small space requirement, low pressure, and high separation selectivity	High operational cost due to membrane fouling
Ion exchange	High treatment capacity, high removal efficiency, and fast kinetics	Cannot handle high concentration of heavy metal because the matrix easily fouled by other species in wastewater, nonselective and highly sensitive to pH, consume expensive expensive electrical supply, and electrode easily corroded and need to be replaced frequently
Electrochemical	Solid metal recovery	Expensive electrical supply
Photocatalytic reduction	Removal of metals and organic pollutant simultaneously, and less harmful by-products	Long duration time and limited applications

Source: Barakat, M., *J. Hazard. Mater.*, 223, 1–12, 2011; Fu, F. and Wang, Q., *J. Environ. Manage.*, 92, 407–418, 2011.

Photocatalytic reduction can be performed on semiconductor surface under light irradiation as shown in Figure 6.1 (Barakat, 2011). If the photon energy is greater than the band gap of the semiconductor ($h\nu > E_g$), the electron is excited from the VB to the CB of semiconductor forming electron-hole ($e^- - h^+$) pairs (eq. 6.1). These charge carriers can migrate to the semiconductor surface to perform a redox reaction of species present on the surface of the photocatalyst (Barakat, 2011). The photogenerated electrons are used to reduce the Cr(VI) to Cr(III) (eq. 6.2), whereas photogenerated holes can oxidize water into O_2 (eq. 6.3) (Barrera-Díaz et al., 2012).

$$\text{Semiconductor} \rightarrow e^- + h^+ \tag{6.1}$$

$$Cr_2O_7^{2-} + 14H^+ + 6e^- \rightarrow 2Cr^{3+} + 7H_2O \tag{6.2}$$

$$2H_2O + 4h^+ \rightarrow O_2 + 4H^+ \tag{6.3}$$

Oxides like ceria (CeO_2) and hematite α-Fe_2O_3 can also be used for photoreduction of Cr(VI) either under ultraviolet (UV) or visible light irradiations. Wu et al. (2015) reported on the reduction of 78.9% Cr(VI) after 60 min with addition of

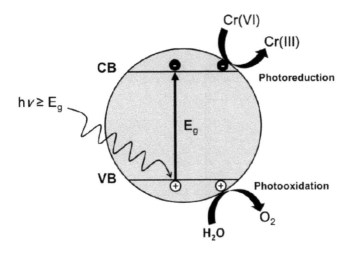

FIGURE 6.1   Mechanism of photocatalytic reduction of Cr(VI) on semiconductor surface.

isopropanol as a hole scavenger and 99.6% reduction in the presence of oxalic acid due to slow oxidation of water. Cr(VI) reduction on core-shell type magnetite (Fe$_3$O$_4$)@Fe$_2$O$_3$ dispersed over nano alumina has also be reported (Nagarjuna et al., 2017). 100% removal can be achieved after 35 min at pH 3. There are several parameters that affect the photoreduction of Cr(VI): (1) catalysts loading, (2) presence of scavengers, (3) light intensity, (4) pH of Cr(VI) solution, and (5) initial concentration of Cr(VI). Nevertheless, the most important is the selection of the right photocatalyst.

As already mentioned, ZrO$_2$ is a wide band gap semiconductor ~5.0 eV (Hernández-Ramírez and Medina-Ramírez, 2015) but it has a highly negative CB potential of −1.0 eV versus standard hydrogen electrode (SHE) at pH 0 and the corresponding VB potential is +4.0 V versus SHE (Jafari et al., 2016), shown in Figure 6.2. The high negative value of its CB edge potential (−1.0 eV versus SHE at pH = 0) has made ZrO$_2$ as suitable photocatalyst to reduce Cr(VI) and other inorganic pollutants (Botta et al., 1999, Karunakaran et al., 2009). Water splitting on ZrO$_2$ has been reported by several authors (Reddy et al., 2003, Soaid et al., 2017) indicating the possible generation of electrons under illumination. The photocatalytic ability of ZrO$_2$ to degrade organic compounds has also been reported (Stojadinović et al., 2015, Rozana et al., 2017). Table 6.5 summarizes previous works reported on photocatalytic activity of ZrO$_2$ from year 2012 to 2017.

However, due to its large band gap, ZrO$_2$ can only be activated under UV light that accounts for only 4% of the total solar energy (Pathakoti et al., 2013). Nevertheless, several attempts have been made to extend the light absorption of ZrO$_2$ under visible light as well as generally improving the properties of ZrO$_2$ including doping (Du et al., 2013, Agorku et al., 2015, Gurushantha et al., 2015, Xiao et al., 2015, Poungchan et al., 2016, Renuka et al., 2016), or coupling with other lower band gap oxides such as CeO$_2$ (Wang et al., 2013, Hao et al., 2017), ZnO (Sherly et al., 2014, Ibrahim, 2015), PbO$_2$ (Kaviyarasu et al., 2017), Bi$_2$O$_3$

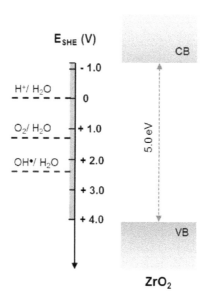

FIGURE 6.2 Illustration of over-simplified energy band diagram for ZrO_2 relatives to redox levels for several systems.

(Vignesh et al., 2013), and TiO_2 (Li et al., 2015, Zhang et al., 2015, Ji et al., 2017). Defective and disordered oxide semiconductors have been recently attracted great attention for photocatalysis. For example, Sinhamahapatra et al. (2016) synthe-sized a black oxygen-deficient ZrO_2 to enable sunlight activation. The defective ZrO_2 is said to have a lower band gap energy of ~1.5 eV. The defective black ZrO_2 can split water under light irradiation with high H_2 production than the crystalline white ZrO_2 due to large amount of oxygen vacancies and surface defects (Sinhamahapatra et al., 2016). Oxygen vacancies and surface defects are therefore required to induce activation under visible light for ZrO_2. Control over defects formation is however not as straight forward as on the nature of the defects. Defects can act as recombination centers which may impede efficient photocatalytic process.

Formation of oxygen vacancy as defects in ZrO_2 can improve the light absorp-tion via formation of new energy levels within the band gap of the oxide (Fang et al., 2013). These oxygen vacancies create localized states within the band gap that overlap with the VB edge of semiconductor (Figure 6.3b) (Wang et al., 2012a). Hence, the position of the VB edge is raised leading to a narrower band gap for excitation of the electron from the VB to the CB of the semiconductor (Suresh et al., 2014, Kumar and Ojha, 2015, Stojadinović et al., 2015, Sinhamahapatra et al., 2016, Rozana et al., 2017).

TABLE 6.5

Photocatalytic Activities of ZrO$_2$-Based Photocatalysts from Literature (2012–2017)

Photo Catalyst	Fabrication Method	Application	Light Source	Results	Modification Effects	Ref.
Flower-like ZrO$_2$ nanostructures	Hydrothermal	Degradation of RhB	UV	100% degradation after 40 min	Special exposed (100) facet of the petals and strong absorption in UV region	Shu et al. (2012)
ZrO$_2$ nanoparticles	Thermal plasma route	Degradation of MB	UV	100% degradation after 30 min	High defect states in ZrO$_2$ and blue shift in monoclinic phase	Nawale et al. (2012)
Tetragonal star-like ZrO$_2$ nanostructures	Hydrothermal	Degradation of MO, Congo red, and RhB	UV	95% degradation after 40 min	Selective acetate anions adsorption on different ZrO$_2$ facets due to surface charge and acid centers of the product	Shu et al. (2013)
ZrO$_2$ nanostructures	Microwave combustion	Degradation of 4-Chlorophenol (4-CP)	UV	62.5% degradation after 240 min		Selvam et al. (2013)
Mesoporous assembled ZrO$_2$ nanoparticles	Surfactant-aided sol-gel	Degradation of MO	UV	40.8% degradation after 90 min	High crystallinity, uniform particle size, and narrow pore size distribution	Sreethawong et al. (2013)
RE/ZrO$_2$ (RE = Sm, Eu) nanocomposite	Sol-gel	Degradation of MB	Visible	100% degradation after 90 min	Incorporation of rare earth ions create structural defects that reduce the band-gap and increase absorption of visible light	Du et al. (2013)

(Continued)

TABLE 6.5 (*Continued*)
Photocatalytic Activities of ZrO$_2$-Based Photocatalysts from Literature (2012–2017)

Photo Catalyst	Fabrication Method	Application	Light Source	Results	Modification Effects	Ref.
ZrO$_2$/CeO$_2$ nanocomposite	Hydrothermal	Degradation of RhB	Visible	40% degradation after 150 min	Generation of new active sites resulted from interactions between ZrO$_2$–CeO$_2$	Wang et al. (2013)
ZrO$_2$ nanotubes	Anodization	Degradation of MO	UV	20% degradation after 120 min	High adsorption at low pH and high removal efficiency at larger tube diameter	Fang et al. (2013)
Bi$_2$O$_3$–ZrO$_2$ nanocomposite	Chemical precipitation	Photoreduction of Cr(VI)	Visible	92.3% removal after 180 min	Efficient electron–hole separation and increased surface area	Vignesh et al. (2013)
ZrO$_2$ nanoparticles supported activated carbon	Microwave irradiation	Degradation of textile dye wastewater	UV	68.1% degradation after 120 min	Oxygen vacancies increase photon absorption and improve charge carriers separation	Suresh et al. (2014)
ZnO-ZrO$_2$ nanoparticles	Microwave assisted combustion	Degradation of 2,4-dichlorophenol	UV	90% degradation after 5 h	Enhanced electron–hole separation	Sherly et al. (2014)
HNO$_3$–ZrO$_2$ nanoparticles	Solvothermal	Photoreduction of Cr(VI)	Visible	100% degradation after 5 h	High photon absorption	Zhao et al. (2014)
ZrO$_2$/graphene	Combustion	Degradation of MO	UV	100% degradation after 40 min	Better electron–hole separation	Rani et al. (2014)
Fe-doped C-ZrO$_2$ nanoparticles	Low temperature green combustion	Degradation of AO7	UV and visible	98% degradation after 90 min (UVA) and 24% (visible)	Effective crystallite size, narrow band gap, and enhanced electron–hole separation	Gurushantha et al. (2015)

(Continued)

TABLE 6.5 (*Continued*)
Photocatalytic Activities of ZrO$_2$-Based Photocatalysts from Literature (2012–2017)

Photo Catalyst	Fabrication Method	Application	Light Source	Results	Modification Effects	Ref.
ZrO$_2$ nanoparticles	Sol-gel	Degradation of textile azo-dye (Acid Blue 25)	UV	20% degradation after 75 min	–	Sultana et al. (2015)
Fe-doped ZrO$_2$ nanostructures	Template method	H$_2$ evolution from aqueous methanol	Visible	Quantum efficiency of H$_2$ production: 0.17% within 15 h	Incorporation of Fe narrowing the band gap	Xiao et al. (2015)
m-ZrO$_2$, t-ZrO$_2$, and c-ZrO$_2$ nanoparticles	Hydrothermal	Degradation of MO	UV	99% degradation after 110 min over m-ZrO$_2$, 90% (t-ZrO$_2$), and 80% (c-ZrO$_2$)	m-ZrO$_2$ contained small amount of oxygen vacancies, high crystallinity, broad pore size distribution, and high density of surface hydroxyl groups	Basahel et al. (2015)
Ferro-magnetic ZrO$_2$ nanostructures	Sol-gel	Degradation of MB	UV	35% degradation after 60 min	High surface to volume ratio of small particle size and high density of oxygen vacancies	Kumar and Ojha (2015)
Macroporous TiO$_2$-ZrO$_2$ composite	Template method	Degradation of MO	UV	100% degradation after 40 min	Better charge carriers separation due to substitution of Ti with Zr ions created a space charge region	Zhang et al. (2015)
Macro–meso porous ZrO$_2$–TiO$_2$ composites	Sol-gel	Degradation of RhB	UV	86.9% degradation after 3 h	Better charge carriers separation due to substitution of Ti with Zr ions created a space charge region	Li et al. (2015)

(Continued)

TABLE 6.5 (Continued)
Photocatalytic Activities of ZrO_2-Based Photocatalysts from Literature (2012–2017)

Photo Catalyst	Fabrication Method	Application	Light Source	Results	Modification Effects	Ref.
ZrO_2 films	Plasma electrolytic oxidation	Degradation of MO	UV	23% degradation after 8 h	High concentration of oxygen vacancies induced new energy levels in the band gap	Stojadinović et al. (2015)
C, N, S-doped ZrO_2 and Eu doped C, N, S-ZrO_2	Coprecipitation	Degradation of Indigo carmine (IC)	Visible	100% degradation after 150 min	Enhanced charge carriers separation and narrowed band gap	Agorku et al. (2015)
ZnO-ZrO_2	Impregnation	Conversion of coumarin to 7-hydroxy coumarin	UV	50% degradation after 120 min	Enhanced charge carriers separation due to heterojunction	Ibrahim (2015)
Oxygen-deficient black ZrO_{2-x}	Magnesio-thermic reduction	H_2 production from methanol-water and degradation of RhB	Visible	H_2 generation: 505 μmolg-1 per hour. Degradation of RhB: not stated.	Black ZrO_2 more photoactive than white ZrO_2 due to high concentration of oxygen vacancies and Zr^{3+}	Sinhamahapatra et al. (2016)
Hollow micro-spheres Mg-doped ZrO_2 nanoparticles	Green assisted route	Degradation of RhB	UV	90% degradation after 1 h	Suitable m-ZrO_2: t-ZrO_2 ratio and enhanced charge carriers separation due to Mg dopant	Renuka et al. (2016)
Flower-like carbon-doped ZrO_2	Sol-gel-hydrothermal	Degradation of MB	Visible	96% degradation after 1 h	Nanopetals provide good intercrystalline connections and carbon doping increase photon absorption due to band gap narrowing	Poungchan et al. (2016)
ZrO_2 nanotubes (our work)	Anodization	Degradation of MO and water splitting	UV	55% degradation after 5 h	Long tubes and enhanced charge carriers separation due to hole trapping by carbonate ions	Soaid et al. (2017)

(Continued)

TABLE 6.5 (Continued)
Photocatalytic Activities of ZrO$_2$-Based Photocatalysts from Literature (2012–2017)

Photo Catalyst	Fabrication Method	Application	Light Source	Results	Modification Effects	Ref.
Freestanding t-ZrO$_2$ nanotubes (our work)	Anodization	Degradation of MO	UV	t-ZrO$_2$: 82% degradation after 5 h; t-ZrO$_2$ + m-ZrO$_2$: 70% degradation after 5 h	High concentration of oxygen vacancies in t-ZrO$_2$ and less photoactive of m-ZrO$_2$	Rozana et al. (2017)
Freestanding amorphous ZrO$_2$ nanotubes (our work)	Anodization	Photoreduction of Cr(VI)	Visible	100% degradation after 5 h	Enhanced adsorption due to high concentration of hydroxyl and carbonate groups and oxygen vacancies	Bashirom et al. (2017)
Multilayer and open structure of dendritic crosslinked CeO$_2$-ZrO$_2$ composite	Sol-gel decompression filling method	Degradation of Congo red and water splitting	UV visible	55% (UV) and 70% (visible) after 120 min	Enhanced charge carriers separation and higher specific surface area	Hao et al. (2017)
ZrO$_2$ doped PbO$_2$	Hydrothermal	Degradation of RhB	UV visible	93.32% degradation after 180 min	–	Kaviyarasu et al. (2017)
ZrO$_2$(Er^{3+})/TiO$_2$	Co-precipitation	Degradation of diesel pollutants in seawater	Visible	87.74% degradation after 2.5 h	–	Ji et al. (2017)

*MO = methyl orange, MB = methylene blue, RhB = rhodamine B, AO7 = acid orange 7.

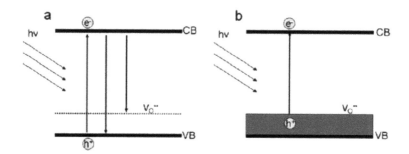

FIGURE 6.3 Influence of oxygen vacancy concentration on band gap narrowing. (a) low concentration, and (b) high concentration. (Reprinted with permission from Wang, J. et al., *ACS Appl. Mater. Interfaces*, 4, 4024–4030, 2012a. Copyright 2012 American Chemical Society.)

6.3 PHOTOREDUCTION OF Cr(VI) ON ZrO_2 PHOTOCATALYST

Despite the extremely negative CB electrons, not many researchers have reported on the use of ZrO_2 as a photocatalyst to photo reduce Cr(VI) as portrayed in Table 6.6. Karunakaran et al. (2009) and Botta et al. (1999) reported on Cr(VI) photoreduction over pure ZrO_2 photocatalyst under UV illumination. Karunakaran et al. irradiated ZrO_2 with UV-A (365 nm) and UV-C (254 nm) and from the diffuse reflectance spectra (DRS), they realized that ZrO_2 did not absorb UV-A and hence, the photo reduction path of Cr(VI) to Cr(III) was proposed to be different than that of ZrO_2 illuminated under UV-C. As shown in Figure 6.4, chromates show a strong absorption band near 365 nm due to metal-ligand charge transfer transition. The VB electron of ZrO_2 can jump to the photoexcited chromate forming a transient Cr(V) species, a hypochromate ion, CrO_4^{3-}, thus creating a hole in the VB of ZrO_2. The hole in the semiconductor oxidizes the adsorbed water molecules to OH^* which subsequently yields H_2O_2. Formed H_2O_2 reduces chromate to Cr(III) with the liberation of O_2. This could be an alternative path for reduction of Cr(VI) on ZrO_2 along with the electron transfer process.

Modifying ZrO_2 photocatalysts has also been done either by doping with Fe (Botta et al., 1999), or by coupling with other smaller band gap oxide semiconductors such as TiO_2 (Smirnova et al., 2006), Bi_2O_3 (Vignesh et al., 2013), and Fe_3O_4 (Kumar et al., 2016) in order to improve the light harvesting capability thus photocatalytic efficiency. The most outstanding result is shown by Botta et al. (1999) whereby the photocatalytic performance of undoped-ZrO_2 is reported to be comparable with TiO_2 Degussa P-25 in the presence of Ethylenediaminetetraacetic acid (EDTA). Without EDTA, only ~10% reduction was achieved. In our recent effort in applying ZrO_2 as photocatalysts for Cr(VI) we discovered that the use of ZrO_2 nanotube (ZNT) arrays fabricated by anodic process has yielded high reduction efficiency for Cr(VI) removal even without the presence of scavengers or any other organic compound. Redox reaction that occur would be the oxidation of water and reduction of Cr(VI) to Cr(III).

As shown in Figure 6.5, 80% of Cr(VI) can be removed over as-anodized ZNTs fabricated by anodization process at 80 V after 5 h. Whereas, only 63% can be removed

TABLE 6.6
Previous Researches on Cr(VI) Photoreduction over ZrO$_2$-Based Photocatalyst

Photocatalyst	Fabrication Method	Light Source	Co (mg/L)	Catalyst Dosage (g/L)	pH	Hole Scavenger	Removal Efficiency (%)	Time (min)
TiO$_2$ Degussa P-25 (Botta et al., 1999)	Commercial sample	UV (150 W Xe lamp)	12	1	2	–	90	120
TiO$_2$ Degussa P-25 (Botta et al., 1999)	Commercial sample	UV (150 W Xe lamp)	12	1	2	EDTA	100	120
Pure ZrO$_2$ (Botta et al., 1999)	Sol-gel	UV (150 W Xe lamp)	12	4	2	–	17	120
Pure ZrO$_2$ (Botta et al., 1999)	Sol-gel	UV (150 W Xe lamp)	12	4	2	EDTA	100	120
5 wt% Fe-doped ZrO$_2$ (Botta et al., 1999)	Sol-gel	UV (150 W Xe lamp)	12	4	2	–	22	120
5 wt% Fe-doped ZrO$_2$ (Botta et al., 1999)	Sol-gel	UV (150 W Xe lamp)	12	4	2	EDTA	100	120
Nonporous TiO$_2$ film (Smirnova et al., 2006)	Sol-gel	UV (Hg lamp)	59	0.05	2	EDTA	60	80
Mesoporous TiO$_2$ film (Smirnova et al., 2006)	Sol-gel	UV (Hg lamp)	59	0.05	2	EDTA	80	80
Mesoporous TiO$_2$-30% ZrO$_2$ (Smirnova et al., 2006)	Sol-gel	UV (Hg lamp)	59	0.05	2	EDTA	90	80
Mesoporous TiO$_2$-5% ZrO$_2$ (Smirnova et al., 2006)	Sol-gel	UV (Hg lamp)	59	0.05	2	EDTA	70	80

(Continued)

TABLE 6.6 (Continued)
Previous Researches on Cr(VI) Photoreduction over ZrO_2-Based Photocatalyst

Photocatalyst	Fabrication Method	Light Source	C_0 (mg/L)	Catalyst Dosage (g/L)	pH	Hole Scavenger	Removal Efficiency (%)	Time (min)
ZrO_2 nanoparticles (Karunakaran et al., 2009)	Commercial powder	UV ($\lambda = 365$ nm; 8 W Hg lamp)	20	20	7	–	50	120
Bi_2O_3-ZrO_2 nanoparticles (Vignesh et al., 2013)	Chemical precipitation	Visible (300 W Xe lamp)	30	1.25	2	–	92.3	180
ZrO_2 nanoparticles (Vignesh et al., 2013)	Chemical precipitation	Visible (300 W Xe lamp)	30	1.25	2	–	40	180
TiO_2 nanoparticles (Vignesh et al., 2013)	Chemical precipitation	Visible (300 W Xe lamp)	30	1.25	2	–	48	180
N-modified ZrO_2 nanoparticles (Zhao et al., 2014)	Solvothermal	Visible	50	1	N/A	–	100	5 h
ZrO_2/Fe_3O_4 (Kumar et al., 2016)	Co-precipitation	Sunlight	70	2	0.5	–	71.2	120
ZrO_2/Fe_3O_4/chitosan (Kumar et al., 2016)	Co-precipitation	Sunlight	70	2	0.5	–	84.7	120
ZrO_2/Fe_3O_4/chitosan (Kumar et al., 2016)	Co-precipitation	Sunlight	70	2	0.5	Ethanol	98.5	120
Freestanding ZrO_2 nanotubes (our work) (Bashirom et al., 2017)	Anodization	Sunlight	5	1	2	–	100	5 h
Freestanding ZrO_2 nanotubes (our work) (Bashirom et al., 2018)	Anodization	Sunlight	10	1	2	–	95	5 h

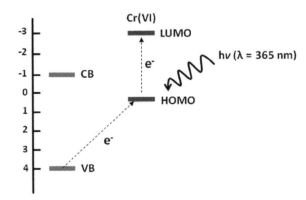

FIGURE 6.4 Excitation of chromate under UV from excited electron in ZrO$_2$.

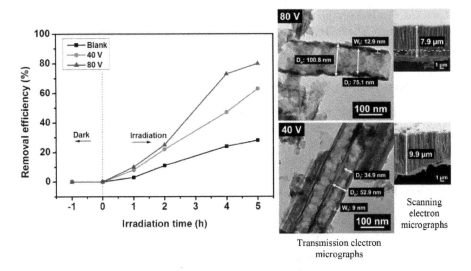

FIGURE 6.5 Photoreduction of Cr(VI) ions over ZNTs under formed by anodic oxidation at 80 and 40 V under sunlight.

over the 40 V sample. In the figure, field-emission scanning electron and transmission electron micrographs of the ZNTs are shown. Seeing the interesting properties displayed by ZNTs, the topic of ZNTs formation is reviewed next. This is followed by anodic processes for ZNTs formation; our own experiences and from literature.

6.4 ZrO$_2$ NANOTUBES FORMATION AND CHARACTERIZATIONS

There are arrays of techniques used in fabricating the ZNTs. In the 1990s and early 2000s, these techniques revolved around a hard-templated technique whereby suitable templates are needed as molds to grow the nanotubes. In recent years, formation of ZNTs has been done by anodizing Zr in suitable electrolyte which can be considered as a semi-templated method. The process is different than the typical process to produce other types of 1-D nanomaterials (nanowires or nanorods) in which a spontaneous growth from vapor or liquid phase occurred, resulting in the formation of single crystal nanowires or nanorods.

6.4.1 TEMPLATING AGAINST EXISTING NANOSTRUCTURES

ZNTs can be produced by templating against existing nanostructures. The templated technique can be further classified as (1) nanofibrous template, and (2) cylindrical nanowells template. The former requires fibrous template in nanoscale onto which Zr-precursor solution or Zr-vapor will be deposited onto. Once the oxide is solidified, the inner fiber will be removed, yielding a nanotubular structure. This method allows for the control of the diameter and length of the ZNTs by controlling the dimensions on the fibrous template used. The latter requires cylindrical wells inside on which the Zr-precursor or Zr-vapor will be deposited. Growth on the walls of the wells will result in the formation of tubular oxide structure. The cylindrical wells will then be removed by chemical processes or pyrolysis producing ZNTs.

6.4.2 NANOFIBROUS TEMPLATED METHOD

Two main ingredients are needed; nanofiber and a Zr-precursor to produce the ZNTs by this method. The precursor can be Zr salt solution (zirconium nitrate, zirconium acetate, zirconium perpoxide, zirconium tert-butoxide, and zirconium acetylacetonate). The deposition process on the fiber can be done by immersing nanofibers in the precursor solution for a pre-determined time followed by heat treatment. During heat treatment, nucleation and crystallization of ZrO_2 on the template will result in the formation of nanofiber-coated ZrO_2. The nanofibers must then be removed to produce hollow fibers of ZrO_2: nanotubes. The nanofibers used are often carbon-base (organic) like carbon nanotubes (CNTs) (Rao et al., 1997), cellulose (Huang and Kunitake, 2003) or carbon nanofibers (Ogihara et al., 2006). Figure 6.6 shows the scanning electron microscope (SEM) images of ZNTs prepared by a templated method against cellulose and a single ZNT is seen from the transmission electron microscope (TEM) image in Figure 6.6b. The TEM image of CNTs is shown in Figure 6.7a. After deposition and removal of CNTs, the resulting ZNTs are shown in Figure 6.7b. The template used does not have to be straight; in Figure 6.7a, a TEM image of carbon nanocoils is shown. Figure 6.7d shows the TEM image of ZNTs formed using carbon nanocoils as a template (Figure 6.7c). Clearly, formed ZNTs had a helical structure, reflecting the shape of the nanocoils. Polymeric material has also been used as a template for instance polyacrylonitrile (PAN) fibers produced by electrospinning. The resulting ZNTs are shown in Figure 6.8. The PAN fibers can either be woven or non-woven and hence, the resulting ZNTs can also be in a form of loose powder or a mat since they replicate the structure of the electrospun fibers.

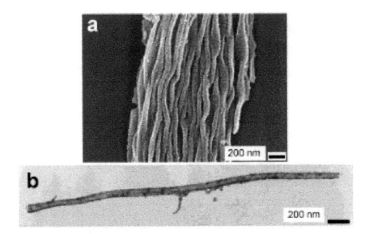

FIGURE 6.6 (a) SEM images of ZrO$_2$ nanotubes prepared by templated method against cellulose. (b) Transmission electron microscope image of an individual ZrO$_2$ nanotube as prepared in (a). (Reprinted with permission from Huang, J. and Kunitake, T., *J. Am. Chem. Soc.*, 125, 11834–11835, 2003. Copyright 2003 American Chemical Society.)

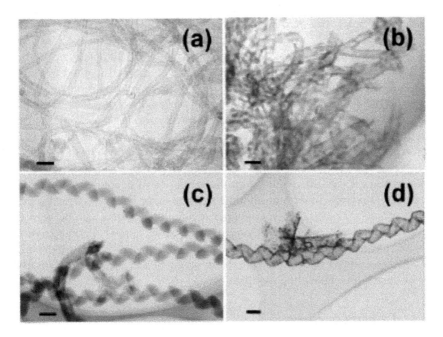

FIGURE 6.7 TEM images of (a) CNTs, (b) ZNTs synthesized using CNTs, (c) carbon nanocoils, (d) ZNTs synthesized using carbon nanocoils. The scale bar for (a) and (b) is 30 nm and (c) and (d) is 1 μm. (Reprinted with permission from Ogihara, H. et al., *Chem. Mater.*, 18, 4981–4983, 2006. Copyright 2006 American Chemical Society.)

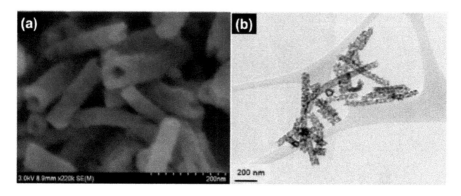

FIGURE 6.8 (a) SEM image and (b) TEM image of ZNTs formed on PAN fibers. (Reprinted from *J. Power Sources*, 337, Aziz, M.A. and Shanmugam, S., Zirconium oxide nanotube–Nafion composite as high-performance membrane for all vanadium redox flow battery, 36–44, Copyright 2017, with permission from Elsevier.)

Through the nanofibrous templated technique, the inner diameter of the obtained ZNTs is similar to the outer diameter of the nanofibers used, and their wall thicknesses could be controlled by the number of coating processes. The coating can be done for more than 50 times to produce the several nanometer-thick of ZNT walls. Once the template is removed, the solid ZrO_2 will have a hollow interior forming the nanotubular material. Template removal can be performed by thermal or chemical methods. The thermal method appears to be the one often adopted for ZNTs formation. Calcination can be done in air at 700°C–800°C for several hours. Ideally, the process of removal must be done at fairly low temperatures and should not leave any residue that would interfere with the final product. Subsequent heating is needed to crystallize the ZNTs to a desired phase. Nonetheless, by combusting the carbon core, one would expect the ZNTs formed are subjected to unintentional doping. Another major disadvantage of this process is the time required for the formation of nanotubes and the complexity of the process in term of multiple steps needed. The ZNTs are also prone to agglomeration. Nevertheless, nanofiber template techniques can be seen as a rather versatile way to produce the multilayered oxide, core-shell, composite and branched nanotubes. Controlled doping can also be achieved, either by intentionally adding in dopant material during the impregnation of the Zr-precursor to the fiber or during post-annealing treatment.

6.4.3 CYLINDRICAL WELL

In this process, a micron-sized template with uniform, elongated, cylindrical pores is required. Similar to the previous process, zirconium-precursor is also needed. Typically, templates like track-etched polymer (TEP) membrane, nanoporous polymer and anodic alumina (AAO) are used. Tsai et al. (2008) used an AAO template where a Zr precursor solution was deposited into the elongated channels of the AAO, then the sol-gel process was encouraged for oxide nucleation and growth.

AAO contains two-dimensional hexagonal pores and the diameter and depth of these pores can be precisely controlled by the anodization conditions; electrolyte, voltage, time, and temperature.

The choice of precursor used would depend on the method of filling pores. For ZNTs formation, the sol-gel method is the most used. Other typical pores-filling methods are electrodeposition, electrophoretic and vapor deposition despite rarely applied for ZNTs formation. Another more adopted method is atomic layer deposition (ALD) or layer-by-layer deposition whereby ZrO_2 precursor solution was deposited by ALD in a nanoporous polycarbonate (PC) template (Shin et al., 2004).

Similar to the nanofibrous method, the template ought to be removed once ZrO_2 has solidified and crystallized. One advantage of this method is the uniform diameter and length of the nanotubes formed unless post process is done. Other advantages are similar to nanofibrous methods including the possibility of performing controlled doping of the ZNTs formed and formation of multilayered oxide, despite branches and other complex nanostructures that may not be able to be formed. It is known that yttria-stabilized zirconia (YSZ) is an extremely important ceramic oxide with applications ranging from medical to energy devices. Through a templated method, YSZ nanotubes (Meng et al., 2008) can be produced as shown in Figure 6.9. As can be seen, the nanotubes are rather aligned with uniform diameter and length.

In addition to creating uniform nanotubes, it is also possible to modify the morphology of the nanotubes along the length by this approach. Segmented nanotubes and core-shell nanotubes can also be produced by appropriately depositing various different layers of precursor solution to the ZNTs. Moreover, post treatment can be done to produce branched and other hierarchical ZNTs structure. Similar to the fibrous method, metal-ZrO_2 junction can easily be formed by using the ZNTs powder prepared by this technique. Reduction of metal ions can be done on the surface of the ZNTs as to produce metal nanoparticles-ZNTs nanostructure. Disadvantages of this process are similar to the previous method; moreover, with the use of polymeric

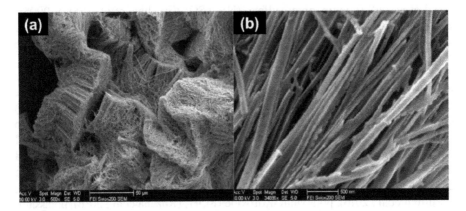

FIGURE 6.9 SEM micrographs of the yttria-stabilized ZNTs. (a) low magnification, and (b) high magnification. (Reprinted from *Mater. Chem. Phys.*, 111, Meng, X. et al., Preparation and characterization of yttria-stabilized zirconia nanotubes, 275–278, Copyright 2008, with permission from Elsevier.)

membrane oxide deposition process, it must also be done at low temperature and hence another step is required for crystallization of the oxide material.

As can be seen through this technique, the length and the inner diameter of the nanotubes will correspond to the dimension of the track-etched polymer/nanoporous polymer or AAO template. The wall thickness can be varied according to the number of coatings applied. There is a possibility that once the ZNTs are free from the TEP/ nanoporous polymer or AAO, due to their high surface energy, they bundle up and agglomerate. This indeed will add to the complexity of integrating the ZNTs to functional devices. Nonetheless, the ZNTs can be compacted and sintered on a functional substrate, for instance in the formation of anode material for an electrochemical cell (such as for solar cell), but by doing so the advantages of the tubular geometry may be much reduced.

The soft template method has been reported for oxide nanotubes formation; however, the resulting tubes that are dispersed in solution typically have a considerably wide distribution of dimensions; i.e. the length and diameter cannot be controlled as easily as the hard template method. Nonetheless this route has been less explored for ZNTs formation compared to the hard-templated technique as described above.

6.5 ANODIC ZNTs

Electrochemical anodization can be defined as a controlled electrochemical growth of an oxide film on a metal substrate by polarizing the metal anodically in an electrochemical cell. There are four main components as shown in Figure 6.10a that are needed in the process of anodization of Zr for ZNTs formation: (1) anode: Zr substrate, (2) cathode: carbon or platinum, (3) power supply, and (4) electrolyte.

The electrolyte can be aqueous or non-aqueous and for the formation of nanotubes, fluoride ions (F^-) must be added in. Typically, ethylene glycol (EG) (Shin and Lee, 2009, Li et al., 2011, Bashirom et al., 2016, 2016a, 2017, Rozana et al., 2016, 2017, Soaid et al., 2017), ammonium sulfate, $(NH_4)_2SO_4$ (Tsuchiya et al., 2005a, 2005b, Zhao et al., 2008b, Guo et al., 2009b, Zhang and Han, 2011, Wang et al., 2012b) or sodium sulfate, Na_2SO_4 (Ismail et al., 2011) and glycerol (Zhao et al., 2008a, Guo et al., 2009a, Wang and Luo, 2010, 2011, 2012, Muratore et al., 2010, 2011, Fang et al., 2011, 2012, 2013, Vacandio et al., 2011, Stępień et al., 2014, Hosseini et al., 2015) are used as the electrolyte. Often glycerol is added with formamide (FA) (Zhao et al., 2008a, Guo et al., 2009a, Fang et al., 2011, 2012, 2013) or dimethylformamide (DMF) (Hosseini et al., 2015) to increase the rate of ZNTs formation. It is apparent, due to limited water content in organic electrolytes, a small volume of additive was added into the electrolyte such as water (Zhao et al., 2008a, Guo et al., 2009a, Shin and Lee, 2009, Wang and Luo, 2010, 2011, 2012, Fang et al., 2011, 2012, 2013, Muratore et al., 2010, 2011, Vacandio et al., 2011, Stępień et al., 2014, Hosseini et al., 2015, Bashirom et al., 2016), potassium hydroxide (KOH) (Rozana et al., 2016), potassium carbonate (K_2CO_3) (Bashirom et al., 2016a, 2017, Soaid et al., 2017), or hydrogen peroxide (H_2O_2) (Bashirom et al., 2016, Rozana et al., 2017).

Figure 6.10b shows typical ZNTs formed by anodic process. The SEM is used to view both the surface and the cross section of the anodic film. Normally the surface

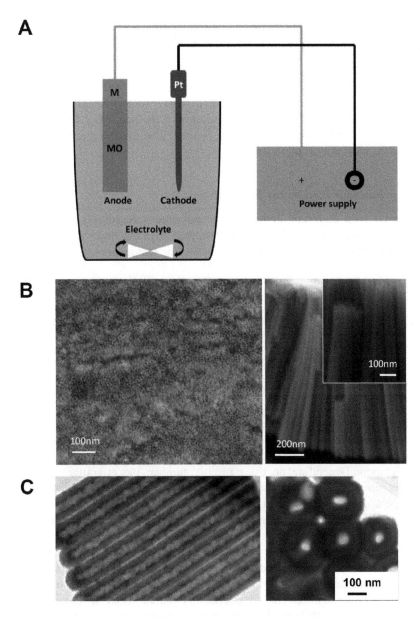

FIGURE 6.10 (a) Schematic diagram of anodization process (M = metal, MO = metal oxide, and Pt = platinum), (b) Typical FESEM micrographs of ZNTs formed by anodic process, and (c) TEM micrographs of nanotubular structure of ZNTs.

of anodic film consisted of ring structures indicative the success of nanotubes formation whereas the cross section the anodic film is comprised of well-aligned nanotubes (Figure 6.10b). These samples were prepared in fluoride containing glycerol electrolyte. TEM can be used to verify the existence of nanotubes. For instance, in fluoride containing EG electrolyte, TEM shows the presence of hollow fibrous (nanotube) structure (Figure 6.10c). From the micrograph, the inner and outer diameter of the nanotubes can be measured as well as the walls thickness of the nanotubes. It is obvious that the barrier layer is scalloped in shape, making the nanotubes open only at the top, but not at the bottom.

The growth of the oxide can then be monitored by recording the current–time characteristics during the course of the anodization process. There are several factors that influence the ZNTs; either the success of the nanotubes formation or the geometry and composition of the ZNTs formed. The following section describes some of the most important factors.

6.5.1   FLUORIDE IONS IN THE ELECTROLYTE

Perhaps the most important ingredient that determines the success of ZNTs formation by anodization is fluoride ions in the electrolyte. In the absence of fluoride ions, a compact anodic film is developed on Zr according to equations (6.4) to (6.7):

$$Zr \rightarrow Zr^{4+} + 4e^- \qquad (6.4)$$

$$Zr + 2H_2O \rightarrow ZrO_2 + 4H^+ + 4e \qquad (6.5)$$

$$Zr^{4+} + 4H_2O \rightarrow Zr(OH)_4 + 4H^+ \qquad (6.6)$$

$$Zr(OH)_4 \rightarrow ZrO_2 + H_2O \qquad (6.7)$$

Anodic film formation involves an ion formation (eq. 6.4), reaction with O^{2-} (due to deprotonation of H_2O) or OH^- to form ZrO_2 or $Zr(OH)_4$, respectively. $Zr(OH)_4$ then will develop into ZrO_2. The thickening of the anodic film will occur by high field ion migration. Depending on the transport number, the migration rate of Zr^{4+} and O^{2-} across the film will result in formation of a new oxide layer either at the oxide|electrolyte interface or at oxide|metal interface. Regardless of which ions dominate the growth process and hence the growth front of the anodic film, the new oxide layer will result in a thicker film. When the thickness, d, of the oxide is increased, then according to F = U/d, at a constant voltage, U, the field, F, developed across the oxide will be reduced and the driving force for solid state migration of ions is reduced. Because of this the anodic film will have a finite thickness. Usually, anodic film at the oxide|electrolyte interface is contaminated by anions or cations from the electrolyte while the layer at the metal|oxide interface consists of a more compact and purer ZrO_2. Depending on the nature of the electrolyte (pH, temperature and composition), dissolution of ZrO_2 can also occur. Moreover, dissolution can occur when the surface of the anodic film is saturated by H^+ from the anodic oxide formation as seen in equations (6.5) and (6.6).

The presence of fluoride ions in the electrolyte indeed affects the anodic process, which obviously alter the morphology of the anodic film, whereby fluoride can form water soluble $[ZrF_6]^{2-}$ species by complexation with Zr^{4+} ions that are ejected at the oxide–electrolyte across the anodic film driven by high field effect (eq. 6.5) and by chemical etching of the ZrO_2 (eq. 6.6). Dissolution of anodic ZrO_2 was studied by El-Mahdy et al. (1996) whereby film growth is restricted with the increase of F^- ion concentrations and the temperature of electrolyte. They attributed to the dissolution process of anodic ZrO_2 to follow equation (6.10).

$$Zr^{4+} + F^- \rightarrow [ZrF_6]^{2-} + 4H^+ \tag{6.8}$$

$$ZrO_2 + 6F^- \rightarrow [ZrO_6]^{2-} + H_2O \tag{6.9}$$

$$ZrO_2 + 6HF \rightarrow H_2ZrF_6 + 2H_2O \tag{6.10}$$

Typical SEM micrographs for anodic ZrO_2 formed in electrolytes with 0.07 wt% NH_4F are shown in Figure 6.11a whereby the nanotubes are obvious. With excess NH_4F content in the electrolyte, the structure deteriorates. An optimum amount of fluoride ions required is often in the range of 0.05–0.5 wt% NH_4F to ensure the success in the formation of ZNTs.

When the fluoride ions are replaced by chloride ions (Cl^-), instead of nanotubular structure the anodic film appears to have a layered structure. This was observed by Bashirom et al. (2016) as shown in Figure 6.12a and b (Bashirom et al., 2016). The anodic process appears to be very fast as well as corroding the Zr foil.

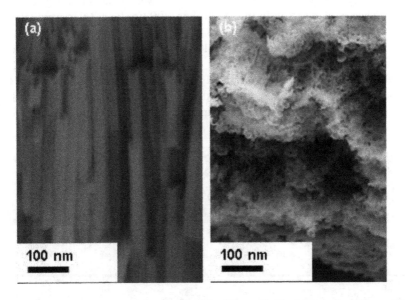

FIGURE 6.11 Micrographs of Zr anodized in EG at 60 V added to it different NH_4F contents: (a) 0.07 wt% and (b) 0.7 wt% for 1 h.

FIGURE 6.12 SEM micrographs (a) surface and (b) cross-sectional morphologies of anodic ZrO_2 film grown in $EG/H_2O_2/NH_4Cl$ at 20 V for 10 min. (Reprinted from *Procedia Chem.*, 19, Bashirom, N. et al., Effect of fluoride or chloride ions on the morphology of ZrO_2 thin film grown in ethylene glycol electrolyte by anodization, 611–618, Copyright 2016, with permission from Elsevier.)

6.6 ANODIZATION TIME

ZNTs are characterized by their length and diameters in which the aspect ratio of 1-D structure can be determined. The length can be directly measured from the cross-section SEM image of the anodic film whereas the diameters mentioned can be measured from TEM. The longest ZNTs reported were formed in FA/glycerol electrolyte (Zhao et al., 2008a). In this electrolyte, 190 μm long ZNTs were fabricated after 24 h of oxidation duration. Figure 6.13 shows the ZNTs formed in FA/glycerol/NH_4F/H_2O (Amer et al., 2014). Figure 6.14 shows the SEM images of ZNTs formed at four different anodization times from our laboratory. The formation of ZNTs is quite fast; ~2 μm long nanotubes formed after 10 min of anodization (Figure 6.14a) and increased to ~3 μm after 30 min, ~4 μm after 60 min, and >10 μm after 180 min of anodization time (Figure 6.14b, c, and d, respectively). The diameter of the

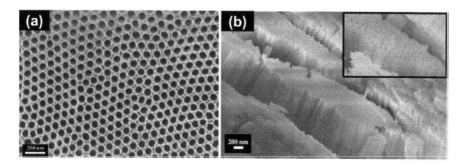

FIGURE 6.13 Field emission SEM images of ZNTs formed in glycerol/FA/NH_4F/H_2O, (a) top-view, and (b) side-view. (From Amer, A.W. et al., *RSC Adv.*, 4, 36336–36343, 2014. Reproduced by permission of The Royal Society of Chemistry.)

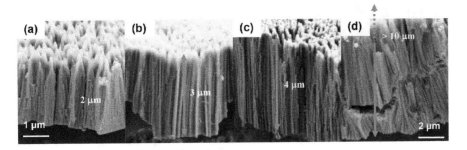

FIGURE 6.14 Field emission SEM images of anodized Zr in fluoride EG at 40 V at different anodization times: (a) 10 min, (b) 30 min, (c) 60 min, and (d) 180 min.

nanotubes is not affected too much with anodization time, but the length increases significantly with anodization times. Nonetheless, when the ZNTs are too long the anodic film is found to be easily detached from the substrate.

6.7 ANODIZATION VOLTAGE

Diameter of ZNTs is dependent on the anodization voltage. Diameter changes with voltage are not sensitive to the electrolyte used whereby regardless of aqueous or organic, the diameter of anodic ZNTs appear to be larger when anodized at higher voltage. Surface microstructures of anodic ZNTs formed at 40, 50, and 60 V are shown in Figure 6.15a–c, respectively. Nevertheless, under the same anodization voltage, ZNTs formed in different electrolytes will have different diameters. Normally anodization in organic electrolyte will resulted in larger diameter compared to anodization in aqueous electrolyte.

Dark-field TEM micrographs of ZNTs formed at 20 and 40 V are shown in Figure 6.16a and b, respectively. Several conclusions can be derived from these micrographs: (1) the barrier layers of the films are ~18 and ~36 nm thick at 20 and 40 V respectively, indicating that barrier layer thickness is voltage dependent and (2) the external diameters of the ZNTs are ~30 nm at 20 V and 65 nm at 40 V

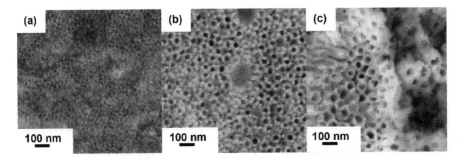

FIGURE 6.15 Surface SEM micrographs of Zr anodized in glycerol with 0.07 wt% NH$_4$F at (a) 40 V, (b) 50 V, and (c) 60 V for 1 h.

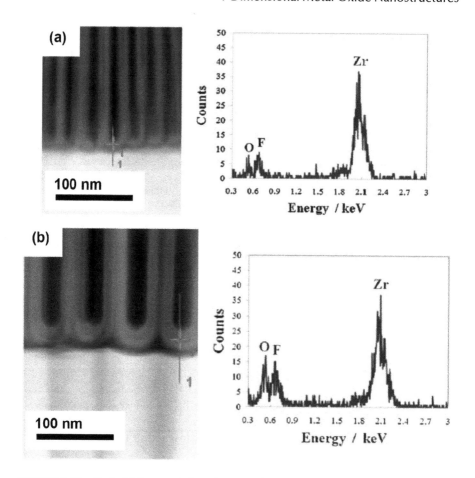

FIGURE 6.16 Dark-field transmission electron micrographs of FIB sections, with coupled EDX analyses at the locations of the crosses, of the bottom region of the anodic film formed by anodizing zirconium at (a) 20 V and (b) 40 V in 0.35 M NH₄F in glycerol, containing 5 vol.% added water, at room temperature. (Reprinted from *Electrochima Acta*, 58, Muratore, F. et al., Comparison of nanotube formation on zirconium in fluoride/glycerol electrolytes at different anodizing potentials, 389–398, Copyright 2011, with permission from Elsevier.)

and the diameter increases with anodization voltage (Muratore et al., 2011). When energy dispersive X-ray (EDX) analysis was done at the bottom of the nanotube (at the marked region as seen in the micrographs), the concentration of fluorine is also seen to be influenced by the anodic voltage. More fluorine is detected at 40 V sample indicating more anion insertion and contamination occur at high voltage especially at the bottom of the nanotubes. Fluorine insertion within the oxide is important in elucidating the development of nanotubular structure, not ordered porous anodic film. As shown, fluoride ions are essential in nanotube formation and the incorporation and importance can be described by considering the oxide growth model.

Similar with other high oxidation state metals, growth of anodic film on Zr follows the Point Defect Model (PDM) as put forward by Macdonald (1992) and revised as PDM II and III in 2011. According to this model, physicochemical processes that occur within a passive film will include constant generation and annihilation of cation interstitials and oxygen vacancies. During film growth, Zr^{4+} vacancies are produced at the oxide|electrolyte interface and are consumed at the oxide|metal interface and oxygen vacancies are formed at the oxide|metal interface and are consumed at the oxide|electrolyte interface. The model predicts or rather explains the formation of bi-layer passive film with a defective and anions-contaminated barrier layer and an outer layer consisting of some precipitates. It is known that the migration rate of fluoride ions is higher than O^{2-} ions, therefore, fluoride ions can be expected to accumulate at the barrier layer near the oxide|metal interface. As already mentioned, the existence of fluorine-rich regions at the scalloped-shaped barrier layer has already been reported and hence, it is safe to assume that there exists a rich layer of fluoride; or termed fluoride-rich layer (FRL) at the bottom of ZNTs.

Anodic film at the oxide|electrolyte interface can be subjected to electric field dissolution as described by Amer et al. (2014) resulting in surface irregularities, which will then be attacked by fluoride ions to induce chemical dissolution. This leads to the formation of pits and pits are considered as pore nucleation centers, which with time will lead to the formation of ordered porous structure (Ismail et al., 2011). There is a possibility of further field-assisted dissolution process occurring within each pore resulting in pore deepening and enlargement. The size (diameter) of the pore then is determined by electric field distribution at the pore walls. Apart from pores deepening and widening, cells separation is an important step in the transforming nanopores to nanotubes.

In anodic alumina formation, the oxide flow concept, which is assumed to originate from the plasticity of the barrier oxide layer generated by compressive stresses, has been used to describe circular pores formation. To minimize stress, the oxide behaves viscoplastically and it is continually being pushed upwards from the metal|oxide interface. Due to the compressive stress developed within the pore wall, the pore walls will be pushed outwards along with the FRL. It is rational to assume that such an FRL is a common phenomenon in ordered nanostructured anodic oxides grown in fluoride containing electrolytes hence, similar growth mechanisms are operative here as well. The transition from pore to tube structures can occur by dissolving the pore boundaries which consists of the FRL. The fluoride compound must be soluble in the electrolyte water (or any other solvent that can dissolve the FRL) which is required in the electrolyte. Often water is added to dissolve the FRL in organic electrolytes as summarized in Table 6.7.

TABLE 6.7

Literature Surveys on Fabrication of ZNTs by Anodization Method

Electrolyte	Additive	Voltage (V)	Time	Length (μm)	D_o (nm)	D_i (nm)	Wall Thickness (nm)	Ref.
1 M H_2SO_4/0.1 wt% NH_4F	–	10–50	5 h	30	10	–	–	Tsuchiya and Schmuki (2004)
1 M $(NH_4)_2SO_4$/0.5 wt% NH_4F	–	20	1 h	17	50	–	–	Tsuchiya et al. (2005b)
1 M $(NH_4)_2SO_4$/0.5 wt% NH_4F	–	20	–	16.6	50	–	–	Tsuchiya et al. (2005a)
1 M $(NH_4)_2SO_4$/0.5 wt% NH_4F	–	10–50	3 h	12	40	–	–	Zhao et al. (2008b)
1 M $(NH_4)_2SO_4$/0.5 wt% NH_4F	–	20	3 h	20	50	–	–	Guo et al. (2009b)
1 M Na_2SO_4/0.01–5 wt% of NH_4F	–	20–50	1 h	6	40	–	10	Ismail et al. (2011)
1 M $(NH_4)_2SO_4$/0.15 M NH_4F	–	20	1 h	7	65	–	–	Zhang and Han (2011)
1M $(NH_4)_2SO_4$/1 wt% NH_4F	–	3–15	2.5 h	13	70	–	–	Wang et al. (2012b)
Glycerol/0.05 M $(NH_4)_2HPO_4$/0.35 M NH_4F	5 vol.% H_2O	30	1 h	12	75	–	–	Wang and Luo (2010)
Glycerol/0.35 M NH_4F	0, 1, 5 vol.% H_2O	40	–	6.90 / 7.10	58 / 75	22 / 40	–	Muratore et al. (2010)
Glycerol/0.35 M NH_4F	0,1,5 vol.% H_2O	20–40	–	40–65	28	15	–	Muratore et al. (2011)
Glycerol/5 wt% NH_4F	5 vol.% H_2O	15–30	–	1–5	20–50	–	–	Wang and Luo (2011)
Glycerol/0.35 M NH_4F	5% vol. H_2O	40	10 min	5	25–30	–	–	Vacandio et al. (2011)
Glycerol/0.35M NH_4F	5 vol.% H_2O	20	1 h	7.74	75	–	–	Wang and Luo (2012)
Glycerol/0.1 M HF	–	10–60	1–24 h	5–45	40–80	15–55	16	Stępień et al. (2014)
Glycerol/DMF (1:1)/1 wt% NH_4F	3 wt% H_2O	20 / 50 / 60	3 h	–	40 / 150 / 160	10 / 40 / 53	–	Hosseini et al. (2015)
FA + Glycerol (1:1)/1 wt% NH_4F	3 wt% H_2O	50	24 h	190	130	66	32	Zhao et al. (2008a)

(Continued)

TABLE 6.7 (Continued)
Literature Surveys on Fabrication of ZNTs by Anodization Method

Electrolyte	Additive	Voltage (V)	Time	Length (μm)	D$_o$ (nm)	D$_i$ (nm)	Wall Thickness (nm)	Ref.
FA + Glycerol (1 : 1)/0.5–5 wt% HCl	3.5 wt% H$_2$O	20	5 h	33	250–300	–	–	Guo et al. (2009a)
FA + Glycerol (1 : 1)/1 wt% NH$_4$F	3 wt% H$_2$O	50	3 h	30.93	60	–	–	Fang et al. (2011)
FA + Glycerol (1 : 1)/1 wt% NH$_4$F	3 wt% H$_2$O	50	3 h	23.2	130	–	–	Fang et al. (2012)
FA + Glycerol (1 : 1)/1 wt% NH$_4$F	3 wt% H$_2$O	20–50	–	58–115	–	–	–	Fang et al. (2013)
EG/17.5 wt% NH$_4$F	–	30	30 min	2.3	26–36	–	–	Li et al. (2011)
EG/0.38 wt% NH$_4$F	1.79 wt% H$_2$O	30	30 min	–	16	–	–	Shin and Lee (2009)
EG/0.3 wt% NH$_4$F	1 vol.% K$_2$CO$_3$	20–60	1 h	9.5	71.4	–	–	Bashirom et al. (2016a)
EG/0.3 wt% NH$_4$F	5 vol.% H$_2$O$_2$ or H$_2$O	20	10 min	–	28.8	–	6.3	Bashirom et al. (2016)
					25.1		4.9	
EG/0.3 wt% NH$_4$F	1 M KOH	60	1 h	10	40	–	–	Rozana et al. (2016)
EG/0.3 wt% NH$_4$F	1 M K$_2$CO$_3$	20–60	1 h	3.9–9.4	72	–	14.8	Soaid et al. (2017)
EG/0.3 wt% NH$_4$F	1 M K$_2$CO$_3$	60	1 h	13.6	84.0	41.8	22.9	Bashirom et al. (2017)
EG/0.27 wt% NH$_4$F	3 wt% H$_2$O$_2$	20	1 h	7	35	–	6	Rozana et al. (2017)
		40		12	40		8	
		60		14	46		10	
		80		7.5	60		11	
		100		7	80		9	

6.8 CRYSTALLINITY OF ANODIC ZNTs

ZrO_2 can exist in three different crystalline phases under atmospheric pressure: monoclinic (m), tetragonal (t) and cubic (c). The monoclinic phase is thermodynamically the most stable phase at ambient temperature therefore, ZNTs are thought to be comprised of m-ZrO_2. Nevertheless, anodic ZrO_2 is found to be comprised of t or c-ZrO_2. The presence of crystalline t- or c-ZrO_2 in as-anodized ZNTs was reported in aqueous electrolytes with fluoride addition (Tsuchiya et al., 2005a, 2005b, Zhao et al., 2008b, Ismail et al., 2011). Apparently, the growth of oxygen vacancies which occurs at metal/oxide interfaces induces crystalline ZNTs and F$^-$ ions content in the electrolyte affect the crystallization of anodized ZNTs (Ismail et al., 2011). However, it is generally reported that the as-anodized ZNTs in organic electrolytes are amorphous (Zhao et al., 2008a, Guo et al., 2009b, Fang et al., 2011, 2013), although nanocrystalline regions may also exist.

As mentioned, oxide film growth results from the competition of oxide formation and oxide dissolutions. In the oxidation process, H_2O from the electrolyte solution acted as the oxygen source. The diffusion of electroneutral H_2O molecules is rather difficult across the oxide layer. Therefore, H_2O dissociation on the anode will result in the formation of free O^{2-} or OH$^-$ ions for oxide formation. These ions will travel to the metal/oxide interface under the influence of an electric field. It is obvious that the process depends on various parameters; concentration of diffusing ions, field applied and oxide nucleation. It will be inevitable for oxide nearer to the metal to suffer from lack of oxygen. Simultaneously, oxygen vacancies are produced at the metal/oxide interface. Therefore, it is expected that there exists a concentration gradient across the oxide with higher concentrations of oxygen vacancies at the inner layer near the metal/oxide interface (Ismail et al., 2011). According to Ismail et al., the stabilization of c-ZrO_2/t-ZrO_2 in ZNTs may be due to the combination of the presence of oxygen vacancies and the size effect (Ismail et al., 2011). Experimental evidence on the structure of anodic ZNTs has concluded that upon annealing at relatively low temperature 400°C–500°C, t- or c-ZrO_2 formed as evidenced from high resolution TEM (HRTEM) shown in Figure 6.17. As for the use of ZNTs for Cr(VI) reduction, we have recently reported on the fast reduction of amorphous ZNTs as opposed to crystalline ZNTs under sunlight. Based on the results, 95% of 10 ppm Cr(VI) can be removed over the amorphous ZNTs compared to crystalline ZNTs (33%) after 5 h. High photocatalytic activity of the amorphous ZNTs was attributed to enhanced Cr(VI) adsorption due to its high density of –OH groups, and high concentration of oxygen vacancies leading to enhanced light absorption under visible spectrum (Bashirom et al., 2018). Therefore, ZrO_2 photocatalyst can be a promising candidate for Cr(VI) removal especially in minute concentrations. Besides Cr(VI), there were also other attempts made to remove other heavy metal ions such as As(III) and As(V) (Cui et al., 2012), Pb(II) (Sengupta et al., 2017), and Cu(II) (Kwon et al., 2016) on ZrO_2.

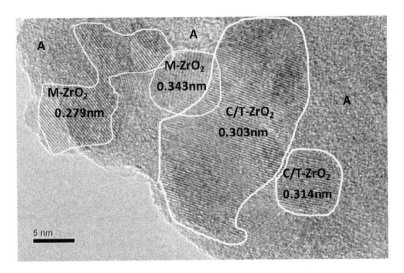

FIGURE 6.17 A high resolution transmission electron microscopy (HRTEM) image of a wall of ZNTs after annealing showing t- and c-ZrO$_2$.

6.9 CONCLUSION

Heavy metals pollution in several states in Malaysia in recent years is reviewed from recent literatures. It appears that the distribution of heavy metal ions in Malaysia can be related to human activities including industries, plantations, agricultural, mining, and shipping. Contamination from chromium is the highest near agricultural sites and electronic and automotive industries. Cr(VI) is carcinogenic and hence removal of Cr(VI) from industrial waste is required. One method discussed here is by photocatalytic reduction of Cr(VI) on ZNTs as the heterogenous photocatalyst. Formation of nanotubes has been discussed in great detail. Templated techniques have been concluded as the main technique in early days, and only until recently anodization has been applied for nanotubes formation. ZNTs derived from anodization are well-aligned, perfectly arranged and have diameter and length that can be controlled. Literature survey on the anodic ZNTs is presented along with all the important characteristics of ZNTs.

ACKNOWLEDGMENT

Heavy Metal Mitigation project is supported by USM-Research University Grant for Toyohashi University of Technology, Japan-USM collaboration; 1001/PBAHAN/870048. Crystallization of ZrO$_2$ work is supported by Fundamental Research Grant Scheme, Ministry of Higher Education Malaysia 1001/PBAHAN/6071363. KPT-UniMAP's SLAB Scholarship for postgraduate scholar is also acknowledged.

REFERENCES

Agency for Toxic Substances and Disease Registry. *Toxicological profile for nickel*. Available: https://www.atsdr.cdc.gov/toxprofiles/tp60-c6.pdf [Accessed February 13, 2018].

Agency for Toxic Substances and Disease Registry. 2004. *Public health statement for copper.* Agency for Toxic Substances and Disease Registry. Available: https://www.atsdr.cdc. gov/phs/phs.asp?id=204&tid=37 [Accessed February 15, 2018].

Agency for Toxic Substances and Disease Registry. 2005. *Public health statement for nickel.* Agency for Toxic Substances and Disease Registry. Available: https://www.atsdr.cdc. gov/phs/phs.asp?id=243&tid=44 [Accessed February 15, 2018].

Agency for Toxic Substances and Disease Registry. 2013a. *Cadmium toxicity. Where is cadmium found?* Available: https://www.atsdr.cdc.gov/csem/csem.asp?csem=6&po=5 [Accessed February 13, 2018].

Agency for Toxic Substances and Disease Registry. 2013b. *Chromium Toxicity. Where Is Chromium Found?* Available: https://www.atsdr.cdc.gov/csem/csem. asp?csem=10&po=5 [Accessed February 13, 2018].

Agency for Toxic Substances and Disease Registry. 2015. *Public health statement for cadmium.* Available: https://www.atsdr.cdc.gov/phs/phs.asp?id=46&tid=15 [Accessed February 15, 2018].

Agency for Toxic Substances and Disease Registry. N/A. *Potential for human exposure: Copper.* Available: https://www.atsdr.cdc.gov/toxprofiles/tp132-c6.pdf [Accessed February 15, 2018].

Agorku, E., Kuvarega, A., Mamba, B., Pandey, A. and Mishra, A. (2015) Enhanced visible-light photocatalytic activity of multi-elements-doped ZrO_2 for degradation of indigo carmine. *Journal of Rare Earths*, 33(5), p.498–506.

Amer, A. W., Mohamed, S. M., Hafez, A. M., Alqaradawi, S. Y., Aljaber, A. S. and Allam, N. K. (2014) Self-assembled zirconia nanotube arrays: Fabrication mechanism, energy consideration and optical activity. *RSC Advances*, 4(68), p.36336–36343.

Aziz, M. A. and Shanmugam, S. (2017) Zirconium oxide nanotube–Nafion composite as high-performance membrane for all vanadium redox flow battery. *Journal of Power Sources*, 337, p.36–44.

Barakat, M. (2011) New trends in removing heavy metals from industrial wastewater. *Arabian Journal of Chemistry*, 4(4), p.361–377.

Barrera-Díaz, C. E., Lugo-Lugo, V. and Bilyeu, B. (2012) A review of chemical, electrochemical and biological methods for aqueous Cr(VI) reduction. *Journal of Hazardous Materials*, 223, p.1–12.

Basahel, S. N., Ali, T. T., Mokhtar, M. and Narasimharao, K. (2015) Influence of crystal structure of nanosized ZrO_2 on photocatalytic degradation of methyl orange. *Nanoscale Research Letters*, 10(1), p.1–13.

Bashirom, N., Kian, T. W., Kawamura, G., Matsuda, A., Razak, K. A. and Lockman, Z (2018) Sunlight activated anodic freestanding ZrO_2 nanotube arrays for Cr(VI) photoreduction. *Nanotechnology*. doi:10.1088/1361-6528/aaccbd.

Bashirom, N., Razak, K. A. and Lockman, Z. (2017) Synthesis of freestanding amorphous ZrO_2 nanotubes by anodization and their application in photoreduction of Cr(VI) under visible light. *Surface and Coatings Technology*, 320, p.371–376.

Bashirom, N., Razak, K. A., Yew, C. K. and Lockman, Z. (2016) Effect of fluoride or chloride ions on the morphology of ZrO_2 thin film grown in ethylene glycol electrolyte by anodization. *Procedia Chemistry*, 19, p.611–618.

Bashirom, N., Ye, B. C., Razak, K. A. and Lockman, Z. (2016a) Formation of freestanding ZrO_2 nanotubes for Cr(VI) removal. *AIP Conference Proceedings*, 1733, p.020029.

Botta, S. G., Navío, J. A., Hidalgo, M. C., Restrepo, G. M. and Litter, M. I. (1999) Photocatalytic properties of ZrO$_2$ and Fe/ZrO$_2$ semiconductors prepared by a sol–gel technique. *Journal of Photochemistry and Photobiology A: Chemistry*, 129, p.89–99.

Cui, H., Li, Q., Gao, S. and Shang, J. K. (2012) Strong adsorption of arsenic species by amorphous zirconium oxide nanoparticles. *Journal of Industrial and Engineering Chemistry*, 18(4), p.1418–1427.

Division, E. S. (2010) *Drinking water quality standard.* Ministry of Health Malaysia. Available: http://kmam.moh.gov.my/public-user/drinking-water-quality-standard.html [Accessed February 15, 2018].

Du, W., Zhu, Z., Zhang, X., Wang, D., Liu, D., Qian, X. and Du, J. (2013) RE/ZrO$_2$ (RE= Sm, Eu) composite oxide nano-materials: Synthesis and applications in photocatalysis. *Materials Research Bulletin*, 48(10), p.3735–3742.

El-Mahdy, G. A., Mahmoud, S. S. and El-Dahan, H. A. (1996) Effect of halide ions on the formation and dissolution behaviour of zirconium oxide. *Thin Solid Films*, 286(1), p.289–294.

Fang, D., Huang, K., Luo, Z., Wang, Y., Liu, S. and Zhang, Q. (2011) Freestanding ZrO$_2$ nanotube membranes made by anodic oxidation and effect of heat treatment on their morphology and crystalline structure. *Journal of Materials Chemistry*, 21(13), p.4989–4994.

Fang, D., Luo, Z., Liu, S., Zeng, T., Liu, L., Xu, J., Bai, Z. and Xu, W. (2013) Photoluminescence properties and photocatalytic activities of zirconia nanotube arrays fabricated by anodization. *Optical Materials*, 35(7), p.1461–1466.

Fang, D., Yu, J., Luo, Z., Liu, S., Huang, K. and Xu, W. (2012) Fabrication parameter-dependent morphologies of self-organized ZrO$_2$ nanotubes during anodization. *Journal of Solid State Electrochemistry*, 16(3), p.1219–1228.

Fu, F. and Wang, Q. (2011) Removal of heavy metal ions from wastewaters: A review. *Journal of Environmental Management*, 92(3), p.407–418.

Gehle, K. (2009) Arsenic toxicity. ATSDR Case Studies in Environmental Medicine, Agency for Toxic Substances and Disease Registry (ATSDR), Course WBCBDV1576.

Guo, L., Zhao, J., Wang, X., Xu, R. and Li, Y. (2009a) Synthesis and growth mechanism of zirconia nanotubes by anodization in electrolyte containing Cl$^-$. *Journal of Solid State Electrochemistry*, 13, p.1321–1326.

Guo, L., Zhao, J., Wang, X., Xu, R., Lu, Z. and Li, Y. (2009b) Bioactivity of zirconia nanotube arrays fabricated by electrochemical anodization. *Materials Science and Engineering C*, 29, p.1174–1177.

Gurushantha, K., Anantharaju, K., Nagabhushana, H., Sharma, S., Vidya, Y., Shivakumara, C., Nagaswarupa, H., Prashantha, S. and Anilkumar, M. (2015) Facile green fabrication of iron-doped cubic ZrO$_2$ nanoparticles by Phyllanthus acidus: Structural, photocatalytic and photoluminescent properties. *Journal of Molecular Catalysis A: Chemical*, 397, p.36–47.

Han, N. M. I. M., Latif, M. T., Othman, M., Dominick, D., Mohamad, N., Juahir, H. and Tahir, N. M. (2013) Composition of selected heavy metals in road dust from Kuala Lumpur city centre. *Environmental Earth Sciences*, 72(3), p.849–859.

Hao, Y., Li, L., Zhang, J., Luo, H., Zhang, X. and Chen, E. (2017) Multilayer and open structure of dendritic crosslinked CeO$_2$-ZrO$_2$ composite: Enhanced photocatalytic degradation and water splitting performance. *International Journal of Hydrogen Energy*, 42(9), p.5916–5929.

Harasim, P. and Filipek, T. (2015) Nickel in the environment. *Journal of Elementology*, 20(2), p.525–534.

Hernández-Ramírez, A. and Medina-Ramírez, I. 2015. *Photocatalytic Semiconductors: Synthesis, Characterization, and Environmental Applications*, Basel, Switzerland, Springer.

Hossain, M. A., Ali, N. M., Islam, M. S. and Hossain, H. Z. (2014) Spatial distribution and source apportionment of heavy metals in soils of Gebeng industrial city, Malaysia. *Environmental Earth Sciences*, 73(1), p.115–126.

Hosseini, M., Daneshvari-Esfahlan, V. and Ordikhani-Seyedlar, R. (2015) Fabrication, characterisation and investigation of zirconium oxide corrosion behaviour on resistance of zirconium oxide nanotubes in artificial saliva as biological environment. *Corrosion Engineering, Science and Technology*, 50(7), p.533–537.

Huang, J. and Kunitake, T. (2003) Nano-precision replication of natural cellulosic substances by metal oxides. *Journal of the American Chemical Society*, 125(39), p.11834–11835.

Ibrahim, M. (2015) Photocatalytic activity of nanostructured $ZnO–ZrO_2$ binary oxide using fluorometric method. *Spectrochimica Acta Part A: Molecular and Biomolecular Spectroscopy*, 145, p.487–492.

Ismail, S., Ahmad, Z. A., Berenov, A. and Lockman, Z. (2011) Effect of applied voltage and fluoride ion content on the formation of zirconia nanotube arrays by anodic oxidation of zirconium. *Corrosion Science*, 53(4), p.1156–1164.

Jafari, T., Moharreri, E., Amin, A., Miao, R., Song, W. and Suib, S. (2016) Photocatalytic water splitting—The untamed dream: A review of recent advances. *Molecules*, 21(7), p.900.

Järup, L. (2003) Hazards of heavy metal contamination. *British Medical Bulletin*, 68(1), p.167–182.

Ji, Q., Yu, X., Zhang, J., Liu, Y., Shang, X. and Qi, X. (2017) Photocatalytic degradation of diesel pollutants in seawater by using $ZrO_2(Er^{3+})/TiO_2$ under visible light. *Journal of Environmental Chemical Engineering*, 5(2), p.1423–1428.

Kadhum, S. A., Ishak, M. Y., Zulkifli, S. Z. and Binti Hashim, R. (2015) Evaluation of the status and distributions of heavy metal pollution in surface sediments of the Langat River Basin in Selangor Malaysia. *Marine Pollution Bulletin*, 101(1), p.391–396.

Karunakaran, C., Sujatha, M. P. and Gomathisankar, P. (2009) Photoreduction of chromium(VI) on ZrO_2 and ZnS surfaces. *Monatshefte für Chemie-Chemical Monthly*, 140(11), p.1269–1274.

Kaviyarasu, K., Kotsedi, L., Simo, A., Fuku, X., Mola, G. T., Kennedy, J. and Maaza, M. (2017) Photocatalytic activity of ZrO_2 doped lead dioxide nanocomposites: Investigation of structural and optical microscopy of RhB organic dye. *Applied Surface Science*, 421, p.234–239.

Khodami, S., Surif, M., Wo, W. M. and Daryanabard, R. (2016) Assessment of heavy metal pollution in surface sediments of the Bayan Lepas area, Penang, Malaysia. *Marine Pollution Bulletin*, 114(1), p.615–622.

Kumar, A., Guo, C., Sharma, G., Pathania, D., Naushad, M., Kalia, S. and Dhiman, P. (2016) Magnetically recoverable ZrO_2/Fe_3O_4/chitosan nanomaterials for enhanced sunlight driven photoreduction of carcinogenic Cr(VI) and dechlorination & mineralization of 4-chlorophenol from simulated waste water. *RSC Advances*, 6(16), p.13251–13263.

Kumar, S. and Ojha, A. K. (2015) Oxygen vacancy induced photoluminescence properties and enhanced photocatalytic activity of ferromagnetic ZrO_2 nanostructures on methylene blue dye under ultra-violet radiation. *Journal of Alloys and Compounds*, 644, p.654–662.

Kwon, O.-H., Kim, J.-O., Cho, D.-W., Kumar, R., Baek, S. H., Kurade, M. B. and Jeon, B.-H. (2016) Adsorption of As(III), As(V) and Cu(II) on zirconium oxide immobilized alginate beads in aqueous phase. *Chemosphere*, 160, p.126–133.

Latif, M. T., Yong, S. M., Saad, A., Mohamad, N., Baharudin, N. H., Mokhtar, M. B. and Tahir, N. M. (2013) Composition of heavy metals in indoor dust and their possible exposure: A case study of preschool children in Malaysia. *Air Quality, Atmosphere & Health*, 7(2), p.181–193.

Lenntech. 2018a. *Chemical properties of chromium - Health effects of chromium - Environmental effects of chromium.* Available: https://www.lenntech.com/periodic/elements/cr.htm [Accessed June 7, 2018].

Lenntech. 2018b. *Chemical properties of zinc - Health effects of zinc - Environmental effects of zinc* [Online]. Available: https://www.lenntech.com/periodic/elements/zn.htm [Accessed June 7, 2018].

Li, L., Yan, D., Lei, J., He, J., Wu, S. and Pan, F. (2011) Fast fabrication of highly regular and ordered ZrO$_2$ nanotubes. *Materials Letters*, 65(9), p.1434–1437.

Li, M., Li, X., Jiang, G. and He, G. (2015) Hierarchically macro–mesoporous ZrO$_2$–TiO$_2$ composites with enhanced photocatalytic activity. *Ceramics International*, 41(4), p.5749–5757.

Lias, K., Jamil, T. and Aliaa, S. (2013) A preliminary study on heavy metal concentration in the marine bivalves Marcia marmorata species and sediments collected from the coastal area of Kuala Perlis, North of Malaysia. *IOSR Journal of Applied Chemistry*, 4(1), p.48–54.

Lim, W. Y., Aris, A. Z. and Tengku Ismail, T. H. (2013) Spatial geochemical distribution and sources of heavy metals in the sediment of Langat River, Western Peninsular Malaysia. Environmental Forensics, 14(2), p.133–145.

Macdonald, D. D. (1992) The point defect model for the passive state. *Journal of the Electrochemical Society*, 139(12), p.3434–3449.

Meng, X., Tan, X., Meng, B., Yang, N. and Ma, Z.-F. (2008) Preparation and characterization of yttria-stabilized zirconia nanotubes. *Materials Chemistry and Physics*, 111(2), p.275–278.

Mohajernia, S., Mazare, A., Gongadze, E., Kralj-Iglič, V., Iglič, A. and Schmuki, P. (2017) Self-organized, free-standing TiO$_2$ nanotube membranes: Effect of surface electrokinetic properties on flow-through membranes. *Electrochimica Acta*, 245, p.25–31.

Mohamed, A., Osman, T. A., Toprak, M. S., Muhammed, M., Yilmaz, E. and Uheida, A. (2016) Visible light photocatalytic reduction of Cr(VI) by surface modified CNT/titanium dioxide composites nanofibers. *Journal of Molecular Catalysis A: Chemical*, 424(Supplement C), p.45–53.

Muratore, F., Baron-Wiecheć, A., Gholinia, A., Hashimoto, T., Skeldon, P. and Thompson, G. (2011) Comparison of nanotube formation on zirconium in fluoride/glycerol electrolytes at different anodizing potentials. *Electrochimica Acta*, 58, p.389–398.

Muratore, F., Baron-Wiecheć, A., Hashimoto, T., Skeldon, P. and Thompson, G. (2010) Anodic zirconia nanotubes: Composition and growth mechanism. *Electrochemistry Communications*, 12(12), p.1727–1730.

Nagarjuna, R., Challagulla, S., Ganesan, R. and Roy, S. (2017) High rates of Cr(VI) photoreduction with magnetically recoverable nano-Fe$_3$O$_4$@Fe$_2$O$_3$/Al$_2$O$_3$ catalyst under visible light. *Chemical Engineering Journal*, 308(Supplement C), p.59–66.

Nawale, A. B., Kanhe, N. S., Bhoraskar, S. V., Mathe, V. L. and Das, A. K. (2012) Influence of crystalline phase and defects in the ZrO$_2$ nanoparticles synthesized by thermal plasma route on its photocatalytic properties. *Materials Research Bulletin*, 47(11), p.3432–3439.

Ogihara, H., Sadakane, M., Nodasaka, Y. and Ueda, W. (2006) Shape-controlled synthesis of ZrO$_2$, Al$_2$O$_3$, and SiO$_2$ nanotubes using carbon nanofibers as templates. *Chemistry of Materials*, 18(21), p.4981–4983.

Owlad, M., Aroua, M. K., Daud, W. A. W. and Baroutian, S. (2009) Removal of hexavalent chromium-contaminated water and wastewater: A review. *Water, Air, and Soil Pollution*, 200(1–4), p.59–77.

Pathakoti, K., Morrow, S., Han, C., Pelaez, M., He, X., Dionysiou, D. D. and Hwang, H.-M. (2013) Photoinactivation of Escherichia coli by sulfur-doped and nitrogen–fluorine-codoped TiO$_2$ nanoparticles under solar simulated light and visible light irradiation. *Environmental Science & Technology*, 47(17), p.9988–9996.

Poungchan, G., Ksapabutr, B. and Panapoy, M. (2016) One-step synthesis of flower-like carbon-doped ZrO_2 for visible-light-responsive photocatalyst. *Materials & Design*, 89, p.137–145.

Praveena, S. and Aris, A. (2017) Status, source identification, and health risks of potentially toxic element concentrations in road dust in a medium-sized city in a developing country. *Environmental Geochemistry and Health*, p.1–14.

Rani, S., Kumar, M., Sharma, S., Kumar, D. and Tyagi, S. (2014) Effect of graphene in enhancing the photo catalytic activity of zirconium oxide. *Catalysis Letters*, 144(2), p.301–307.

Rao, C., Satishkumar, B. and Govindaraj, A. (1997) Zirconia nanotubes. *Chemical Communications*, 16, p.1581–1582.

Reddy, V. R., Hwang, D. W. and Lee, J. S. (2003) Photocatalytic water splitting over ZrO_2 prepared by precipitation method. *The Korean Journal of Chemical Engineering*, 20(6), p.1026–1029.

Renuka, L., Anantharaju, K., Sharma, S., Nagaswarupa, H., Prashantha, S., Nagabhushana, H. and Vidya, Y. (2016) Hollow microspheres Mg-doped ZrO_2 nanoparticles: Green assisted synthesis and applications in photocatalysis and photoluminescence. *Journal of Alloys and Compounds*, 672, p.609–622.

Rozana, M., Soaid, N. I., Kawamura, G., Kian, T. W., Matsuda, A. and Lockman, Z. (2016) Effect of KOH added to ethylene glycol electrolyte on the self-organization of anodic ZrO_2 nanotubes. *AIP Conference Proceedings*. AIP Publishing, 020024.

Rozana, M., Soaid, N. I., Kian, T. W., Kawamura, G., Matsuda, A. and Lockman, Z. (2017) Photocatalytic performance of freestanding tetragonal zirconia nanotubes formed in H_2O_2/NH_4F/ethylene glycol electrolyte by anodisation of zirconium. *Nanotechnology*, 28(15), p.155604.

Sany, S. B. T., Salleh, A., Sulaiman, A. H., Sasekumar, A., Rezayi, M. and Tehrani, G. M. (2013) Heavy metal contamination in water and sediment of the Port Klang coastal area, Selangor, Malaysia. *Environmental Earth Sciences*, 69(6).

Selvam, N. C. S., Manikandan, A., Kennedy, L. J. and Vijaya, J. J. (2013) Comparative investigation of zirconium oxide (ZrO_2) nano and microstructures for structural, optical and photocatalytic properties. *Journal of Colloid and Interface Science*, 389(1), p.91–98.

Sengupta, A., Mallick, S. and Bahadur, D. (2017) Tetragonal nanostructured zirconia modified hematite mesoporous composite for efficient adsorption of toxic cations from wastewater. *Journal of Environmental Chemical Engineering*, 5(5), p.5285–5292.

Shafie, N. A., Aris, A. Z., Zakaria, M. P., Haris, H., Lim, W. Y. and Isa, N. M. (2013) Application of geoaccumulation index and enrichment factors on the assessment of heavy metal pollution in the sediments. *Journal of Environmental Science and Health, Part A*, 48(2), p.182–190.

Sherly, E., Vijaya, J. J., Selvam, N. C. S. and Kennedy, L. J. (2014) Microwave assisted combustion synthesis of coupled $ZnO–ZrO_2$ nanoparticles and their role in the photocatalytic degradation of 2, 4-dichlorophenol. *Ceramics International*, 40(4), p.5681–5691.

Shin, H., Jeong, D. K., Lee, J., Sung, M. M. and Kim, J. (2004) Formation of TiO_2 and ZrO_2 nanotubes using atomic layer deposition with ultraprecise control of the wall thickness. *Advanced Materials*, 16(14), p.1197–1200.

Shin, Y. and Lee, S. (2009) A freestanding membrane of highly ordered anodic ZrO_2 nanotube arrays. *Nanotechnology*, 20(10), p.105301.

Shu, Z., Jiao, X. and Chen, D. (2012) Synthesis and photocatalytic properties of flower-like zirconia nanostructures. *CrystEngComm*, 14(3), p.1122–1127.

Shu, Z., Jiao, X. and Chen, D. (2013) Hydrothermal synthesis and selective photocatalytic properties of tetragonal star-like ZrO_2 nanostructures. *CrystEngComm*, 15(21), p.4288–4294.

Sinhamahapatra, A., Jeon, J.-P., Kang, J., Han, B. and Yu, J.-S. (2016) Oxygen-deficient zirconia (ZrO$_{2-x}$): A new material for solar light absorption. *Scientific Reports*, 6, p.27218.

Smirnova, N., Gnatyuk, Y., Eremenko, A., Kolbasov, G., Vorobetz, V., Kolbasova, I. and Linyucheva, O. (2006) Photoelectrochemical characterization and photocatalytic properties of mesoporous TiO$_2$/ZrO$_2$ films. *International Journal of Photoenergy*, 85469, p.1–6.

Soaid, N. I., Bashirom, N., Rozana, M. and Lockman, Z. (2017) Formation of anodic zirconia nanotubes in fluorinated ethylene glycol electrolyte with K$_2$CO$_3$ addition. *Surface and Coatings Technology*, 320, p.86–90.

Sreethawong, T., Ngamsinlapasathian, S. and Yoshikawa, S. (2013) Synthesis of crystalline mesoporous-assembled ZrO$_2$ nanoparticles via a facile surfactant-aided sol–gel process and their photocatalytic dye degradation activity. *Chemical Engineering Journal*, 228, 256–262.

Stępień, M., Handzlik, P. and Fitzner, K. (2014) Synthesis of ZrO$_2$ nanotubes in inorganic and organic electrolytes by anodic oxidation of zirconium. *Journal of Solid State Electrochemistry*, 18(11), p.3081–3090.

Stojadinović, S., Vasilic, R., Radic, N. and Grbic, B. (2015) Zirconia films formed by plasma electrolytic oxidation: Photoluminescent and photocatalytic properties. *Optical Materials*, 40, p.20–25.

Sultana, S., Khan, M. Z. and Shahadat, M. (2015) Development of ZnO and ZrO$_2$ nanoparticles: Their photocatalytic and bactericidal activity. *Journal of Environmental Chemical Engineering*, 3(2), p.886–891.

Suresh, P., Vijaya, J. J. and Kennedy, L. J. (2014) Photocatalytic degradation of textile-dyeing wastewater by using a microwave combustion-synthesized zirconium oxide supported activated carbon. *Materials Science in Semiconductor Processing*, 27, p.482–493.

Taib, M. A. A., Razak, K. A., Jaafar, M. and Lockman, Z. (2017) Initial growth study of TiO$_2$ nanotube arrays anodised in KOH/fluoride/ethylene glycol electrolyte. *Materials & Design*, 128, p.195–205.

Tashakor, M., Yaacob, W. Z. W., Mohamad, H. and Ghani, A. A. (2014) Geochemical characteristics of serpentinite soils from Malaysia. *Malaysian Journal of Soil Science*, 18, p.35–49.

Tsai, M. C., Te Lin, G., Chiu, H. T. and Lee, C. Y. (2008) Synthesis of zirconium dioxide nanotubes, nanowires, and nanocables by concentration dependent solution deposition. *Journal of Nanoparticle Research*, 10(5), p.863–869.

Tsuchiya, H., Macak, J. M., Ghicov, A., Taveira, L. and Schmuki, P. (2005a) Self-organized porous TiO$_2$ and ZrO$_2$ produced by anodization. *Corrosion Science*, 47(12), p.3324–3335.

Tsuchiya, H., Macak, J. M., Taveira, L. and Schmuki, P. (2005b) Fabrication and characterization of smooth high aspect ratio zirconia nanotubes. *Chemical Physics Letters*, 410(4), p.188–191.

Tsuchiya, H. and Schmuki, P. (2004) Thick self-organized porous zirconium oxide formed in H$_2$SO$_4$/NH$_4$F electrolytes. *Electrochemistry Communications*, 6(11), p.1131–1134.

United States Environmental Protection Agency. (2017) *Mercury Emissions: The Global Context*. United States Environmental Protection Agency. Available: https://www.epa.gov/international-cooperation/mercury-emissions-global-context [Accessed February 2, 2018].

US-EPA. (2017) *Health effects of exposures to mercury*. Available: https://www.epa.gov/mercury/health-effects-exposures-mercury.

US-EPA. (2018) *Table of regulated drinking water contaminants*. Available: https://www.epa.gov/ground-water-and-drinking-water/national-primary-drinking-water-regulations#Inorganic [Accessed February 15, 2018].

Vacandio, F., Eyraud, M., Knauth, P. and Djenizian, T. (2011) Tunable electrical properties of self-organized zirconia nanotubes. *Electrochemistry Communications*, 13(10), p.1060–1062.

Vignesh, K., Priyanka, R., Rajarajan, M. and Suganthi, A. (2013) Photoreduction of Cr(VI) in water using Bi_2O_3–ZrO_2 nanocomposite under visible light irradiation. *Materials Science and Engineering: B*, 178(2), p.149–157.

Wang, A.-J., Bong, C. W., Xu, Y.-H., Hassan, M. H. A., Ye, X., Bakar, A. F. A., Li, Y.-H., Lai, Z.-K., Xu, J. and Loh, K. H. (2017) Assessment of heavy metal pollution in surficial sediments from a tropical river-estuary-shelf system: A case study of Kelantan River, Malaysia. *Marine Pollution Bulletin*, 125(1–2), p.492–500.

Wang, L.-N. and Luo, J.-L. (2010) Enhancing the bioactivity of zirconium with the coating of anodized ZrO_2 nanotubular arrays prepared in phosphate containing electrolyte. *Electrochemistry Communications*, 12(11), p.1559–1562.

Wang, L.-N. and Luo, J.-L. (2011) Fabrication and formation of bioactive anodic zirconium oxide nanotubes containing presynthesized hydroxyapatite via alternative immersion method. *Materials Science and Engineering: C*, 31(4), p.748–754.

Wang, L.-N. and Luo, J.-L. (2012) Electrochemical behaviour of anodic zirconium oxide nanotubes in simulated body fluid. *Applied Surface Science*, 258(10), p.4830–4833.

Wang, X., Zhai, B., Yang, M., Han, W. and Shao, X. (2013) ZrO_2/CeO_2 nanocomposite: Two step synthesis, microstructure, and visible-light photocatalytic activity. *Materials Letters*, 112, p.90–93.

Wang, J., Wang, Z., Huang, B., Ma, Y., Liu, Y., Qin, X., Zhang, X. and Dai, Y. (2012a) Oxygen vacancy induced band-gap narrowing and enhanced visible light photocatalytic activity of ZnO. *ACS Applied Materials & Interfaces*, 4, p.4024–4030.

Wang, X., Zhao, J., Hou, X., He, Q. and Tang, C. (2012b) Catalytic activity of ZrO_2 nanotube arrays prepared by anodization method. *Journal of Nanomaterials*.

WHO 2011. *Guidelines for Drinking-Water Quality*. Geneva, Switzerland: World Health Organization.

Wu, J., Wang, J., Du, Y., Li, H., Yang, Y. and Jia, X. (2015) Chemically controlled growth of porous CeO_2 nanotubes for Cr(VI) photoreduction. *Applied Catalysis B: Environmental*, 174, p.435–444.

Xiao, M., Li, Y., Lu, Y. and Ye, Z. (2015) Synthesis of ZrO_2: Fe nanostructures with visible-light driven H_2 evolution activity. *Journal of Materials Chemistry A*, 3(6), p.2701–2706.

Xu, Y. and Schoonen, M. A. (2000) The absolute energy positions of conduction and valence bands of selected semiconducting minerals. *American Mineralogist*, 85(3–4), p.543–556.

Yuswir, N. S., Praveena, S. M., Aris, A. Z., Ismail, S. N. S. and Hashim, Z. (2015) Health risk assessment of heavy metal in urban surface soil (Klang District, Malaysia). *Bulletin of Environmental Contamination and Toxicology*, 95(1), p.80–89.

Zhang, J., Li, L., Liu, D., Zhang, J., Hao, Y. and Zhang, W. (2015) Multi-layer and open three-dimensionally ordered macroporous TiO_2–ZrO_2 composite: Diversified design and the comparison of multiple mode photocatalytic performance. *Materials & Design*, 86, p.818–828.

Zhang, L. and Han, Y. (2011) Enhanced bioactivity of self-organized ZrO_2 nanotube layer by annealing and UV irradiation. *Materials Science and Engineering: C*, 31(5), p.1104–1110.

Zhao, J., Wang, X., Xu, R., Meng, F., Guo, L. and Li, Y. (2008a) Fabrication of high aspect ratio zirconia nanotube arrays by anodization of zirconium foils. *Materials Letters*, 62(29), p.4428–4430.

Zhao, J., Xu, R., Wang, X. and Li, Y. (2008b) In situ synthesis of zirconia nanotube crystallines by direct anodization. *Corrosion Science*, 50(6), p.1593–1597.

Zhao, Y., Zhang, Y., Li, J. and Du, X. (2014) Solvothermal synthesis of visible-light-active N-modified ZrO_2 nanoparticles. *Materials Letters*, 130, p.139–142.

7 One-Dimensional Metal Oxide Nanostructures in Sensor Applications

Ahalapitiya H. Jayatissa and Bharat R. Pant

CONTENTS

7.1 INTRODUCTION

Metal oxides have been extensively investigated for applications in gas sensors, biosensors, photodetectors and humidity sensors over the past two decades (Seiyama et al., 1962, Kolmakov and Moskovits, 2004, Kolmakov et al., 2005, Chen et al., 2010, Kumar et al., 2011, Devan et al., 2012, Arafat et al., 2012, Li et al., 2012, Jeong et al., 2014, Zhang et al., 2017). Metal oxides are composed of positive metallic ions and negative oxygen ions, and the strong bond between metal and oxygen made them solid and firm (Devan et al., 2012). Some of the unique properties that make metal oxides suitable for sensing applications include stability, high surface to volume ratio, easy fabrication, good selectivity and ability to sense a wide range of gases. Metal oxides are thermally and chemically stable because their s-shells are totally filled, but their d-shells are not completely filled, which makes them useful for electronic device applications. Other attractive properties of 1-D metal oxides are good thermal and chemical stabilities and low weight of electronic devices fabricated with

1-D materials. There are two types of metal oxides, n-type and p-type. For n-type metal oxides, the majority of charge carriers are electron whereas, for p-type metal oxides, the majority of charge carrier are holes (Zhang et al., 2017). The dramatic change in resistance of semiconductor metal oxide due to adsorption and desorption of gases was first reported by Seiyama et al. in 1962. The 1-D metal oxides have a large surface to volume ratio due to the high ratio of length to lateral dimension, which makes them highly sensitive towards target gases (Kolmakov et al., 2005).

There are many forms of 1-D metal oxides such as nanobelts, nanorods, nanowires and nanotubes. These nanostructures are considered one-dimensional because their width is negligible compared with length, which makes the length to width ratio very large. One can ask for the reason why 1-D nanostructures can have a more improved performance than other nanostructures, such as nanoparticles or nanosheets, as shown in Figure 7.1. A more specific answer can be that the lateral dimension of 1-D structure can be fabricated smaller than or equal to the Debye length of the 1-D structure. Therefore, the electrical transport phenomena can be better guided by the localized defect (doping) centers along the Debye length region throughout the 1-D nanostructure. Thus, the uniform sensitivity can be achieved throughout the nanostructure along

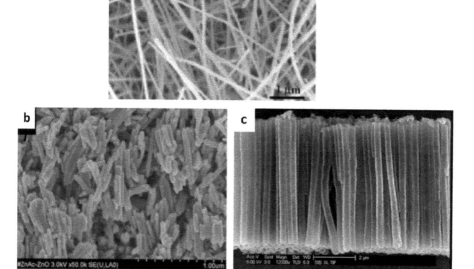

FIGURE 7.1   (a) Titanium Oxide nanowires with 10 nm diameter and 10 μm length (Reprinted with permission from Chen, P.-C. et al., *ACS Nano*, 4, 4403–4411, 2010. Copyright 2010 American Chemical Society.), (b) SEM image of ZnO nanorods with diameter ~60 nm and length ~500 nm (Reprinted with permission from Jeong, H.W. et al., *J. Phys. Chem. C.*, 118, 21331–21338, 2014. Copyright 2014 American Chemical Society.), and (c) SEM image of TiO$_2$ nanotubes with diameter 295 nm and length 6.1 μm. (Reprinted with permission from Kang et al., *Nano Lett.*, 9, 601–606, 2009. Copyright 2009: American Chemical Society.)

the length direction rather than the lateral directions. Metal oxides that form 1-D nanostructures are ZnO, SnO_2, TiO_2, and WO_3. These metal oxides can be fabricated with various sizes and configurations depending upon the requirements and applications (Arafat et al., 2012). Nanowires, nanorods and nanotubes are distinguished by their aspect ratio (length to width ratio). The aspect ratio of nanowires is higher than 20 whereas the aspect ratio of nanorods is between 2 and 20 (Arafat et al., 2012). The nanotubes are hollow structures and can be single-walled or multi-walled.

7.2   APPLICATIONS IN BIOSENSORS

The biosensor is a device that can be used to detect biomolecules such as protein, enzymes, body chemicals and antibodies by converting their presence or activities into an electrical signal (Kumar et al., 2011). Figure 7.2 shows the schematic of the working principle of a biosensor. A biosensor usually consists of a sensing part and transducer part. The sensing part detects the target materials and the transducer submits the information to the output system. The analytes or body chemicals are bound to the active surface of the sensor, and the output signal is received in the form of change in capacitance, conductance, current, or voltage of the active surface (Kolmakov and Moskovits, 2004). The biosensors are used to examine the function of body materials or organs as well as for the detection of diseases in human or animals. People have been using complex, expensive and time-consuming equipment and processes for the detection of a disease-causing agent such as a virus. The invention of metal oxide biosensors promises the development of low-cost, fast and hassle-free detection of diseases or pathogens (Guo et al., 2013).

1-D metal oxides have shown promising results for their applications in biosensors. In 2-D materials, the charge accumulation occurs on the surface of the film, but in 1-D materials the charge accumulation occurs on the bulk of the materials (Kumar et al., 2011). This difference makes a better sensing phenomenon of 1-D materials than 2-D nanomaterials. 1-D metal oxides can be deposited using many techniques such as chemical vapor deposition (CVD), sol-gel, vapor phase growth, laser ablation, and gaseous-disperse synthesis. The nanorods of ZnO were deposited on quartz substrates using the gaseous-disperse synthesis (GDS) technique (Viter et al., 2014).

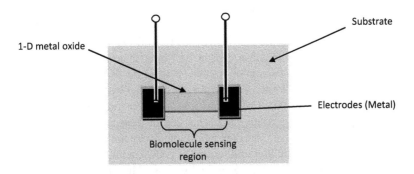

FIGURE 7.2   Schematic of two terminal biosensor that can be used to measure the interaction between biomolecules and nanostructure using impedance, temperature or voltage.

TABLE 7.1

Summary of the Literature of Biosensors Using 1-D Metal Oxides

Materials	1-D Structure	Synthesis Technique	Analyte	LDL Sensitivity (S)		References
ZnO	Nonorods	Gaseous-disperse synthesis (GDS)	Salmonella antigens	10^2–10^6		Viter et al. (2014)
ZnO	Nanorods	Hydrothermal decomposition	Enzymatic glucose	0.01–3.45 mM		Wei et al. (2006)
In$_2$O$_3$, FET	Nanowires	Laser ablation	Lipoprotein	—		Rouhanizadeh et al. (2006)
TiO$_2$:G	Nanowires	Electrospinning	Cholesterol	LDL = 6 µM S = 3.82 µA/cm^2		Komathi et al. (2016)
ZnO	Nanowires	Vapor-solid	Streptavidin	50 ng/mL S = 20%		Liu et al. (2008)
TiO$_2$	Nanotubes	Anodic oxidation	Body glucose	LDL = 3.8 µM S = 199.6 µAmM^{-1}cm^{-2}		Wang et al. (2014)
SnO$_2$	Nanorods	Hydrothermal	H$_2$O$_2$	LDL = 0.2 µM S = 379 µAmM^{-1}cm^{-2}		Liu et al. (2010)

Note: LDL = lower detection limit, FET = field-effect transistor, G = graphene, S = sensitivity.

A laser ablation method was also employed to synthesized In$_2$O$_3$ nanowires and carbon nanotube network-based field-effect transistors (FETs) on a silicon wafer for the detection of lipoproteins (Rouhanizadeh et al., 2006). Also, TiO$_2$ nanowires embedded with graphene were synthesized using an electrospinning method for the application in cholesterol biosensor (Komathi et al., 2016). Primarily, passive biosensors such as electrochemical sensors and FET-based sensors have been developed to detect the target analyte (Table 7.1).

7.2.1 ELECTROCHEMICAL BIOSENSORS

Sensing behavior of electrochemical biosensors basically depends upon the enzymatic catalysis of reaction, which produces or consumes electrons. In this type of sensor, there are three electrodes: (1) reference electrode, (2) working electrode, and (3) counter electrode. The reaction caused by target material on a working electrode produces a change of current or voltage (Wikipedia, 2018a). Figure 7.3a shows the SEM image of the ZnO nanorods synthesized by the hydrothermal decomposition method (Wei et al., 2006). The nanorods have a hexagonal cross-section, which indicates morphology of wurtzite ZnO. The size of nanorods are uniform with a dimension of 300 nm × 4 µm and they are aligned perpendicular to the substrates. Figure 7.3b shows the cyclic voltammogram of GO$_x$/ZnO nanorods with and without glucose. With the addition of 5 nM of glucose, the voltammogram changed

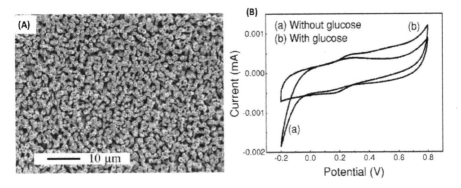

FIGURE 7.3 (a) SEM image of as-grown ZnO nanorod array, and (b) cyclic voltammograms of GO_x/ZnO nanorods/Au electrodes in the absence and presence of 5 mM glucose in 0.01M PBS buffer. (Reprinted with permission from Wei, A. et al., *Appl. Phys. Lett.*, 89, 123902, 2006. Copyright 2006 by the American Institute of Physics.)

dramatically and increments of current were observed starting from 0.2 V. This clearly indicates that the presence of glucose affects the I-V characteristics of the ZnO nanorods.

The cyclic voltammetry and chronoamperometry were performed to understand the electrochemical performance of sensors. Figure 7.4a shows the cyclic voltammetry of TiO_2 nanotubes with and without H_2O_2 in 0.2 M phosphate buffer solution (Wang et al., 2014). The input voltage was in the range −0.6 V to 0.6 V at the scan rate of 100 mV/s. The peak was reduced at about −0.5 V after addition of H_2O_2. For this reason, −0.5 V was chosen to be an optimal detection potential. The highest performance of 540.97 μA $mM^{-1}cm^{-2}$ was found for bare TiO_2 nanotube array, which is much higher than bare Ti sheet and bare TiO_2 film. This might be attributed to the high surface area and electro-catalytic sites of the TiO_2 nanotubes (Benvenuto et al., 2009). Figure 7.4b shows the cyclic voltammetry of an unhybridized nanotube array (curve a), after addition of 0.5 mM glucose (curve b) and 1 mM glucose (curve c). The peak potential was found at −0.5 V, which is related to H_2O_2 reduction at the time of GO_x oxidation. This clearly indicates that TiO_2 nanotubes have good sensing ability for glucose.

Figure 7.5a displays the CVs of HRP/SnO_2 array at the scan rate of 100, 300 and 500 mV/s. The peak of HRP is displayed by all the CV traces by the Fe^{3+}/Fe^{2+} conversion reaction (−0.47 V) (Kafi et al., 2008). Figure 7.5b exhibits the CV of electrocatalytic diminution of HRP towards H_2O_2 by a nanorods array electrode. When H_2O_2 was added to the phosphate buffer solution (PBS) solution the reduction current increases greatly with the decrease in oxidation current because of the presence of more oxidative HRP arising from the reaction between HRP_{red} and H_2O_2 (Kafi et al., 2008). This clearly indicates that the addition of H_2O_2 decreases current, which indicated that modified HRP can act as a catalyst in the reduction of H_2O_2 (Liu et al., 2010).

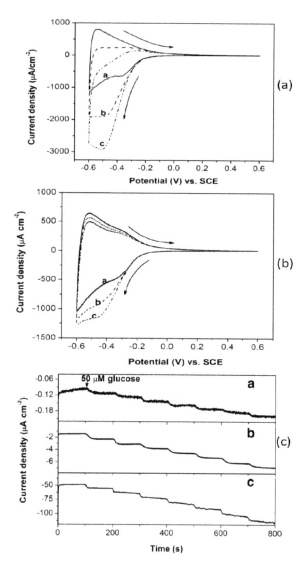

FIGURE 7.4 (a) Cyclic voltammograms of the TiO_2 nanotube in 0.2 M phosphate buffer with and without H_2O_2 solution pH = 6.8 (**a** without H_2O_2, **b** with 0.5 mM H_2O_2, **c** with 1 mM H_2O_2). (b) Cyclic voltammograms of TiO_2 nanotube array enzyme electrode in 0.2 M phosphate buffer solution (pH = 6.8) at different concentration of glucose (curve **a** without glucose, curve **b** 0.5 mM glucose, curve **c** 1 mM glucose) (scan rate = 100 mV/s). (c) The time-current response curve of three enzyme electrodes when 50 μM glucose was added to 0.2 M phosphate buffer solution and −0.5 V was applied. (Curve **a** I-t response of Ti sheet enzyme electrode, curve **b** I-t response of TiO_2 film enzyme electrode, curve **c** I-t response of TiO_2 nanotube array enzyme electrode). (Reprinted with permission from Wang et al., *Microchimica Acta*, 181, 381–387, 2014. Copyright 2014: Springer Nature.)

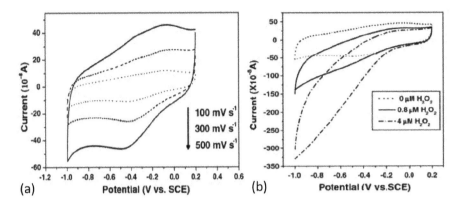

FIGURE 7.5 (a) Cyclic voltammetry (CV) of horseradish peroxides (HRP)/SnO_2 nanorod array electrode at different scan rates in N_2- saturated PBS solution. (b) CV of HRP/SnO_2 electrodes in N_2- saturated solution consisting of 0, 0.8 and 4 μM H_2O_2. (Reprinted with permission from Liu, J. et al., *Nanoscale Res. Lett.*, 5, 1177, 2010. Copyright 2010.)

7.2.2 FIELD-EFFECT TRANSISTOR (FET)-BASED BIOSENSORS

The most common type of transistor used in the analysis of analyte using biosensor is the field-effect transistor (FET). The device consists of three terminals in FET, namely source, drain, and gate. FET, also known as a unipolar transistor, exhibits a high input impedance at low frequencies, and the electric field generated by the voltage difference between body and gate controls the conductance between the source and drain (Wikipedia, 2018b). By applying a correct voltage onto its gate, properties of FET can be controlled. The 1-D metal oxide nanostructures act as a channel connecting source and drain, through which current flows and the conductivity of the channel depends upon the diameter of the channel (WhatIs.com, 2018) (Figure 7.6).

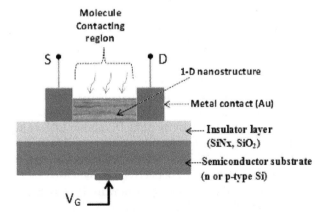

FIGURE 7.6 Schematic of FET based sensors. The most of the sensors are fabricated on Si/SiO_2 by placing a 1-D nanostructure (channel). The metallization is carried out with gold layer. The current (I_{SD}) between source (S) and drain (D) is measured and see the change of I_{SD} upon interaction of molecules on the channel region. The gate voltage (Vg) is applied to amplify the I_{SD}.

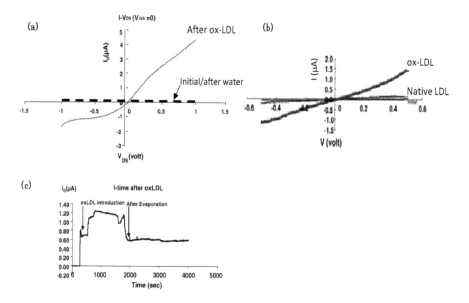

FIGURE 7.7 (a) I-V_{DS} curve of In$_2$O$_3$ shows the increase in conductivity (initially water) before and after addition of oxidized low-density lipoprotein (oxLDL). (initial and after water lines are coincided with V_{DS} axis). (b) I-V_{DS} selectivity curve of oxLDL and native LDL under dynamic condition. (c) Current-time dynamic response of FET when exposed to oxLDL sample (15.1% of oxLDL). (Reprinted with permission from Rouhanizadeh et al., *Sens. Actuators B Chem.*, 133, 638–643, 2006. Copyright 2006: Elsevier.)

Figure 7.7 shows the testing results of a FET biosensor fabricated using In$_2$O$_3$ nanowires. This transistor-based sensor is fabricated for sensing oxidized and low-density lipoprotein (oxLDL). Ti/Au electrodes act as the source and drain and In$_2$O$_3$ nanowire acts as a channel through which current flows. The source and drain electrodes are patterned by photolithography and successive Ti/Au deposition. Figure 7.7a shows an I-V_{DS} trends curve of initially water and after addition of oxLDL. It can be seen that the conductivity increased when oxLDL is added. Figure 7.7b displays the I-V_{DS} curve of FET for 15.1% oxidized LDL and 4.4% oxidized LDL. It is clear that conductivity of nLDL and oxLDL is higher than the initial conductivity (in black). The conductivity is increased by five-fold of magnitude with the addition of 15.1% oxLDL. Figure 7.7c shows the current versus time, dynamic response, of FET. The FET conductivity is increased by the addition of 15.1% LDL. These results indicated that the nanowire FET is highly sensitive to LDL particles. The graph shows a decrease in conductivity after 1800 s because of the evaporation of water. This example indicates that if appropriate matching occurs between the channel region of FET and functionality of target materials, practical sensors and sensing methods can be designed with 1-D nanostructures of semiconducting metal oxides.

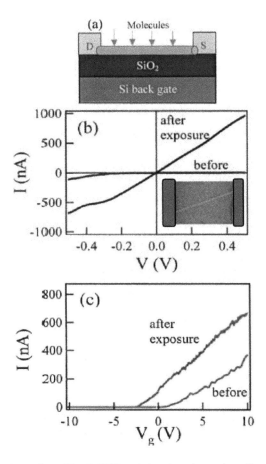

FIGURE 7.8   (a) Illustration of In_2O_3 FET and showing absorption of molecules on nanowires. (b) I-V curve before and after exposure to butylamine. (c) I-Vg curve before and after absorption of butylamine at drain-source bias voltage of 0.1 V. (Reprinted with permission from Li, C. et al., *Appl. Phys. Lett.*, 82, 1613–1615, 2003. Copyright 2003 by the American Institute of Physics.)

Also, In_2O_3 nanowire-based FETs have been investigated for sensing of organic and biomolecules (Li et al., 2003). The nanowires were single crystalline with a diameter of approximately 10 nm. Figure 7.8a shows that the analyte molecules are being absorbed to the surface of the nanowires. Figure 7.8b shows the I-V curves of FET at gate bias voltage $V_g = 0$ before and after the exposition of butylamine. It can be seen that the conductivity increased after nanowires are exposed to butylamine. The amine groups have electron-donating capability and electrons are being donated to the nanowire. As a result, the conductivity of nanowires increased and there is a shift at gate threshold voltage (Li et al., 2003). Figure 7.8c shows the dependence of current on gate bias. The I-V_g curve is recorded at a constant drain-source voltage of 0.1 V before and after exposure to butylamine. There is a −3 V shift in the threshold voltage after the exposure to amine due to the increase in electron concentration in In_2O_3 nanowires (Li et al., 2003).

7.3   ULTRAVIOLET LIGHT SENSORS

Ultraviolet (UV) light helps in the production of vitamin D in the body, it can be used to treat some skin diseases, and it also reduces hypertension and increases serotonin level (Wikipedia, 2018c). Despite some benefits, UV light can also cause many health problems. Sunburn is the major cause of skin cancer and sun rays constitute about 10% UV light. The UV light sensors based on 1-D metal oxides have attracted much attention because of their unique features and ability to detect photons effectively (Tian et al., 2015). Other properties of 1-D metal oxides that can be used as UV detector include high optical bandgap and high sensitivity. Various 1-D metal oxides have been investigated for their applications in photodetectors. Among possible UV photonic materials, ZnO, SnO_2, TiO_2, ITO and In_2O_3 have attracted much attention in recent years. Table 7.2 summaries the most important literature in this field.

There are several parameters for the evaluation of metal oxides as an efficient photodetector. Some of the common parameters are sensitivity, responsivity and external quantum efficiency (EQE). Sensitivity is defined as the ratio of change in current to the original dark current and expressed in percentage (Tian et al., 2015).

$$S = \frac{(I_{light} - I_{dark})}{I_{dark}}, \qquad (7.1)$$

where I_{dark} and I_{light} are the currents before and after exposure to UV light, respectively. The responsivity (R_λ) is defined as the photocurrent per unit power (A/W) of irradiation per active area (Tian et al., 2015).

$$R_\lambda = \frac{(I_{light} - I_{dark})}{I_{dark}}, \qquad (7.2)$$

where, P is the intensity of light, and S is the active area exposed to light.

TABLE 7.2
Summary of 1-D Metal Oxides as UV Photodetector in Literature

Material	Structure	Fabrication Method	Type of Light	Wavelength (nm)	Bias (V)	Maximum Sensitivity	References
ZnO/Cds	Nanorods	RF sputtering	UV	350	5	12.8 A/w	Lam et al. (2017)
ZnO	Nanorods	Hydrothermal	UV	365	10	–	Ko et al. (2015)
ZnO	Nanorods	Hydrothermal	UV	370	5	2 A/W	Humayun et al. (2014)
ZnO	Nanowires	Hydrothermal	Near UV	400	1	$I_p/I_d = 29$	Swanwick et al. (2012)
TiO_2	Nanorods	Hydrothermal	UV	320–400	0	0.025 A/w	Xie et al. (2013)
ITO	Nanowires	CVD	UV/Vis	355/543	3	0.07–0.2 A/W	Zhao et al. (2014)
SnO_2, ZnO	Nanobelts	Thermal evaporation	UV/gas	350	0		Arnold et al. (2003)
In_2O_3	Nanowires	Vapor-liquid-solid	UV	254/365	0.32	$I_{uv}/I_d = 10^4$	Zhang et al. (2003)

The mechanism of UV light sensing by a zinc oxide (ZnO) surface has been described based on equations (7.3) to (7.5).

$$O_2 + e^- \rightarrow O_2^-. \tag{7.3}$$

$$hv \rightarrow h^+ + e^- \text{(Conduction Band)}. \tag{7.4}$$

$$O_2^- + h^+ \rightarrow O_2. \tag{7.5}$$

During the illumination of UV light on the ZnO surface, the free oxygen molecules (O_2) absorb electrons from the conduction band and become O_2^-, resulting in increased resistance of ZnO nanorods, as shown by equation (7.3). Electron-hole pair is produced when photon energy of incident radiation is greater than band gap energy of ZnO, as shown by equation (7.4). The holes produced neutralize-adsorbed oxygen, which is responsible for the increase in resistance of ZnO, as a result photo current increases, as shown by equation (7.5).

7.3.1 OPTICAL PROPERTIES

The UV light sensor is fabricated using TiO_2 nanorods array synthesized by a hydrothermal process (Xie et al., 2013). The self-powered TiO_2 nanorods/water, solid-liquid heterojunction, is created for the efficient detection of UV light. Figure 7.9a shows the I-V curve in dark and under 1.25 mW/cm² UV light illumination. The UV light detector has an outstanding performance producing a short-circuit current of 4.67 µA and open circuit voltage of 0.48 V. The forward turn-on voltage of 0.4 V and rectification ratio of about 0.44 V are reported for this device. Figure 7.9b shows the dynamic response of a UV light sensor at the 1.25 mW/cm² illumination (Xie et al., 2013). The sensor showed an excellent sensitivity and repeatability with fast response and recovery of 0.15 s and 0.05 s, respectively. Figure 7.9c shows the responsivity of TiO_2 NRs/water at a different wavelength of UV light. The wavelength selective ability of a UV light detector has been performed in the wavelength range of 260–550 nm in unbiased conditions. It can be seen that the sensor displayed excellent selectivity in the 310–420 nm range with a maximum responsivity of 0.025 A/W at 350 nm.

SnO₂ and ZnO have also been used for the development of an FET-based gas sensor and UV light sensor. SnO_2 and ZnO nanobelts are fabricated on the SiO_2 substrates using a thermal evaporation technique (Arnold et al., 2003). The size of the nanobelts are in the range of 10–30 nm. The SnO_2 nanobelt-based FET is used as an oxygen sensor while ZnO nanobelts are used to detect UV light. Figure 7.10a shows the relation between source-drain current and gate potential of the ZnO FET. The ZnO FET had a threshold voltage of –15 V, switching ratio of about 100 and peak conductivity of $1.25 \times 10^{-3} (\Omega \text{ cm})^{-1}$. Figure 7.10b shows a different regime of a steady state of current decay after exposure to UV was ceased. The first regime shows a sharp decrease in current immediately after UV was turned off. This is attributed to the reunification of photo-generated electron-hole pairs. The second

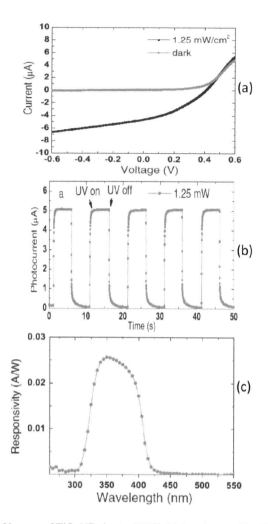

FIGURE 7.9 (a) I-V curve of TiO_2 NRs/water UV light detector at a bias from −0.6 to 0.6 V under dark and 365 nm illumination. (b) Current-time dynamic response of TiO_2 NRs/water UV light detector at 1.25 mW/cm² of UV radiation. (c) The responsivity of TiO_2/water from 260 to 550 nm under 0.0 V bias. (Reprinted with permission from Xie, Y. et al., *Nanoscale Res. Lett.*, 8, 188, 2013. Copyright 2013.)

regime is not as sharp as the first regime. The slow decrease in current might be due to a surface chemical reaction (Arnold et al., 2003). The inset of Figure 7.10b shows a dynamic response of ZnO-based FET when exposure to UV is ON and OFF, alternatively. There is a sharp increase in current when nanotubes are exposed to the UV light, and sharp decrease in current when UV exposure is ceased. The authors concluded that ZnO nanobelts can be successfully used to detect UV light with good response.

Vapor phase synthesized In_2O_3 nanowires has been used for applications in UV photodetectors (Zhang et al., 2003). The In_2O_3 individual nanowire-based

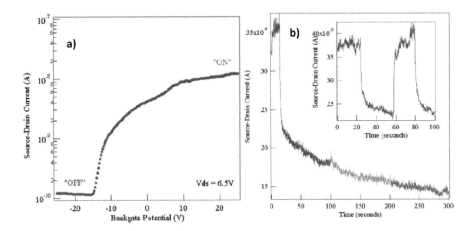

FIGURE 7.10   (a) Source drain current Vs back-gate voltage of ZnO nanobelt FET. (b) Response of ZnO nanobelt FET when exposed to the 350 nm UV light. Insert: time response of ZnO FET when UV light was turned on and off, alternatively. (Reprinted with permission from Arnold, M.S. et al., *J. Phys. Chem. B*, 107, 659–663, 2003. Copyright 2003 American Chemical Society.)

sensor showed sensitivities up to 10^4 when illuminated by UV light with a significant shift in threshold voltage. Figure 7.11a shows the I-V curve of the nanowire before and after exposure to the UV light. Before exposure to the UV light, the resistance was found to be very high with the differential conductance of 3.6×10^{-1} nS for V = 0 V. It can be seen that the conductance increased significantly after exposure to the UV light. The conductivity was higher at 254 nm UV light than that of 365 nm UV light with the conductance of 5.0×10^3 and 1.7×10^3, respectively. The I-V curve before exposure to the UV light is asymmetric in nature, which can be attributed to the local gaining effect. Figure 7.11b shows the current-gate voltage curve at a bias of 0.32 V under various conditions. It can be seen that slope of the curve and threshold gate voltage were remarkably regulated by UV light. Before the exposure to the UV light the threshold gate voltage was +5 V, but after exposure, it shifted to −2 V and −25 V at 365 and 254 nm, respectively.

Figure 7.11c displays the current-time response and repeatability curve of the In_2O_3 sensor at 254 nm UV light illumination. The drain-source voltage was kept constant at 0.3 V and gate bias was 0 V during measurement. It can be seen that the conductance increased rapidly when UV light was turned on and decayed gradually when the UV light was turned off. Figure 7.11d shows the current-time response of the sensor at 254 and 365 nm of the UV light. At 254 nm UV light, the current reached up to 290 nA and at 354 nm the current reached up to 33 nA. Hence, it is clear that the sensor is more sensitive to 254 nm UV light than 365 nm UV light, which also indicates that the sensor is selective to a different wavelength of light. It can also be seen that the response time is very low, but recovery time is very high.

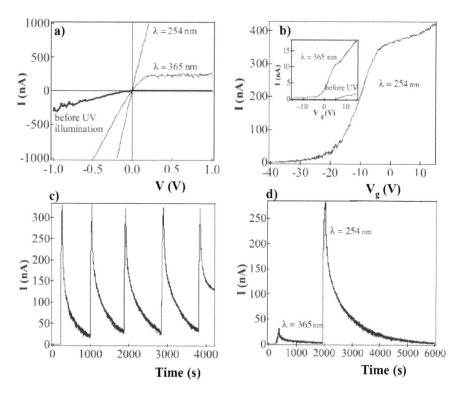

FIGURE 7.11 (a) I-V curve of the In_2O_3 nanowires before and after illumination with UV light. (b) I-V_g characteristics curve at a bias of 0.32 V, under various conditions. (c) The current-time response of In_2O_3 nanowires when the 254-nm UV light is switched on and off repeatedly. (d) Photoresponse of In_2O_3 nanowires at 254 and 365 nm UV light. (Reprinted by permission from Macmillan Publishers Ltd. *Appl. Phys. A*, Zhang, D. et al., 2003, Copyright 2003.)

7.4   APPLICATIONS IN HUMIDITY AND GAS SENSORS

7.4.1   HUMIDITY SENSOR

Humidity is the presence of water vapor in the air. Humidity plays very important roles in some industries, agriculture, human health and medical devices. There are two types of humidity: absolute humidity and relative humidity. Absolute humidity is defined as the amount of water vapor (in gram) per cubic meter of air. The unit of absolute humidity is g/m^3 and it does not depend on temperature. Relative humidity is the ratio of the amount of water vapor in air at a certain temperature to the amount of water vapor needed for the saturation at the same temperature. Relative humidity is expressed as a percentage and it is temperature dependent. Humidity affects human health: too low or too high humidity is not good for a human. Low humidity can cause eye irritation, dryness of skin, nose, and throat, and affects mucous membranes. High humidity augments the growth of micro-organisms such as fungi, bacteria and viruses in the atmosphere and hence increases the chance of infection caused by micro-organisms (Arundel et al., 1986). High humidity also causes allergies, sweating and feelings of discomfort.

Due to several effects of humidity to the human health and other industrial processing, it is very important to monitor and control it. Humidity sensors play important roles in monitoring and controlling humidity. It is very important to have humidity sensors with high sensitivity and good stability. The successful humidity sensor requires high sensitivity, thermal and chemical stability, low cost, durability and low operation temperature (Zhang et al., 2005). 1-D metal oxides have been used in the fabrication of humidity sensors. The disadvantages of conventional humidity sensors are high cost and large volume. But 1-D metal oxides are cheap, and their size varies from micrometre to nanometre.

7.4.1.1 Working Mechanism of Humidity Sensor

The sensing of humidity can be observed in many ways such as a change in resistance, capacitance, I-V characteristics and impedance. The change in resistance is due to physisorption and chemisorption of water molecules on the surface. When the metal oxide interacts with humidity, the chemisorption of water molecules takes place and water molecules diffuse quickly into H^+ and OH^- ions due to self-ionization of water (Farahani et al., 2014). The released H^+ ion combines with other water molecule and creates an H_3O^+ ion as shown in equations (7.6) and (7.7). This process causes the singly bonded water molecules to move from a continuous layer of dipoles and electrolyte between electrodes. Hence the conductivity and dielectric are increased.

$$H_2O <\text{---------------}> H^+ + OH^- \tag{7.6}$$

$$H_2O + H^+ <\text{--------------}> H_3O^+ \tag{7.7}$$

Proton (H^+) transfer is another cause of change in conductivity of the metal oxide. The proton transfer on the surface of metal oxide is explained by the Grotthuss mechanism. The proton (H^+) jumps from one water molecule to another molecule and there is an exchange of an H_2 bond with a covalent bond (Takeo and Marco, 2016). The proton hopping results in a change in conductivity of metal oxide. The interaction of water molecules with the metal oxide surface also creates free electrons. The release of these free electrons to the surface of metal oxides causes a change in conductivity of metal oxides (Shankar and Rayappan, 2015). It has been reported of the fabrication of a highly sensitive humidity sensor based on ZnO nanorods (Mohseni Kiasari et al., 2012). The sensor is tested for a relative humidity concentration of 0%–60% and R_{dry}/R_{humid} at 60% RH and the sensitivity is in 10^5 order of magnitude. The result also showed a fast response and recovery time of 60 s and 3 s, respectively.

Figure 7.12a shows the I-V curve at different relative humidity concentrations. The sweeping voltage is applied from −5 V to +5 V with steps of 0.25 V. It can be seen that the current increases when the RH increases. The inset of Figure 7.12a is the semi-logarithmic plot, which shows that the NWs resistance decreases with an increase in relative humidity. The current-voltage characteristics curve showed mostly ohmic behavior. There is an exponential increase in current when exposed to relative humidity due to the operation of nanowires in subthreshold process. Figure 7.12b shows the dynamic response of the sensor when exposed to 30% RH

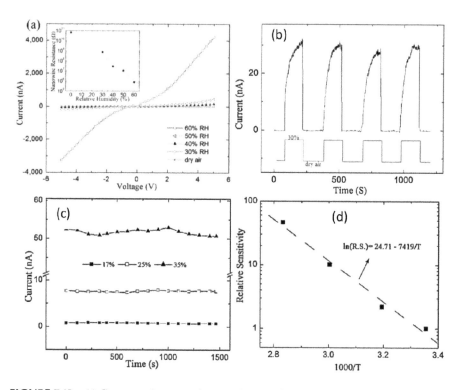

FIGURE 7.12   (a) Current-voltage steady state characteristics curve of ZnO nanowire at different RH values. (b) Dynamic response of the ZnO nanowires at 30% relative humidity shows a change in current when exposed to humid air and dry air, alternatively. (c) Sensor current at 17%, 25% and 35% relative humidity with respect to time. (d) Arrhenius plot of relative sensitivity to absolute temperature. (Reprinted from *Sens. Actuators A Phys.*, 182, Mohseni Kiasari, N. et al., Room temperature ultra-sensitive resistive humidity sensor based on single zinc oxide nanowire, 101–105, Copyright 2012, with permission from Elsevier.)

and dry air, alternatively. When exposed to humidity, the current increases with a response time of 60 s. When the humid air is turned off and dry air is passed, the current decreased with the recovery time of 3 s. This sensor exhibited a high repeatability. Figure 7.12c shows the sensor's current level at different RH concentrations over time. The test is performed after 1-day, 1-week, and 2-weeks and found similar results over time. The graph also shows that current increases with increasing humidity concentrations. Figure 7.12d illustrates the Arrhenius plot of relative sensitivity with respect to the absolute temperature. The sensitivity increases with increasing temperature, which might be attributed to the increase in surface activity with respect to adsorption/desorption process. The activation energy at 30% RH is reported to be 0.6 eV.

A humidity sensor has been developed using SnO_2 nanowires that are synthesized by a CVD process (Kuang et al., 2007). The sensor is tested for different humidity concentrations (5%–85%), and it is found that $R_{RH}/R_{dry\ air}$ is 32 at 85% RH. Figure 7.13a shows the SEM images of SnO_2 nanowires with high yield.

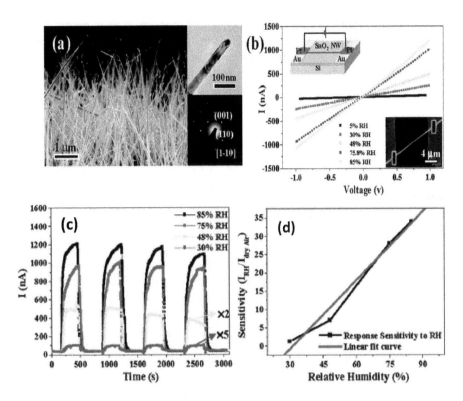

FIGURE 7.13 (a) SEM image of SnO_2 nanowires. (b) I-V curve of single SnO_2 nanowires at different relative humidity at 30°C. (c) Dynamic response of SnO_2 nanowires sensor when switched between dry air and different relative humidity (30%–85%) at 30°C. (d) A Linear relationship between current sensitivity and relative humidity. (Reprinted with permission from Kuang, Q. et al., *J. Am. Chem. Soc.*, 129, 6070–6071, 2007. Copyright 2007 American Chemical Society.)

The NW has rutile structure and the diameter varies from 50 to 300 nm, and the length is a few micrometres. The TEM image (lower inset) reveals that the SnO_2 nanowire is single crystalline and grows along (001) direction. Figure 7.13b shows the I-V curve of the single SnO_2 nanowire at different humidity concentrations. The sensor is based on the field-effect transistor (FET). A single nanowire is connected between two Au electrodes and the current signal is measured at 30°C for RH of ranges 5%–85%. The I-V curve exhibited a linear behavior, which indicates that there is ohmic contact between the nanowire and electrodes. The resistance is decreased promptly when the concentration of humidity increases. The resistance at 5% RH, 48% RH and 85% RH is found to be 28, 2 and 0.85 MΩ, respectively. Figure 7.13c shows the dynamic response of the SnO_2 nanowire at different humidity concentrations at 30°C. The dynamic response gives information about sensitivity, response time, recovery time and reproducibility. The current increased promptly when the dry air (5%) was switched with 30%–85% RH air and the current decreased promptly to the original state when the humid air was switched with dry air. The response and recovery time is reported to be 120–170 s and 20–60 s, respectively. The sensor also displayed good reproducibility although the response decreased slightly in the fourth cycle. Figure 7.13d shows the relation between the relative humidity and response sensitivity (R_{RH}/R_{dry} or I_{RH}/I_{dry}). It can be seen that there is a good linear relationship between sensitivity and RH.

7.4.2 HAZARDOUS GAS SENSORS

Semiconductor metal oxides are considered the most promising materials for chemical sensors because of low cost, small size, low power consumption and huge compatibility with many devices (Soleimanpour et al., 2013b). The metal oxides are cheap, abundant, highly stable, exhibit high response and recovery time and easy to fabricate. The application of a metal oxide gas sensor includes automotive industries, environmental pollution monitoring, aircraft and spacecraft and other industries. 1-D metal oxides such as nanowires, nanorods, and nanobelts have been shown promising results in gas sensors. High surface to volume ratio increases interaction of gas molecules with metal oxides and hence performance is increased. In addition, grain size, porosity and film thickness also play an important role in the performance of gas sensors. There are some drawbacks of metal oxide semiconductor sensors such as poor selectivity and high operating temperature. Researchers have been trying to overcome these drawbacks by introducing modifications to the metal oxide such as doping with noble metals, laser treatment, and synthesis of mixtures of metal oxides. Table 7.3 lists few examples for humidity and gas sensors made with 1-D nanostructure of metal oxides.

The gas sensor works on the principle of adsorption and desorption of gas molecules on the surface of the films. When the film is heated, the film adsorbs oxygen molecules from the air and oxygen molecules gain electrons from the conduction band to produce the following oxygen species:

$$O_2(gas) \rightarrow O_2(absorbed) \tag{7.8}$$

$$O_2(absorbed) + e^- \rightarrow O_2^- \tag{7.9}$$

$$O_2^- + e^- \rightarrow 2O^- \tag{7.10}$$

$$O^- + e^- \rightarrow O^{2-} \tag{7.11}$$

The stabilities of these oxygen species depend on temperature. O_2^- is stable below 100°C, O^- is stable between 100°C to 300°C and O^{2-} is stable above 300°C (Barsan and Weimar, 2001). When the target gases such as NH_3, H_2 and CH_4 are exposed to the heated films, the following reaction occurs at the surface of the films (Soleimanpour et al., 2013a):

$$H_2 + O^-(absorbed) \rightarrow H_2O + e^- \tag{7.12}$$

TABLE 7.3
Summary of 1-D Metal Oxide Based Gas and Humidity Sensors Reported in the Literature

Material	Structure	Synthesis Method	Types of Analyte	Analyte Concentration	Operating Temp (°C)	Sensitivity	References
ZnO	Nanorods	Hydrothermal	H_2S C_2H_5OH	1 ppm- C_2H_5OH 0.05 ppm- H_2S	25, 350	$R_a/R_g = 1.7$ for H_2S $R_a/R_g = 2.6$ for C_2H_5OH	Wang et al. (2006)
TiO$_2$	Nanotubes	Anodization	H_2	1000 ppm	290	$(R_g/R_0)^{-1} = 10^2$	Varghese et al. (2003)
SnO$_2$	Nanowires	Thermal evaporation	C_2H_5OH, CO, C_3H_8, H_2S	5 ppm H_2S 100 ppm other	300–500	$R_a/R_g = 26.2$ for C_2H_5OH and 1.3-2 for other gases	Hwang et al. (2011)
TiO$_2$	Nanotubes	Anodic oxidation	H_2	L- 20 ppm H- 1000 ppm	RT	5E10%	Maggie et al. (2006)
In$_2$O$_3$	Nanorods	Sol-gel	H_2	50–5000 ppm	340	$R_a/R_g = 6$	Lu and Yin (2011)
ZnO/SnO$_2$	Nanorods	CVD RF sputtering	Humidity	40%–90%	—	11-265	Md Sin et al. (2014)
ZnO	Nanowire	CVD	Humidity	0%–60%	RT	$R_{dry}/R_{60\%} = 10^5$	Mohseni Kiasari et al. (2012)
SnO$_2$	Nanowires	CVD	Humidity	5%–85%	30	$R_{5\%}/R_{85\%} = 32.59$	Kuang et al. (2007)
ZnO	Nanorods	Wet chemical/ Screen printing	Humidity	11%–95%	25	$R_{11\%}/R_{95\%} = 10^5$	Qi et al. (2008)
TiO$_2$/nafion	Nanowires	hydrothermal	Humidity	12%–97%	—	$R_{12\%}/R_{97\%} = 10^3$	Wu et al. (2006)

Note: CVD = chemical vapor deposition, L = lower, H = higher.

$$2NH_3 + 3O^{2-} \rightarrow 3H_2O + N_2 + 3e^- \tag{7.13}$$

$$CH_4 + 4O^- \rightarrow CO_2 + 2H_2O + 4e^- \tag{7.14}$$

From equations (7.12) through (7.14) it can be seen that after interaction of target gases with different oxygen species, free electrons are produced on the surface. These free electrons are injected back to grain and hence the resistance of the film is changed. When electrons are injected to the grain of p-type films, the resistance is increased whereas the resistance of the n-type film is decreased when electrons are injected back into the grains. The sensitivity of the sensor is defined as the ratio of change in resistance after exposure to the target gas to the original resistance in the air $[(R_g - R_a)/R_a]$. Another definition of sensitivity is the ratio of resistance after exposure to target gases to the original resistance in the air (R_g/R_a). Where, R_g is the resistance of the film after exposure to target gas and R_a is the resistance of film in the air.

7.4.2.1   Examples of 1-D Metal Oxide in Hazardous Gas Sensors

ZnO nanorods have been used to detect H_2S and C_2H_5OH (Wang et al., 2006). These nanorods are synthesized in an Al_2O_3 tube using the hydrothermal method. Sodium dodecyl sulfate (SDS) was dissolved in 10.2 mL of heptane and 3.0 mL of ethanol. In a separate beaker $ZnAc_2$ (0.5 M) and NaOH (5.0 M) are mixed in different proportions to produce $Zn(OH)_4^{2-}$. From this solution, 2 mL of $Zn(OH)_4^{2-}$ is added to the above mixture. Micro-emulsion thus formed is transferred into 25 mL of the Teflon-lined autoclave and heated at 180°C for 13 hours. Then, after filtration the precipitate is washed with distilled water and ethanol several times and then dried at 70°C–80°C in the air. Finally, the paste is coated onto the Al_2O_3 tube with PVA solution.

Sensing behavior of SnO_2 nanowires is reported for ethanol (C_2H_5OH), carbon monoxide (CO), C_3H_8 and hydrogen sulfide (H_2S) (Hwang et al., 2011). The SnO_2 nanowires are deposited on a Si substrate using the thermal evaporation method. The Sn metal is used in the process and the pressure of the system is maintained at 10^{-2} Torr using a mechanical pump. The nanowires are deposited at 750°C, the deposition time was 20 min and the flow rate of the oxygen inside the chamber is kept at 0.5 sccm.

The hydrogen sensor behavior of In_2O_3 nanorods synthesized in a sol-gel process has also been reported (Lu and Yin, 2011). For the synthesis process, 0.1 g of $InCl_3.4H_2O$ and 0.3 g of sodium dodecyl sulfate are dissolved in 10 mL of water and kept stirred for 20 min at 60°C. Then sodium hydroxide (0.1 g/mL) solution is added dropwise and kept stirring at 60°C until alkalescence. The solution is kept stirred for 12 h at room temperature for aging and then the precipitate is separated from the remaining solution by centrifugation. After separation, the precipitate was washed with deionized water until it became neutral and dried at 60°C for 12 h in the air. Then, the $In(OH)_3$ was calcined in a furnace at 500°C for 4 h in air to obtain In_2O_3 nanorods. The nanorods were ground to make a paste. The paste was coated on alumina tube type substrate and then annealed in a furnace at 500°C for 2 h to obtain the final sensing device. Table 7.3 lists the sensor characteristics of this device.

A humidity sensor has been developed using ZnO/SnO_2 nanorod composites on a glass substrate by using a CVD technique for the application of a humidity sensor (Md Sin et al., 2014). The overall performance of the sensors can be improved by combining two metal oxides because of the formation of more porous film and heterogeneous interfaces between them. The porosity increases the interaction of water molecules with metal oxide nanorods. At first, ZnO is coated on a glass substrate using a reactive RF sputtering method. The thin layer of SnO_2 is deposited on the ZnO film by a CVD process. The combined sensing effect is very clear (see Table 7.3) and the technology can be transferred to an IC (integrated circuit) processing line.

7.4.2.2 Applications of 1-D Metal Oxides in Hazardous Gas Sensing

Nanorods of TiO_2 have been widely investigated for the fabrication of oxygen sensors as shown in Table 7.3. TiO_2 nanotubes were fabricated by an anodization method at a potential in the range of 12 to 20 V (Varghese et al., 2003). The sensor thus fabricated is tested for different gases such as hydrogen, ammonia and carbon monoxide. The author found that the TiO_2 nanotubes thus fabricated were highly sensitive to hydrogen, but less sensitive to other gases such as oxygen, ammonia and carbon monoxide.

Titanium oxide nanotubes synthesized by anodization methods have been successfully used for sensing of hydrogen. Anodization potential is varied from 12 to 20 V. The sensor is tested mainly for sensing hydrogen gas as well as other hazardous gases. Figure 7.14a shows the change in resistance with respect to base resistance at different temperatures when the sensor was exposed with 1000 ppm of H_2 gas. The sensor is tested in the environment of N_2 gas. The response of the sensor is increased with increasing operating temperatures with three order of magnitude changes in resistance above 300°C. Figure 7.14b shows response of the sensor under repeated exposure to H_2 gas of the concentration varying from 100 to 500 ppm at a constant temperature of 290°C. The response is found to be consistent and the resistance is recovered to original value after H_2 gas was stopped, and the sensitivity increased with increasing concentration of H_2 gas. Figure 7.14c shows sensitivity of the sensor with respect to different gases such as O_2, NH_3, CO and CO_2. The sensor displayed no response to CO_2 gas, while it is sensitive to other gases. For O_2 gas, the resistance did not recover to original value even after many hours.

In_2O_3 nanorods have also been investigated for the application in H_2 sensors (Lu and Yin, 2011). The preparation technique of these nanorods are discussed in the previous section. Figure 7.15a shows an SEM image of In_2O_3 nanorods. The length and diameter of nanorods are found to be 300–900 nm and 70–100 nm, respectively. The nanorods are produced with uniform size and morphology on a large scale. Figure 7.15b shows the sensitivity of In_2O_3 nanorod sensors under 500 ppm of H_2 in the temperature range of 250°C–450°C. The sensitivity increased with increasing temperature until 340°C;

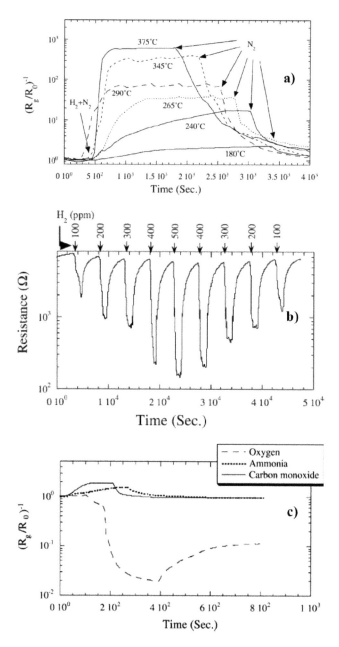

FIGURE 7.14 (a) Response of TiO$_2$ nanotube array for H$_2$ gas at different operating temperatures. The highest sensitivity was found at 375°C. (b) Change in resistance of nanotube array with respect to time when exposed to different concentrations of H$_2$ gases. (c) Response of nanotube array with respect to time for ammonia, oxygen and carbon monoxide. (Reprinted from *Sens. Actuators B Chem.*, 93, Varghese, O.K. et al., Hydrogen sensing using titania nanotubes, 338–344, Copyright 2003, with permission from Elsevier.)

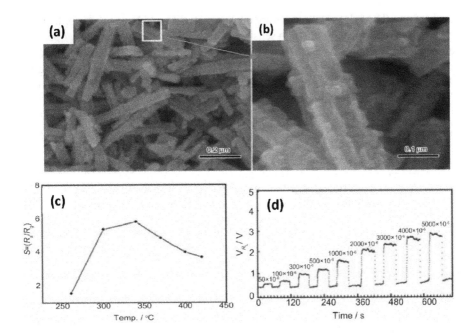

FIGURE 7.15   (a) SEM image of In_2O_3 as prepared nanorods in presence of sodium hydroxide, (b) Magnified SEM image, (c) The sensitivity of In_2O_3 sensor under 500 ppm of H_2 at different operational temperatures, and (d) Voltage dynamic response of In_2O_3 nanorods under various concentrations of H_2 at 340°C. (Reprinted with permission from Lu, X. and Yin, L., *J. Mater. Sci. Technol.*, 27, 680–684, 2011. Copyright 2011: Elsevier.)

the highest sensitivity was observed at 340°C. But the sensitivity started to decrease above 340°C. This phenomenon has been interpreted as due to the increase in desorption rate of gases at a higher temperature. Figure 7.15c shows the dynamic response of voltage for H_2 concentration of 50–500 ppm at 340°C. The sensitivity of the sensor increased with increasing concentration of H_2. The response value recovered to original condition when the H_2 was stopped. The sensor showed fast response and recovery time of 6 s at 50 ppm, which indicates that the sensor has swift response and recovery for H_2.

7.5 CONCLUSION

This chapter describes the application of 1-D nanostructures of metal oxides in four different sensors: biosensors, optical, humidity sensors and hazardous gas sensors. The materials were carefully selected from many research papers published in the literature. The attempts were made to discuss important facts whenever necessary including their unique physical and chemical properties, synthesis techniques, surface morphology and sensing properties. This report, however, is unable to cover all the work and progress that has been done in this field because of a very wide range of studies including material preparation, post treatments, functionalization methods, and many possible device structures. This article clearly discussed huge aspects of biosensors, UV light sensors, humidity sensors and gas sensors and their working mechanisms. The 1-D metal oxides are unique materials and have broad applications in many fields. They exhibit a high sensitivity and recovery time, and a low operating temperature to a variety of analytes. Because of their unique properties, 1-D nanostructures of metal oxides have attracted lots of attention of researchers around the globe. Next step will be the continuous efforts toward development of commercial sensor systems based on these fundamental investigations.

REFERENCES

Arafat, M., Dinan, B., Akbar, S. A. and Haseeb, A. (2012) Gas sensors based on one dimensional nanostructured metal-oxides: A review. *Sensors*, 12(6), 7207–7258.

Arnold, M. S., Avouris, P., Pan, Z. W. and Wang, Z. L. (2003) Field-effect transistors based on single semiconducting oxide nanobelts. *The Journal of Physical Chemistry B*, 107(3), 659–663.

Arundel, A. V., Sterling, E. M., Biggin, J. H. and Sterling, T. D. (1986) Indirect health effects of relative humidity in indoor environments. *Environmental Health Perspectives*, 65, 351–361.

Barsan, N. and Weimar, U. (2001) Conduction model of metal oxide gas sensors. *Journal of Electroceramics*, 7(3), 143–167.

Benvenuto, P., Kafi, A. and Chen, A. (2009) High performance glucose biosensor based on the immobilization of glucose oxidase onto modified titania nanotube arrays. *Journal of Electroanalytical Chemistry*, 627(1–2), 76–81.

Chen, P.-C., Shen, G., Shi, Y., Chen, H. and Zhou, C. (2010) Preparation and characterization of flexible asymmetric supercapacitors based on transition-metal-oxide nanowire/single-walled carbon nanotube hybrid thin-film electrodes. *ACS Nano*, 4(8), 4403–4411.

Devan, R. S., Patil, R. A., Lin, J. H. and Ma, Y. R. (2012) One-dimensional metal-oxide nanostructures: Recent developments in synthesis, characterization, and applications. *Advanced Functional Materials*, 22(16), 3326–3370.

Farahani, H., Wagiran, R. and Hamidon, M. N. (2014) Humidity sensors principle, mechanism, and fabrication technologies: A comprehensive review. *Sensors*, 14(5), 7881–7939.

Guo, D., Zhuo, M., Zhang, X., Xu, C., Jiang, J., Gao, F., Wan, Q., Li, Q. and Wang, T. (2013) Indium-tin-oxide thin film transistor biosensors for label-free detection of avian influenza virus H5N1. *Analytica Chimica Acta*, 773, 83–88.

Humayun, Q., Kashif, M., Hashim, U. and Qurashi, A. (2014) Selective growth of ZnO nanorods on microgap electrodes and their applications in UV sensors. *Nanoscale Research Letters*, 9(1), 29.

Hwang, I.-S., Lee, E.-B., Kim, S.-J., Choi, J.-K., Cha, J.-H., Lee, H.-J., Ju, B.-K. and Lee, J.-H. (2011) Gas sensing properties of SnO$_2$ nanowires on micro-heater. *Sensors and Actuators B: Chemical*, 154(2), 295–300.

Jeong, H. W., Choi, S.-Y., Hong, S. H., Lim, S. K., Han, D. S., Abdel-Wahab, A. and Park, H. (2014) Shape-dependent charge transfers in crystalline ZnO photocatalysts: Rods versus plates. *The Journal of Physical Chemistry C*, 118(37), 21331–21338.

Kafi, A., Wu, G. and Chen, A. (2008) A novel hydrogen peroxide biosensor based on the immobilization of horseradish peroxidase onto Au-modified titanium dioxide nanotube arrays. *Biosensors and Bioelectronics*, 24(4), 566–571.

Kang, T.-S. et al. (2009) Fabrication of highly-ordered TiO2 nanotube arrays and their use in dye-sensitized solar cells. *Nano Letters* 9(2), 601–606.

Ko, Y. H., Nagaraju, G. and Yu, J. S. (2015) Fabrication and optimization of vertically aligned ZnO nanorod array-based UV photodetectors via selective hydrothermal synthesis. *Nanoscale Research Letters*, 10(1), 323.

Kolmakov, A., Klenov, D., Lilach, Y., Stemmer, S. and Moskovits, M. (2005) Enhanced gas sensing by individual SnO_2 nanowires and nanobelts functionalized with Pd catalyst particles. *Nano Letters*, 5(4), 667–673.

Kolmakov, A. and Moskovits, M. (2004) Chemical sensing and catalysis by one-dimensional metal-oxide nanostructures. *Annual Review of Materials Research*, 34, 151–180.

Komathi, S., Muthuchamy, N., Lee, K. and Gopalan, A. (2016) Fabrication of a novel dual mode cholesterol biosensor using titanium dioxide nanowire bridged 3D graphene nanostacks. *Biosensors and Bioelectronics*, 84, 64–71.

Kuang, Q., Lao, C., Wang, Z. L., Xie, Z. and Zheng, L. (2007) High-sensitivity humidity sensor based on a single SnO_2 nanowire. *Journal of the American Chemical Society*, 129(19), 6070–6071.

Kumar M, A., Jung, S. and Ji, T. (2011) Protein biosensors based on polymer nanowires, carbon nanotubes and zinc oxide nanorods. *Sensors*, 11(5), 5087–5111.

Lam, K.-T., Hsiao, Y.-J., Ji, L.-W., Fang, T.-H., Hsiao, K.-H. and Chu, T.-T. (2017) High-sensitive ultraviolet photodetectors based on ZnO nanorods/CdS heterostructures. *Nanoscale Research Letters*, 12(1), 31.

Li, C., Zhang, D., Liu, X., Han, S., Tang, T., Han, J. and Zhou, C. (2003) In_2O_3 nanowires as chemical sensors. *Applied Physics Letters*, 82(10), 1613–1615.

Li, Y., Yang, X.-Y., Feng, Y., Yuan, Z.-Y. and Su, B.-L. (2012) One-dimensional metal oxide nanotubes, nanowires, nanoribbons, and nanorods: Synthesis, characterizations, properties and applications. *Critical Reviews in Solid State and Materials Sciences*, 37(1), 1–74.

Liu, J., Goud, J., Raj, P. M., Iyer, M., Wang, Z. L. and Tummala, R. R. Real-time protein detection using ZnO nanowire/thin film bio-sensor integrated with microfluidic system. *Electronic Components and Technology Conference*, 2008. ECTC 2008. 58th. IEEE, 1317–1322.

Liu, J., Li, Y., Huang, X. and Zhu, Z. (2010) Tin oxide nanorod array-based electrochemical hydrogen peroxide biosensor. *Nanoscale Research Letters*, 5(7), 1177.

Lu, X. and Yin, L. (2011) Porous indium oxide nanorods: Synthesis, characterization and gas sensing properties. *Journal of Materials Science & Technology*, 27(8), 680–684.

Maggie, P., Oomman, K. V., Gopal, K. M., Craig, A. G. and Keat, G. O. (2006) Unprecedented ultra-high hydrogen gas sensitivity in undoped titania nanotubes. *Nanotechnology*, 17(2), 398.

Md Sin, N. D., Tahar, M. F., Mamat, M. H. and Rusop, M. 2014. Enhancement of nanocomposite for humidity sensor application. In: GAOL, F. L. and WEBB, J. (Eds.) *Recent Trends in Nanotechnology and Materials Science: Selected Review Papers from the 2013 International Conference on Manufacturing, Optimization, Industrial and Material Engineering (MOIME 2013)*. Cham, Switzerland: Springer International Publishing.

Mohseni Kiasari, N., Soltanian, S., Gholamkhass, B. and Servati, P. (2012) Room temperature ultra-sensitive resistive humidity sensor based on single zinc oxide nanowire. *Sensors and Actuators A: Physical*, 182, 101–105.

Qi, Q., Zhang, T., Yu, Q., Wang, R., Zeng, Y., Liu, L. and Yang, H. (2008) Properties of humidity sensing ZnO nanorods-base sensor fabricated by screen-printing. *Sensors and Actuators B: Chemical*, 133(2), 638–643.

Rouhanizadeh, M., Tang, T., Li, C., Hwang, J., Zhou, C. and Hsiai, T. K. (2006) Differentiation of oxidized low-density lipoproteins by nanosensors. *Sensors and Actuators B: Chemical*, 114(2), 788–798.

Seiyama, T., Kato, A., Fujiishi, K. and Nagatani, M. (1962) A new detector for gaseous components using semiconductive thin films. *Analytical Chemistry*, 34(11), 1502–1503.

Shankar, P. and Rayappan, J. B. B. (2015) Gas sensing mechanism of metal oxides: The role of ambient atmosphere, type of semiconductor and gases–A review. *Science Letters Journal*, 4, 126.

Soleimanpour, A. M., Hou, Y. and Jayatissa, A. H. (2013a) Evolution of hydrogen gas sensing properties of sol–gel derived nickel oxide thin film. *Sensors and Actuators B: Chemical*, 182, 125–133.

Soleimanpour, A. M., Jayatissa, A. H. and Sumanasekera, G. (2013b) Surface and gas sensing properties of nanocrystalline nickel oxide thin films. *Applied Surface Science*, 276, 291–297.

Swanwick, M. E., Pfaendler, S. M., Akinwande, A. I. and Flewitt, A. J. (2012) Near-ultraviolet zinc oxide nanowire sensor using low temperature hydrothermal growth. *Nanotechnology*, 23(34), 344009.

Takeo, M. and Marco, R. (2016) Grotthuss mechanisms: from proton transport in proton wires to bioprotonic devices. *Journal of Physics: Condensed Matter*, 28(2), 023001.

Tian, W., Lu, H. and Li, L. (2015) Nanoscale ultraviolet photodetectors based on onedimensional metal oxide nanostructures. *Nano Research*, 8(2), 382–405.

Varghese, O. K., Gong, D., Paulose, M., Ong, K. G. and Grimes, C. A. (2003) Hydrogen sensing using titania nanotubes. *Sensors and Actuators B: Chemical*, 93(1), 338–344.

Viter, R., Khranovskyy, V., Starodub, N., Ogorodniichuk, Y., Gevelyuk, S., Gertnere, Z., Poletaev, N., Yakimova, R., Erts, D. and Smyntyna, V. (2014) Application of room temperature photoluminescence from ZnO nanorods for salmonella detection. *IEEE Sensors Journal*, 14(6), 2028–2034.

Wang, C., Chu, X. and Wu, M. (2006) Detection of H_2S down to ppb levels at room temperature using sensors based on ZnO nanorods. *Sensors and Actuators B: Chemical*, 113(1), 320–323.

Wang, W., Xie, Y., Wang, Y., Du, H., Xia, C. and Tian, F. (2014) Glucose biosensor based on glucose oxidase immobilized on unhybridized titanium dioxide nanotube arrays. *Microchimica Acta*, 181(3–4), 381–387.

Wei, A., Sun, X. W., Wang, J., Lei, Y., Cai, X., Li, C. M., Dong, Z. and Huang, W. (2006) Enzymatic glucose biosensor based on ZnO nanorod array grown by hydrothermal decomposition. *Applied Physics Letters*, 89(12), 123902.

Whatis.com. 2018. Field-effect transistor (FET) [Online]. TechTarget. Available: https://whatis.techtarget.com/definition/field-effect-transistor-FET [Accessed July 20, 2018].

Wikipedia. 2018a. Biosensor [Online]. Wikipedia. Available: https://en.wikipedia.org/wiki/Biosensor [Accessed July 20, 2018].

Wikipedia. 2018b. Field-effect transistor [Online]. Wikipedia. Available: https://en.wikipedia.org/wiki/Field-effect_transistor [Accessed June 20, 2018].

Wikipedia. 2018c. Ultraviolet [Online]. Wikipedia. Available: https://en.wikipedia.org/wiki/Ultraviolet [Accessed July 20, 2018].

Wu, R.-J., Sun, Y.-L., Lin, C.-C., Chen, H.-W. and Chavali, M. (2006) Composite of TiO_2 nanowires and Nafion as humidity sensor material. *Sensors and Actuators B: Chemical*, 115(1), 198–204.

Xie, Y., Wei, L., Wei, G., Li, Q., Wang, D., Chen, Y., Yan, S., Liu, G., Mei, L. and Jiao, J. (2013) A self-powered UV photodetector based on TiO_2 nanorod arrays. *Nanoscale Research Letters*, 8(1), 188.

Zhang, D., Li, C., Han, S., Liu, X., Tang, T., Jin, W. and Zhou, C. (2003) Ultraviolet photodetection properties of indium oxide nanowires. *Applied Physics A*, 77(1), 163–166.

Zhang, J., Qin, Z., Zeng, D. and Xie, C. (2017) Metal-oxide-semiconductor based gas sensors: Screening, preparation, and integration. *Physical Chemistry Chemical Physics*, 19(9), 6313–6329.

Zhang, Y., Yu, K., Jiang, D., Zhu, Z., Geng, H. and Luo, L. (2005) Zinc oxide nanorod and nanowire for humidity sensor. *Applied Surface Science*, 242(1), 212–217.

Zhao, S., Choi, D., Lee, T., Boyd, A. K., Barbara, P., Van Keuren, E. and Hahm, J.-I. (2014) Indium tin oxide nanowire networks as effective UV/Vis photodetection platforms. *The Journal of Physical Chemistry C*, 119(26), 14483–14489.

8 Sensors with 1-Dimensional Metal Oxide

Khairunisak Abdul Razak, Nur Syafinaz Ridhuan,
Noorhashimah Mohamad Nor, Haslinda
Abdul Hamid, and Zainovia Lockman

CONTENTS

8.1 DEFINITION OF SENSORS

A chemical sensor is a device that interprets chemical information about its surroundings. Precisely, a chemical sensor transforms a certain type of response that is related to the amount of a specific chemical species or analytes. All chemical sensors contain a transducer, which converts the response from qualitative or qualitative signal into an analytical signal via instrumentation (Grieshaber et al. 2008). Chemical sensors own a physical transducer which is attached with a chemically sensitive layer or recognition layer, where analyte molecules interact selectively with receptor molecules or sites included in the structure of the recognition element of the sensor. Consequently, a characteristic physical parameter varies and this variation is reported by means of an integrated transducer that generates the output signal as shown in Figure 8.1. The signal from a sensor is usually in electronic mode, which transforms into a useful signal for instance current or voltage change when the input quantity is being measured.

In order to detect analytes in the solid, liquid, or gaseous phase, the direct-reading selective sensors, such as electrochemical sensors can be used. Electrochemical sensors commonly have a reference electrode (e.g., Ag/AgCl), a counter electrode or auxiliary electrode (e.g., Pt, Au, graphite), and a working electrode (e.g., Ag, glassy carbon) also known as the sensing electrode. The counter electrode is an electrode which is used to close the current circuit and does not contribute to the electrochemical cell. The reference electrode is an electrode which has steady potential and used as a mark of reference for the potential controller and measurement in the electrochemical cell. The working electrode is the electrode on which the electrochemical reaction between electrode surface and analytes take place/occur. Typically, it is divided into two types of configurations: two-electrode and, as shown in Figure 8.2, three-electrode systems.

A few electrochemical sensors are assigned to three main types: potentiometric, amperometric, and conductometric. Potentiometric sensors measure the accumulated charge potential at the working electrode in an electrochemical cell as compared with the reference electrode when zero or no substantial current flows between them. Alternatively, it gives information on the ion motion in the reaction process. Amperometric sensors measure the resultant current caused by the oxidation or reduction of an electro-active species from the reaction process by utilizing applied potential between a reference and a working electrode. Typically, the current which is obtained at a constant potential is called amperometry while the current measured in various potentials is called voltammetry. Conductometric

Electronic signal

FIGURE 8.1   The signal acquired based on the analytes interaction with the transducer.

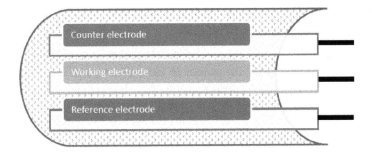

FIGURE 8.2 Schematic view of typical three electrode configuration of electrochemical analysis setup.

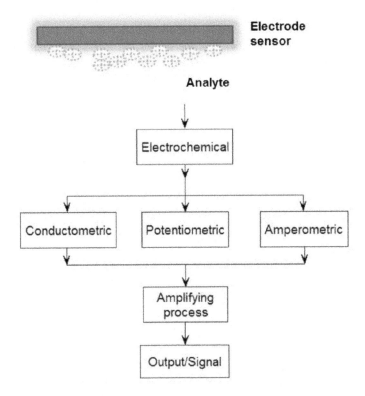

FIGURE 8.3 The flow process of the reaction of analyte and electrode surface which occur at the surface until the signal acquired.

sensors measure the ability of an analyte or electrolyte solution to conduct electrical current between electrodes and reference electrode whose conductivity is affected by the presence of analyte. There are such practical reasons that make conductometric sensor attractive, such as inexpensive and easy as no reference electrodes are required (Stradiotto et al. 2003). The flow process of the reaction of analyte is shown in Figure 8.3.

8.2   1-DIMENSIONAL (1-D) METAL OXIDE IN ELECTROCHEMICAL SENSOR

1-D metal oxide nanostructures are one of the choices for discovering an enormous number of novel phenomena at the nanoscale and exploring size and dimensionality needs of nanostructure properties for potential uses. 1-D metal oxide nanomaterials have inspired great attention due to their significance in basic scientific investigations and potential technological applications (Li et al. 2013). Among nanomaterials of inorganic semiconductor, 1-D metal oxide nanostructures are of interest as they have a unique shape, uniform size, perfect crystalline structure, and homogeneous stoichiometry (Devan et al. 2012). 1-D metal oxide nanostructures have now been widely used in many areas, such as transparent electronics, piezoelectric transducers, ceramics, catalysis, sensors, electro-optical, and electrochromic devices (Devan et al. 2012). 1-D metal oxide nanostructures have fascinated much attention because metal oxides are the most captivating functional materials. The 1-D morphologies can effortlessly enhance the unique properties of the metal oxide nanostructures, which make them appropriate for a wide variety of applications, including gas sensors, liquid sensors, light-emitting diodes, super capacitors, nanoelectronics, and nanogenerators. Therefore, much attention has been made to synthesize and characterize 1-D metal oxide nanostructures in a variety of forms such as nanorods, nanowires, nanotubes, nanobelts, etc. Various physical and chemical deposition techniques and growth mechanisms are demonstrated and developed to monitor the morphology, uniform shape and size, perfect crystalline structure, defects, and homogenous stoichiometry of the 1-D metal-oxide nanostructures. These synthesis methods can be categorized into two: physical deposition and chemical deposition. Each method has its own benefits and strategy for the synthesis of 1-D metal oxide nanostructures. For example, physical deposition methods, such as sputtering techniques, are ideal for the top-down approach while chemical deposition methods are usually used for the bottom-up approach.

8.3   FABRICATION OF 1-D METAL OXIDE

The metal oxide exhibits diversity of structures and 1-D metal oxide nanostructures are the most extensively studied nanostructures. 1-D structure builds up the major group including nanorods (Stradiotto et al. 2003, Devan et al. 2012), nanoneedles (Fioravanti et al. 2014), nanowires (Dar et al. 2012, Fioravanti et al. 2014), nanobelts (Solanki et al. 2011) and nanotubes (Ching et al. 2016). The synthesis methods are numerous as the lateral dimensions of 1-D metal oxide should fall in the range of 1–100 nm with difference shape, size, and structure. Generally, synthesis methods to produce 1-D metal oxide can be classified in two types as shown in Figure 8.4; physical method and chemical method. Typically, the physical methods involve high temperature synthesis and does not require solvent. The chemical method is simple, inexpensive, can be fabricated at low temperature, and has been widely used to engineer the versatility of 1-D nanostructures design with good chemical stability. The methods are discussed below.

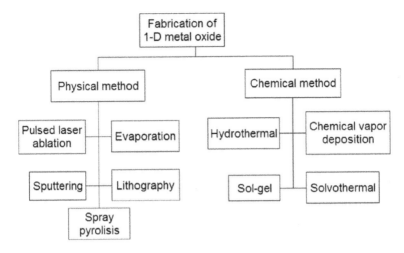

FIGURE 8.4   The variety of methods for 1-D metal oxides synthesis.

8.3.1   PULSED LASER ABLATION

Pulse laser ablation is one of the most popular gas phase methods to obtain nanosize elements with high purity and ultra-fine 1-D nanostructure. Through this method, a high intensity of pulse laser is used to evaporate the material in a high vacuum chamber which is filled with gas such as oxygen. Consequently, the surface of the target is heated up and vaporized as the plasma clouds and settles on the substrate. The process is depicted in Figure 8.5.

8.3.2   EVAPORATION METHOD

Evaporation is a common method to get 1-D nanostructures of metal oxide and is considered the easiest way to form it on a substrate using a vacuum chamber. The vapor pressure of a source material can be obtained by heating the source at a higher temperature. Vacuum is beneficial to reduce the contamination in the chamber and

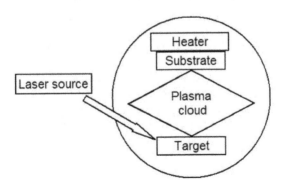

FIGURE 8.5   Schematic diagram of pulsed laser ablation process.

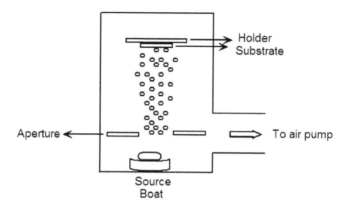

Holder
Substrate

Aperture

To air pump

Source
Boat

FIGURE 8.6   Schematic diagram of evaporation process.

increase the quality of the sample. The process consists of evaporation from melting source material that vaporizes, which is then deposited on the substrate positioned directly above. The process is shown in Figure 8.6. Table 8.1 lists 1-D metal oxide nanostructure reported using the evaporation method.

8.3.3   SPUTTERING METHOD

The sputtering method is one of the methods used to deposit 1-D nanostructure onto a substrate in thin films. The process occurs in a high vacuum chamber with the presence of inert gas such as argon as shown in Figure 8.7. Unlike evaporation, the material to be sputtered does not have to be heated. The gas molecules are ionized in a glow discharge (plasma) and bombard the material of the target. Finally, sputtered atoms deposit onto the substrate.

8.3.4   LITHOGRAPHY METHOD

Lithography is used to produce the self-assembly of nanostructures on different types of substrates. It is also an efficient process for surface patterning (Tiwari et al. 2012). The lithography process offers surface preparation using a patterned mask or photo mask as shown in Figure 8.8. The process begins with preparing the substrate; this example of substrate is silicon. Then, the substrate is coated with photoresist or the addition of an imaging layer such as a light sensitive material. A mask containing the desired pattern is placed on the surface of substrate and exposed to UV radiation. An electron beam is also used as an alternative to describe the pattern. The exposed imaging layer is detached in solution and the final product of 1-D metal oxide is produced.

8.3.5   SPRAY PYROLYSIS

Spray pyrolysis in Figure 8.9 represents a technique which uses liquid solution sprays from an atomizer onto the heating substrate where the constituent reaction takes place to form a chemical compound. Spray pyrolysis signifies a very simple and

TABLE 8.1

Summary of Different Materials with Various Fabrication Methods

Fabrication Method	Material	Structure	Synthesis Condition	References
Pulsed laser ablation	ZnO	Nanorods Diameter: 300–500 nm	• Grown at laser energy density in between 3–4 J/cm^2 • Pulsed laser frequency at 10 Hz with pulse duration of 20 ns • Growth temperature between 500°C and 600°C	Karnati et al. (2018)
	ZnO	Nanorods Diameter: 300 nm	• Grown on sapphire substrates heated at 700°C • Without catalyst	Kawakami et al. (2003)
	ZnO	Nanorods Diameter: 20–60 nm	• Grown on Si substrates at 600°C • No catalyst	Sun et al. (2006)
	ZnO	Nanorods Diameter: 500–700 nm	• Laser energy density 6–10 J cm^{-2} • Laser energy 75–100 mJ and repetition rate of 10 Hz • The deposition time is 60 min with growth temperature of 500°C	Choopun et al. (2005)
	ZnO	Nanorods Diameter: 100–200 nm	• Firstly, ZnS thin films grown on sapphire substrates using the CBD method • The PLD-ZnO layers grown at 500°C, 700°C, and 900°C	Ou et al. (2017)
Evaporation	ZnO	Nanorods Diameter: 50 nm	• Grown on silicon wafers without catalyst free • Low temperature at 550°C–600°C	Pushkariov et al. (2014)
	ZnO	Nanowires Diameter: 300–500 nm	• Use ZnO powders mixed with carbon group elements (C, Si, Ge, Sn, or Pb) as the reducing agent • Growth with and without argon • Si strips were used as substrates • Growth for about 1.5–2 h	Lv et al. (2010)
	ZnO	Nanowires (40–200 nm)	• Grown on the Si substrates • 800°C is the optimum temperature for nanowires growth • Au catalyst	Orvatinia and Imani (2012)

(Continued)

TABLE 8.1 (*Continued*)
Summary of Different Materials with Various Fabrication Methods

Fabrication Method	Material	Structure	Synthesis Condition	References
	ZnO	Nanowires Diameter: (N/A)	• The optimized growth temperature is 700°C • Growth temperature of 650°C and 700°C • No catalyst or additive	Yang et al. (2013)
	TiO$_2$	Nanowires Diameter: 50–200 nm	• Titanium monoxide powder as precursor • Temperature range of 900°C–1050°C	Shang et al. (2012)
Sputtering	SnO$_2$	Nanowires Diameter: 200 nm	• Gas mixture of O$_2$ and Ar were flowed at 900°C for 10 min • Produce Ar plasma, the temperature, deposition time, and DC sputter current were 25°C, 90 s and 65 mA, respectively	Kim et al. (2013)
	ZnO	Nanorods Diameter: 400–500 nm	• Grown on Si and quartz substrates • Argon–oxygen ambient • No seed layer or catalyst • ZnO nanorods formed at substrate temperature of ~750°C • Sputtering pressure of 0.03 mbar • Oxygen to argon ratio of 15%	Nandi (2016)
Spray pyrolysis	ZnO	Nanorods (100–300 nm)	• Growth temperatures at 450°C–550°C • Two substrates: soda-lime glass (SGL) and indium tin oxide-covered glass (ITO/SGL) • The acidity of the solution is at pH 5	Kärber et al. (2011)
	ZnO	Nanorods Diameter: (N/A)	• The precursor used is aerosol • Glass substrate	Febrianti et al. (2017)

(*Continued*)

TABLE 8.1 (*Continued*)
Summary of Different Materials with Various Fabrication Methods

Fabrication Method	Material	Structure	Synthesis Condition	References
	ZnO	Nanorods (100–600 nm)	• Pre-treatment by dip-coating the substrate with a solution of zinc salt to create a seed layer • ITO as substrate • The temperatures were 500°C, 600°C and 850°C • The solution flow rate and the gas pressure were at 2 mL min^{-1} and 2 kgfcm^{-2}. Nitrogen was used as the carrier gas • spray process varied from 5 to 25 min	Dwivedi and Dutta (2012)
	ZnO	Nanorods Diameter: 376 ± 98 nm	• Glass substrate • The solution was sprayed vertically onto substrates at substrate temperature of 300°C ± 5°C	Ikhmayies and Zbib (2017)
	SnO$_2$	Nanorods Diameter: 700 nm	• Precursor was 0.05M of tin tetrachloride diluted in methanol • The process is the atomization of a precursor solution into small discrete droplets	Paraguay-Delgado et al. (2005)
	ZnS	Nanorods Diameter: 80 nm	• Grown on soda-lime glass (SGL) substrate • Air flow rate was 8 L/min and the solution deposition rate was 2.4 mL/min using non-stop spray mode	Dedova et al. (2015)
Hydrothermal	ZnO	Nanorods Diameter: (N/A)	• Using lower precursor concentration (1 μM–9 μM) • preheated oven for 6 hours at 900°C	Mustafa et al. (2017)
	ZnO	Nanorods Diameter: 500 nm	• Glass slides and Si were used as substrates • Temperature from 60°C–95°C • pH value from 10–11	Polsongkram et al. (2008)
	MnO$_2$	Nanorods Diameter: 10–20 nm	• H$_2$O$_2$ solution was added dropwise into 0.01M KMnO$_4$ solution • Kept in the autoclave at 120°C for 12 h	Yang et al. (2008)

(Continued)

TABLE 8.1 (Continued)
Summary of Different Materials with Various Fabrication Methods

Fabrication Method	Material	Structure	Synthesis Condition	References
Chemical vapour deposition	ZnO	Nanorods Diameter: 160 ± 5 nm	• Heat treatment at 95°C for 3 h • Grown on PDMS substrate	Chen et al. (2015)
	TiO_2	Nanorods Diameter: 100 nm	• Grown by using Ti powder and O_2 as precursor • Si/SiO_2 (100) substrate • Growth temperature of 900°C • Growth duration is 3 h	Rizal et al. (2016)
	ZnS	Nanowires Diameter: 50 nm	• Using Zn and S as precursors • $MnCl_2$ or Cd as doping source • Growth of Mn- or Cd-doped ZnS was 15–20 min • Temperature of the furnace was 700°C–720°C	Liu et al. (2006)
	ZnO	Nanorods (50–300 nm)	• Precursor solution was prepared by dissolving Zinc (II)-2-ethylhexanoate in absolute ethanol • Deposition was done at 1050°C–1200°C for 30 minutes	Liu and Liu (2007)
	SiC	Nanorods Diameter: (N/A)	• Catalyst material used is (Me(C_5H_5)$_2$) • Deposition time was 10–60 min • Deposition temperature was 1370–1390K	Leonhardt et al. (2005)
Sol-gel	ZnO	Nanorods Diameter: (N/A)	• Silicon substrate • Solution was stirred for 2 h at 60°C • Spin coater rotated at 3000 rpmf or 30 s	Kashif et al. (2013)
	ZnO	Nanorods Diameter: (N/A)	• Two aqueous solution used; [$Zn(NO_3)_2 \cdot 6H_2O$] solution and hexamethylenetetramine. • Dipped into the solutions at ~100°C for 3.5 h.	Das et al. (2011)

(Continued)

TABLE 8.1 (*Continued*)
Summary of Different Materials with Various Fabrication Methods

Fabrication Method	Material	Structure	Synthesis Condition	References
Solvothermal	CdS/ZnO	Nanorods Diameter: 20 nm	• Used zinc acetate dihydrate sulphur as the source of zinc and thiourea as the source of sulphur • The solution was stirred vigorously for 20 min • The solution sealed in a stainless steel autoclave and heated at 120°C for 20 h • White coloured precipitate was collected • No calcination process	Zou et al. (2014)
	Mn_2O_3	Nanowires Diameter: 75 nm	• PVP and EDTA disodium salt were dissolved into the mixture solution of H_2O and N, N-dime- thylformamide, $Ni(Ac)_2.4H_2O$ and $Co(Ac)_2.4H_2O$ • In autoclave for 21 h at 180°C	Niu et al. (2015)

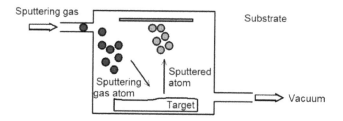

FIGURE 8.7   Representation of sputtering process.

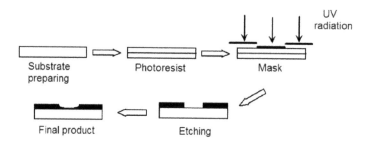

FIGURE 8.8   Lithography flow process.

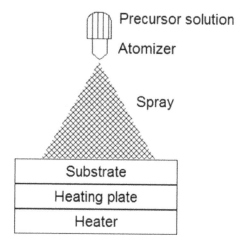

FIGURE 8.9   Spray pyrolysis process illustration.

relatively economical processing method (especially with regard to equipment costs) (Perednis and Gauckler 2005). Spray pyrolysis does not require high-quality substrates or chemicals (Devan et al. 2012). Spray pyrolysis is composed of a precursor solution, atomizer, heating plate, and heater.

8.3.6   HYDROTHERMAL METHOD

Hydrothermal synthesis can be defined as a synthesis method of single crystals that are governed by the solubility of minerals in a hot aqueous solution under

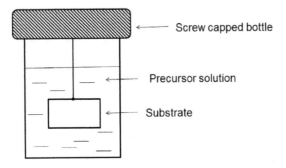

FIGURE 8.10   Schematic diagram of hydrothermal process.

high pressure. The hydrothermal technique maintains good control over the composition for better quality and growth of larger crystals. One of the advantages of hydrothermal is that it does not require high temperatures calcination process. A schematic diagram of the hydrothermal process is represented in Figure 8.10.

8.3.7   CHEMICAL VAPOR DEPOSITION

Chemical vapor deposition (CVD) is a transformation process of gaseous molecules into a solid material on the substrate surface. The substrate is usually exposed to one or more volatile precursors which then reacts or decomposes to produce resultant compounds on the surface of a substrate. Good precursor molecules produce good quality nanostructures with smooth surface. Ideally, the precursor molecule adsorbs on the surface, diffuses, and dissociates into the growth and other volatile molecule products. These products are released from the surface and leave only the growth element on the surface. Figure 8.11 shows the schematic diagram of the typical CVD instrument and its process.

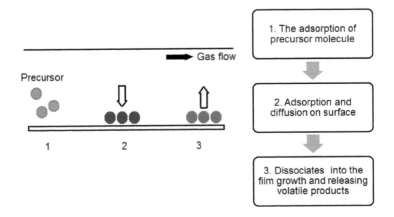

FIGURE 8.11   Schematic diagrams of the typical CVD instrument and its process.

8.3.8   SOL-GEL

The sol-gel method is a wet chemical synthesis which involves a colloidal solution (sol) and gelation colloidal solution (gel) in a liquid phase. The sol-gel method undergoes several steps prior to the final product. The steps are hydrolysis, condensation, and drying. Initially, the precursor such as metal oxide experiences hydrolysis to produce a precursor solution. Then, the condensation process happens to form 3-D gels and are subject to the drying process. Therefore, the final products are readily transformed to xerogel or aerogel. Sol-gel methods can be divided into two routes. There are aqueous sol-gels, where water is used as the medium for reaction. On the other hand, non-aqueous sol-gel routes use organic solvents as the reaction medium. It is noteworthy that the precursor and solvents play important roles in the synthesis of the desired nanostructures. The process of the sol-gel process is shown in Figure 8.12.

8.3.9   SOLVOTHERMAL METHOD

The solvothermal technique is a simple and common synthesis that utilizes solvent at moderate temperatures and high pressures to facilitate the interaction of the precursor during the synthesis. This method is similar to the hydrothermal process where the synthesis or chemical reaction happens in an autoclave, but different a type of precursor is used. The hydrothermal method uses an aqueous solution while the solvothermal method uses a non-aqueous solution such as n-butyl alcohol. Table 8.1 summarizes the 1-D structures of various materials with various fabrication methods.

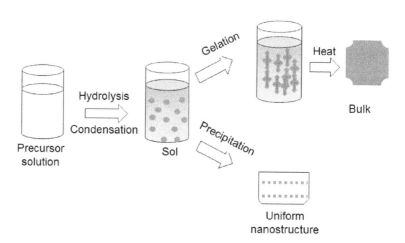

FIGURE 8.12   Illustration of the sol-gel process.

8.4 BIOSENSOR

Biosensor is a self-contained integrated device that is capable of providing specific quantitative or semi-quantitative analytical information using a biological recognition element (biochemical receptor), which is in direct contact with a transducer element as proposed by International Union of Pure and Applied Chemistry (IUPAC). Clark and Lyons (1962), who are the fathers of the biosensor concept, were the first to perform research on glucose quantification through GOx entrapment with a dialysis membrane on the oxygen electrode surface. The glucose amount was estimated based on the reduction in the dissolved oxygen concentration (Turner 2013). Since then, various forms of glucose biosensors have been developed, as well as many other biosensing technologies and biosensing devices. Biosensors often consist of a three-element system; a bioreceptor, a transducer, and a signal-processing unit as shown in Figure 8.13. A bioreceptor is the biological recognition system that translates information from the biochemical domain, usually an analyte concentration, into a chemical or physical output signal (Kissinger 2005). Sensitive biological elements such as enzyme, DNA probe, cells, nucleic acid, and antibodies are immobilized on the electrodes (transducer) in order to recognize the variety of analytes (e.g., body fluid, food samples, cell cultures and environmental samples). The recognition system plays a role in providing a high degree of selectivity to the sensor for the analyte to be measured.

8.4.1 CLASSIFICATION OF BIOSENSOR

Generally, biosensors can be classified into two types; based on the transduction mechanism (transducer) or based on the biorecognization elements (bioreceptor). The classification is shown in Figure 8.14. In the optical transduction method, the

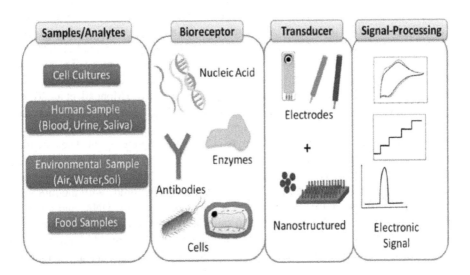

FIGURE 8.13 Elements and selected components in a typical biosensor.

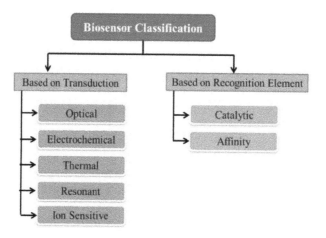

FIGURE 8.14 Classification of biosensors.

physical or chemical change produced by the biorecognition process induces a change in the phase, amplitude, polarization, or frequency of the input light. The biosensor can be based on fluorescence, chemiluminescence, bioluminescence, and surface plasmon resonance optical diffraction (Patel et al. 2010). In electrochemical transduction, chemical reactions between immobilized biomolecules and target analytes that either produce or consume electrons determines the measurable electrical properties of the solution, such as electric current or potential.

Thermal transduction measures the production of heat from the biological interaction, which in turn changes the temperature. The measurement is done by a thermistor known as an "enzyme thermistor" and is most commonly applied for the detection of pesticides and pathogenic bacteria. Biosensors based on ion-selective field-effect transistors (ISFETs) is considered as a category of potentiometric sensors. The surface electrical potential changes due to the interaction between ions and the semiconductor. This change in the potential can be subsequently measured (Monošík et al. 2012).

Biosensors can also be classified based on the biorecognition elements which are catalytic biosensors and affinity biosensors. In catalytic biosensors, enzymes function as the bioreceptor to convert the analyte in order to generate signal. The reaction process of the biocatalysis can be monitored by measuring the formation rate of a product, the disappearance of a reactant, or the inhibition of the reaction (Patel et al. 2010, Goode et al. 2014). In affinity biosensors, the analyte binds with the biological material presence on the biosensor e.g., antibodies, receptors, DNA, and nucleic acids. Thus, affinity biosensors are based on affinity interactions by separating an individual or selected range of components from complex mixtures of biomolecules. There are vast potential application areas for affinity-based biosensor techniques such as clinical/diagnostics, food processing, military/antiterrorism, and environmental monitoring (Rogers 2000, Saber-Tehrani et al. 2013, Pandey et al. 2017).

Among all, the electrochemical biosensor is the most commonly exploited for commercialization devices. Typically, these sensors contain three electrodes: a working electrode, a reference electrode, and a counter/auxiliary electrode. In electrochemical biosensors, analyte is observed through redox reactions of the electrochemical reactions. The analyte is reduced or oxidized at the working electrode thus producing a signal proportional to the concentration of the analyte. The following section focuses on the enzymatic electrochemical biosensors.

8.4.2 Enzymatic Electrochemical Biosensor

The major properties of a biosensor lie on the selectivity and sensitivity, which means that the biosensor can only recognize a certain analyte and does not interact with other interference or contaminant with high accuracy and fast response. In the enzymatic electrochemical biosensor, the enzyme is immobilized to react specifically towards the target analyte molecule by acting as a catalyst. Then, the enzymatic activity is monitored and converted into electrical signal. Enzymatic electrochemical biosensors are commonly applied in health care, food safety, and environmental monitoring. The advantages of enzymatic electrochemical biosensors are their high sensitivity and high specificity. In enzymatic electrochemical biosensors, different types of enzymes are immobilized on the working electrode that are influenced by their target analyte. In order to increase the sensitivity of biosensors, bienzymes are also commonly employed. Table 8.2 lists the types of enzymes and their possible applications. The first step involved in a biosensor fabrication is the selection of biomolecules that form a specific complex between an immobilized biologically active compound and a desired analyte.

There are three categories of the enzymatic electrochemical biosensor that are based on the mode of action; first-generation biosensor, second-generation biosensor, and third-generation biosensor. In first-generation biosensor, the natural mediator (e.g., O_2) is used for monitoring the increase in enzymatic production or decrease of redox enzymes (Figure 8.15a). However, the limitation of the first-generation biosensor is a high potential that leads to interference and oxygen deficit. This occurs due to the limited oxygen solubility in biological fluids. In the second-generation biosensor, an artificial redox mediator is utilized to improve the electron transfer kinetics between an enzyme's active site and electrode surface (Figure 8.15b). However, the major drawbacks of using natural or artificial mediators in glucose biosensor is the difficulty to maintain the presence of the mediator near the electrode and enzyme surface after relatively prolonged use. This is due to the properties of the mediator that are small and highly diffusive thus requires additional and complicated methods to secure them between the two entities.

In third-generation biosensor, direct electrons transferring between an enzyme's active site and electrode surface are introduced without the need for natural or synthetic mediators (Figure 8.15c). The mediator-free biosensor is a more perfect system because it can eliminate the complication from the mediators. Thus, the selectivity and sensitivity of the biosensor are improving. However, the embedded redox active center present in the thick protein needs to be overcome in order to achieve the direct electron transfer between the electrode and enzyme. Figure 8.15 shows the schematic comparison of the three generations of biosensor.

Introducing direct electron transfer in electrochemical biosensors requires modification of the electrode with nanostructured materials such as metal, carbon-based materials, and metal oxide. In this chapter, the focus is mainly in 1-D metal oxide nanostructure for modification of electrode in biosensor. This is due to the properties of the 1-D metal

TABLE 8.2

Types of Enzymes, Properties, and the Potential Applications

Enzymes	Properties	Applications	Reference
Glucose oxidase (GOx)	IEP = 4.22 Active group; FAD	Glucose biosensor	Bankar et al. (2009)
Glucose dehydrogenese (GDH)	IEP = 4.50 Active group; FAD/NAD/PQQ	Glucose biosensor	Deng et al. (2006)
Cholesterol oxidase (ChOx)	IEP = 4.60 Active group; FAD	Cholesterol biosensor	Arya (2008)
Cholesterol esterase (ChEt)	IEP = 5.95 Active group; -	Cholesterol biosensor	Ansari et al. (2009)
Catalase	IEP = 5.42 Active Group; Fe(III/II) prosthetic group	H_2O_2 biosensor	Huang et al. (2011)
HRP	IEP = 3.0–9.0 Active group; -	Glucose biosensor, H_2O_2 biosensor	Xu et al. (2014)
Myoglobin	IEP = 6.8–7.4 Active group; heme	H_2O_2 biosensor	Palanisamy et al. (2014)
Haemoglobin	IEP = 6.87 Active group; heme	H_2O_2 biosensor	Wang, Tang, et al. (2012), Tao et al. (2005)
Urease	IEP = 7.0 Active group; dimeric nickel centre	Urea biosensor	Mahajan et al. (2015)

Note: FAD; flavin adenine dinucleotide, NAD; Nicotinamide adenine dinucleotide, PQQ; pyrroloquinoline quinone.

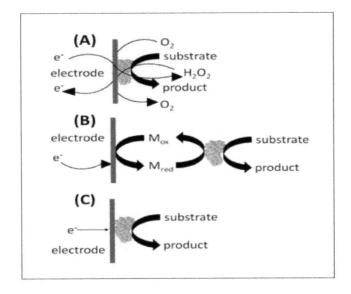

FIGURE 8.15 Schematic comparison of three generations of biosensor where (a) first generation, (b) second generation, and (c) third generation. (Reprinted with permission from Putzbach, W. and Ronkainen, N.J., *Sensors*, 13, 4811, 2013.)

oxide nanostructures that have a high surface-to-volume ratio, excellent electrical conductivity, and able to act as the catalyst in the electrochemical reaction. In electrochemical biosensors, the biomolecules that have an affinity towards the analyte are commonly immobilized on the working electrode to increase specificity and avoid interference. With the presence of 1-D metal oxide on electrodes that have a length up to micron size and a diameter in nanometers aids in increasing the surface area for biomolecule immobilization and provides the biological environment similar to the native redox enzymes. Meanwhile, the high electrical conductivity properties of 1-D metal oxide nanostructure help to transfer the electron transfer between deeply buried enzyme active sites and the electrode. Thus, leading to more rapid current response of the target molecules. The next sub-section focuses on enzymatic electrochemical glucose biosensor, hydrogen peroxide biosensor, and cholesterol biosensor and urea biosensor electrodes modified with 1-D metal oxide nanostructure. The 1-D metal oxide nanostructures that are commonly employed are zinc oxide (ZnO), copper oxide (Cu_2O), ceria (CeO_2), tin oxide (SnO_2), manganese oxide (MnO_2), nickel oxide (NiO), and titania (TiO_2).

8.4.2.1 Glucose Biosensor

Diabetes mellitus has become the major health problem in most developed societies worldwide and has encouraged the development of glucose biosensors. The normal range of glucose concentration in human serum is 4–6 mM. Scholars have focused on producing fast, precise, and low-cost biosensors for monitoring blood glucose at home. Glucose biosensors have been the subject of concern not only in medical science but also in the food industry. A glucose biosensor that is capable of providing fast and quantitative determination is important in the field of clinical chemistry and food analysis (Ozdemir et al. 2012,

Arduini et al. 2016). A glucose biosensor based on commercial strips for self-monitoring of blood glucose have long been in use. The detection signal of glucose biosensors comes directly from glucose or by promoting the conversion of glucose into other determinable electroactive species. So far, both enzymatic and non-enzymatic glucose biosensors have been employed for the determination of glucose. In the enzymatic glucose biosensor, the immobilized GOx catalyzes the oxidation of β-D-glucose utilizing molecular O_2, producing gluconic acid and H_2O_2 as given below (eqs: 8.1–8.3):

$$Glucose + GOx(FAD) \rightarrow Gluconolatone + GOx(FADH_2) \qquad (8.1)$$

$$GOx(FADH_2) + O_2 \rightarrow GOx(FAD) + H_2O_2 \qquad (8.2)$$

$$H_2O_2 \rightarrow O_2 + 2H^+ + 2e^- \qquad (8.3)$$

Glucose is oxidized by oxygen and catalyzed by the immobilized GOx producing gluconic acid and H_2O_2. Then, H_2O_2 is oxidized at the counter electrode which triggers the electron couple proton transfer from active center of the GOx enzyme to the modified electrode. Thus, the number of electron transfers is proportional to the concentration of glucose molecules supplied. In glucose sensing, the amperometric method is commonly applied to measure the amount of H_2O_2 being produced (eq. 8.3).

Commonly, glucose oxidase (GOx) or glucose dehydrogenese (GDH) enzyme is immobilized on the modified electrode for glucose sensing. GOx is preferable in a commercial glucose sensor owing to its benefits such as low cost, high sensitivity, and high specificity for glucose detection. The only limitation of the GOx enzyme is dependency of a glucose sensor to the oxygen concentration presence in the measuring media. Alternatively, the GDH enzyme has been used. The GDH enzyme contains one of three types of enzyme cofactor; flavin adenine dinucleotide (FAD), nicotinamide adenine dinucleotide (NAD), and pyrroloquinoline quinone (PQQ) in which each of them has limitations. FAD-GDH is costly and requires complex preparation process, while PQQ-GDH has poor selectivity due to susceptible interference from variety of saccharides. The NAD-GDH exhibits higher selectivity and stability, but limitations in finding a match with mediator properties.

Recently, scholars reported on improved performance of the enzymatic glucose sensor modified with 1-D metal oxide nanostructure. The modified electrode shows high sensitivity, selectivity, stability, and low limit of detection. Various 1-D metal oxide nanostructures such as nanotubes, nanowires, and nanorods have been applied as the matrices to improve the performance of glucose biosensor. The morphology of nanostructures has influenced the glucose sensor performance. Table 8.3 lists the comparison between types and properties of glucose biosensor modified with 1-D metal oxide nanostructure reported recently. Glucose biosensor based on bienzymes shows improvement in glucose detection performance. Among all, horseradish peroxidase (HRP) is usually chosen to couple with various oxidase enzymes. For a glucose biosensor, H_2O_2 produced by GOx is subsequently reduced by HRP. The oxidized state of HRP is then electrically connected to the electrode surface either by a free diffusing mediator or directly communicating electrons with the electrode under low applied potentials. Higher sensitivity of the glucose sensor is obtained due to signal amplification from the catalytic reaction of the bienzyme molecules (GOx-HRP).

TABLE 8.3

Properties and Performance of Enzymatic Glucose Biosensor Based on 1-D Metal Oxide Nanostructured

Nanostructured		Electrode Modification	Sensitivity ($\mu AmM^{-1}cm^{-2}$)	Linearity	LOD	References
ZnO	Nanotubes	GOx/ZnO/AuCS	2.63	0–6.5 µM	8 µM	Zhou et al. (2016)
	Nanorods	Nafion/GOx/ZnO/Ag/Si	110.76	0.01–23 µM	0.1 µM	Ahmad, Tripathy, et al. (2012)
	Nanorods	Nafion/GOx/ZnO/FcC11SH/Au	27.8	0.05–1.0 µM	20 µM	Ma and Nakazato (2014)
	Nanorods	Nafion-Gluteraldehyde-GOx/ZnO/Au/GCE	1492	0.1 µM –33 µM	10 nM	Wei et al. (2010)
	Nanorods	Nafion/GOx/ZnO Array/Pt	—	0.5–2.5 µM	18.7 mA	Shukla et al. (2017)
	Nanowires	GOx/ZnO/IDE	28.56 ohmµL/mgcm²	1.7–16.7 µM	0.03 mg/dL	Haarindraprasad et al. (2016)
	Nanowires	Nafion/GOx/ZnO/SiNWs	129	1–25 µM	12 µM	Miao et al. (2016)
Co_3O_4	Nanorods	Co_3O_4/PbO_2	460.3	5 µM–1.2 µM	0.31 µM	Chen et al. (2014)
	Nanowires	Co_3O_4/Au	45.8	1 µM–12 µM	0.265 µM	Khun et al. (2015)
CoO	Nanorods	CoO/FTO	571.8	0.2–3.5 µM	0.058 µM	Kung et al. (2011)
NiO	Nanorods	$NiCo_2O/ZnO/SS$	1685	0.3 µM–1.0 µM	0.16 µM	Yang et al. (2016)
	Nanowires	Ag-NiO/GCE	67.51	0–1.28 µM	1 µM	Song et al. (2013)
MnO_x	Nanorods	MnO_x/Au	811.8	0.007–10.6 µM	2.7 µM	Lee et al. (2015)

Note: AuCS; Au cylindrical spiral, FcC11SH; ferrocenyl–alkanethiol, SS; Stainless steel, IDE; interdigitated electrode, GCE; glass carbon electrode, IDE; interdigitated electrode, SiNWs; silicon nanowires.

8.4.2.2 Cholesterol Biosensor

In human serum, normal concentrations of cholesterol are in the range of 1.3–2.6 mg mL^{-1}, consisting of 30% sterol and ~70% fatty acids. Higher ranges of cholesterol present in the human body may lead to other clinical diseases such as hypertension, coronary disease, brain thrombosis, and many others. Therefore, the determination of cholesterol levels in the human blood is of great importance in clinical analysis/diagnosis. Cholesterol biosensors based on commercial strips for self-monitoring of blood cholesterol levels have long been in use (Arya et al. 2008). To obtain high sensitivity and selectivity in cholesterol detection, the enzymatic reaction is commonly employed. Cholesterol oxidase (ChOx), cholesterol esterase (ChEt) and horseradish peroxidase (HRP) are the mainly enzymes immobilized on the electrode. ChOx, in the presence of oxygen (O$_2$) catalyzes the oxidation of cholesterol and produces hydrogen peroxide (H$_2$O$_2$) and cholestenone (Figure 8.16). The cofactor flavin adenine dinucleotide (FAD) of ChOx is responsible for the oxidation of cholesterol. Meanwhile, ChEt converts esterified cholesterol present in the blood into free cholesterol and for HRP, the reaction of HRP with H$_2$O$_2$ produces a colored compound that is detected as a photometry-based biosensor.

In recent years, the electrochemical (amperometric) cholesterol biosensor based on 1-D metal oxides and ChOx enzymes has attracted particular interest. The whole reaction of cholesterol detection catalyze by ChOx are as follows:

$$\text{Cholestrol} + \text{ChOx(FAD)} \rightarrow \text{Cholestenone} + \text{ChOx(FADH}_2) \qquad (8.4)$$

$$\text{ChOx(FADH}_2) + \text{O}_2 \rightarrow \text{ChOx(FAD)} + \text{H}_2\text{O}_2 \qquad (8.5)$$

$$\text{H}_2\text{O}_2 \rightarrow \text{O}_2 + 2\text{H}^+ + 2\text{e}^- \qquad (8.6)$$

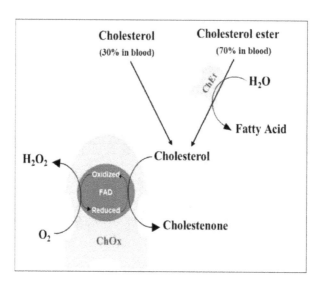

FIGURE 8.16 Schematic diagram of cholesterol reactions catalyst by cholesterol oxidase and cholesterol esterase.

The cholesterol is oxidized by oxygen in the presence of ChOx and H_2O_2 is produced at the same time (Goode et al. 2014). The electro-oxidation current of H_2O_2 is detected to determine the concentration of the cholesterol in the sample. In recent years, scholars reported on the improved performance of enzymatic cholesterol biosensors modified with 1-D metal oxide nanostructures (Table 8.4). The modified electrode shows high sensitivity, selectivity, stability, and a low limit of detection. For cholesterol biosensors based on 1-D metal oxide nanostructure, only ZnO nanostructured and MnO_2 have been reported. Other scholars reported on using metal oxide nanoparticles of NiO (Singh et al. 2011), TiO_2 (Cao et al. 2013), and SnO_2 (Ansari et al. 2009) in modifications of the cholesterol biosensor electrode. Table 8.4 lists the comparison between types and properties of the cholesterol biosensor modified with 1-D metal oxide nanostructure reported recently.

8.4.2.3 Hydrogen Peroxide

Hydrogen peroxide (H_2O_2) is the reactive oxygen species, which is very important in clinical, biological, food, pharmaceutical, and environmental applications. H_2O_2 is generated in several biological processes. Several enzymes that produce H_2O_2 through catalysis are glucose oxidase, cholesterol oxidase, myoglobin, hemoglobin, and horseradish peroxidase (HRP). The electrocatalytic mechanisms of H_2O_2 detection are:

$$\text{Enzymes}_{(reduced)} + H_2O_2 \rightarrow \text{Enzymes}_{(oxidized)} + H_2O \tag{8.7}$$

$$\text{Enzymes}_{(oxidized)} + H_2O_2 \rightarrow \text{Enzymes}_{(reduced)} + H_2O \tag{8.8}$$

$$H_2O_2 + 2e^- + 2H^+ + H_2O \tag{8.9}$$

In the presence of H_2O_2, the enzyme catalyzes the reduction of H_2O_2. The enzyme itself then converts to its oxidized form, which is then reduced at the electrode surface. These reactions produce electrons that are proportional to H_2O_2 concentration.

In the fabrication of highly sensitive and highly specific H_2O_2 biosensors, it is important to ensure good immobilization of enzymes on electrodes and fast electron transfer between enzymes and the electrodes. To do that, scholars reported on modification of the electrode with 1-D metal oxide nanostructures and enzymes (Table 8.5). Most reported works utilized TiO_2 nanotubes to act as the matrix for enzyme immobilization. This could be due to the properties of TiO_2 which are biocompatible, inexpensive, with high chemical and thermal stability. However, in non-enzymatic H_2O_2 biosensor, there are various 1-D metal oxide nanostructure materials that have been utilized such as ZnO, CeO, ZrO, SnO_2, and MnO_2.

8.4.2.4 Urea Biosensor

Real-time monitoring of urea biosensors is of utmost important, because it is the indicator for liver and kidney function. The urea level is also important to be observed in food science and environmental monitoring. In human serum, the normal level of urea concentration is in the range of 15–40 mg/dL (2.5–7.5 μM/L) (Dhawan et al. 2009). Higher urea concentration levels lead to other health issue such as dehydration, shock, renal failure, and gastrointestinal bleeding. Meanwhile, lower urea

TABLE 8.4

Properties and Performance of Enzymatic Cholesterol Biosensor Based on 1-D Metal Oxide Nanostructured

Nanostructured		Electrode Modification	Sensitivity ($\mu AmM^{-1}cm^{-2}$)	Linearity	LOD	References
ZnO	Nanotubes	ChOx/ZnO/Si/Ag	79.4	1.0 μM–13.0 μM	0.5 nM	Ahmad et al. (2014)
	Nanorods	ChOx/Pt–Au@ZnO/MWCNT/MGCE	26.8	0.1–759 μM	0.03 μM	Wang, Tan, et al. (2012)
	Nanorods	ChOx/ZnO-ZnS/CT/GCE	293	0.4–2.0 μM	0.08 μM	Giri et al. (2016)
	Nanorods	ChOx/ZnO/Ag	35.2	1.0 μM–0.1 μM	—	Qadir Israr et al. (2010)
	Nanoflower	ChOx/ZnO/GCE	61.7	1–15 μM	0.012 μM	Umar et al. 2009
MnO_2	Nanorods	ChOx/MnO$_2$/CT/GCE	5.59	0.03–11.7 μM	2.0 μM	Charan and Shahi (2014)
SnO_2	Nanoparticles	ChOx/SnO$_2$-chitosan/ITO	34.7	0.26–10.36 μM	0.13 μM	Ansari et al. (2009)
NiO	Nanoparticles	ChOx/NiO-chitosan/ITO	0.031	0.26–10.34 μM	1.12 μM	Singh et al. (2011)

TABLE 8.5

Properties and Performance of Enzymatic Hydrogen Peroxide Biosensor Based on 1-D Metal Oxide Nanostructured

Nanostructured		Electrode Modification	Sensitivity ($\mu AmM^{-1}\ cm^{-2}$)	Linearity	LOD	References
TiO_2	Nano-needles	Hb/Au-TiO$_2$/ITO	144.5	5 μM–1 μM	0.6 μM	Wei et al. (2011)
	Nanotubes	HRP/Au-TiO$_2$/GCE	0.0018	65 μM–6 μM	5 μM	Liu et al. (2013)
	Nanotubes	HRP-Thionine/TiO$_2$/Ti substrate	—	0.11 μM–2 μM	1.2 μM	Liu and Chen (2005)
	Nanotubes	HRP-chitosan/Au-TiO$_2$/Ti substrate	—	5 μM–40 μM	2 μM	Kafi (2008)
	Nanotubes	Hb/CMC-TiO$_2$/GCE	—	4 μM-64 μM	4.6 μM	Zheng et al. (2008)
ZnO	Nanowire	HRP/Carbon-ZnO/	237.8	0.01–1.6 μM	0.2 μM	Liu et al. (2009)
SnO_2	Nanorod	HRP/SnO$_2$/Fe-alloy	379	0.8–35 μM	0.2 μM	Liu et al. (2010)

Note: CMC; carboxymethyl cellulose, Hb; hemoglobin, HRP; horseradish peroxidase.

concentration causes hepatic failure, nephritic syndrome, and cachexia. In patients suffering from renal problems, urea concentration level in serum vary from 180 to 480 mg/dL and at elevated levels above 180 mg/dL hemodialysis is required.

In the enzymatic urea biosensor, the urease enzyme is commonly used to provide selectivity and improve sensitivity of the biosensor. In recent years, the electrochemical urea biosensor based on 1-D metal oxide and urease enzymes have attracted particular interest. The whole reaction of urea detection catalyze by urease is as follows:

$$NH_2CONH_2 \rightarrow 3H_2 \xrightarrow{\text{urease}} NH_3 + NH_2COOH \qquad (8.10)$$

$$NH_2COOH_2 \rightarrow NH_3 + CO_2 \qquad (8.11)$$

where, urease catalyzes hydrolysis of urea to carbamine acid that further gets hydrolyzed to ammonia (NH_3) and carbon dioxide (CO_2). Electrons generated from these biochemical reactions are transferred to the electrode, which amplifies the electrochemical signal.

In recent years, scholars reported on the improved performance of enzymatic urea biosensors modified with 1-D metal oxide nanostructures. The modified electrode shows high sensitivity, selectivity, stability, and a low limit of detection. For urea biosensors based on 1-D metal oxide nanostructure, only ZnO nanostructured and NiO nanostructured have been reported (Dutta et al. 2014). Other scholars reported on using the metal oxide of ZnO (Solanki et al. 2008, Ali et al. 2009), SnO_2 (Ansari et al. 2008), and NiO (Tyagi et al. 2013), MnO_2 (Mahajan et al. 2015) in nanoparticles shaped for the modification of urea biosensor electrodes. There were also a lot of researches recently in developing non-enzymatic urea biosensors to improve the sensitivity and stability of the biosensors. These were due to enzymes drawbacks of thermal instability and lower reproducibility. Table 8.6 lists the comparison between types and properties of urea biosensors modified with 1-D metal oxide nanostructure.

8.5 ENVIRONMENTAL SENSOR

8.5.1 LIQUID SENSOR

Nanosensors based on 1-D metal oxides have been widely studied in biological science (Dar et al. 2012, Fioravanti et al. 2014), environmental science (Wang et al. 2004, Zhang et al. 2012), and analytical science (Mirzaei et al. 2016). According to the liquid adsorption and desorption kinetics, sensitivity and selectivity of the sensor are dependent on surface properties of the sensor materials and also the target species. Nanostructured 1-D metal oxides have attracted a wide interest due to their outstanding properties of high active surface area that provide efficiency owing to their small particle size. Therefore, the study on 1-D metal oxides to develop such simple, reliable, and cost-effective sensors are crucial for environmental and health monitoring. A schematic view of the reaction mechanism in the presence of analyte is shown in Figure 8.17. The prominent sensitivity of the sensor is attributed to the good adsorption as well as absorption ability of the material. 1-D metal oxide has a large surface area that is beneficial in sensoring for a nano environment where it improves the direct electron movement between the active sites of 1-D metal oxide and the sensor electrode surface.

TABLE 8.6
Properties and Performance of Urea Biosensor Based on 1-D Metal Oxide Nanostructure

Nanostructured	Electrode Modification	Sensitivity	Linearity	LOD	References
ZnO	Urease/ZnO/Ag	41.64 μAmM^{-1}	0.001–24 μM	10 μM	Ahmad et al. (2014)
	Urease/ZnO/Au	52.8 mV/decade	0.1–40 μM	—	Ali et al. (2011)
	Urease/ZnO/ITO	0.4 μAmM^{-1}	1–20 μM	0.13 μM	Palomera et al. (2011)
	Urease-GLDH/ZnO/ITO	1.44 mA/mg/dL	10–80 mg/dL	—	Ali et al. (2009)
NiO	Urease/NiO/ITO	48.0 μAmM^{-1}	0.83–16.65 μM	—	Tyagi et al. (2014)
	Nafion/NiO/GCE	—	0.1–1.1 μM	10 μM	Arain et al. (2016)
SnO$_2$	SnO$_2$/reduced graphene/GCE	1.38 μA/fM	1.6 × 10^{-14} –3.9 × 10^{-12} M	11.7 fM	Dutta et al. (2014)
	Urease/SnO$_2$/Si	—	1–100 μM	—	Ansari et al. (2008)

Note: GLDH; glutamate dehydrogenase, Si; silicon wafer.

Nanorods / Nanowires / Nanorods / Nanoparticles / Nanorods / Nanostructure / Quantum dot / Nanoflakes

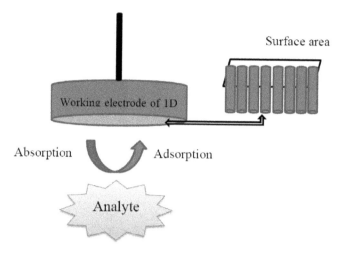

FIGURE 8.17 Schematic view of reaction of mechanism in the presence of analyte.

Redox reaction happens on the surface of the 1-D metal oxide by reacting with the dissolve oxygen ions (O_2^-, O^{2-} and O^-) in bulk solution. These ionized oxygen species enter the lattice of the 1-D metal oxide and subsequently reduce the conductivity. In this context, the formation of 1-D metal oxide facilitates the dissociation of oxygen onto the surface of metal oxide.

Therefore, when the analytes come into contact with the sensing surface, it reacts with the adsorbed oxygen ions, releasing free electrons, and gradually increases the electrons concentration in the conduction band. The possible reaction of mechanism could be expressed in Figure 8.18. Figure 8.18a shows the reaction mechanism of analyte in the presence of 1-D metal oxide whereby the oxygen is adsorbed on the surrounding surface of the sensing surface, while Figure 8.18b shows the reduction process where the analytes release electrons to enhance the sensitivity of the sensor. These produced electrons contribute to a rapid conductivity of the fabricated sensor and thus increase the sensitivity.

8.5.1.1 Volatile Organic Compound Sensor

Generally, any chemical compound that contains at least one carbon and one hydrogen atom in its molecular structure is denoted as an organic compound (Pujol et al. 2014). The term VOC typically refers to organic compounds that exist in the

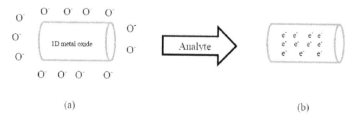

(a)   (b)

FIGURE 8.18 Mechanism of 1-D metal oxide sensor.

atmosphere as gases, but could present in liquids or solids under ordinary conditions of temperature and pressure. Some VOCs are hazardous to human health and cause destruction to the environment depending on the chemical. Volatile organic compounds (VOCs) are discharged from burning fuel, solvents, and paints. Long-term exposure to VOCs can cause harm to the liver, central nervous system, and kidneys. While, short-term contact with VOCs can cause eye and respiratory irritation, headaches, dizziness, visual disorders, fatigue, loss of coordination, allergic skin reactions, nausea, and memory impairment (Faisal et al. 2013). Therefore, nano-structured metal oxide semiconductors are a great emerging technology to improve sensing properties of VOCs in sensitivity, selectivity, and reaction speed. Table 8.7 shows examples of some VOCs and their properties.

Faisal et al. (2011) reported on the acetone sensor utilizing ZnO by using the I–V technique. They synthesized highly crystalline ZnO nanorods with a wurtzite hexagonal structure via the hydrothermal method with an average diameter of ~455 ± 20 nm. The ZnO nanorods exhibited excellent sensitivity and lower detection limit when used as a chemical sensor for the detecting acetone in liquid phase. On the other hand, this group also reported on the detection of trichloromethane by ZnO nanocapsules by the hydrothermal method. From the result, a substantial increase in the current value with applied potential was evidently demonstrated after injection of the target. The limit of detection was 6.67 μM (Lee et al. 2013). Lee et al. (2013) prepared ZnO nanorods for ethanol detection in liquid phase. They reported that the ZnO base sensor displayed excellent sensitivity at room temperature. Moreover, the I–V curves in various ethanol concentrations showed the resistivity increased with increasing concentration of ethanol.

8.5.2 HEAVY METAL SENSOR

Heavy metal ions create a serious environmental problem due to their persistent and non-decomposable nature. Heavy metal ions are poisonous to biological systems even at low concentration and there is an apparent need to determine them at a trace level. Therefore, heavy metals detection and monitoring are crucial due to the potential health and ecological problems. The practice of the chemically modified electrodes enormously improves the efficiency of collecting target analytes and has been established as a captivating and effective way for the heavy metal ions detection.

Devi et al. (2015) synthesized MnO hybrids by one-step electrochemical reduction in Na_2SO_4 electrolyte. They reported on a nanocomposite material of graphene oxide and manganese oxide for sensitive and selective detection of As (III) in water. The synthesized nanocomposite showed good electrical properties of As (III) adsorption properties using MnO_2 electrode. The results showed the sensitivity of the fabricated electrode to As (III) by the presence of stripping peak at ~ 0.26 V in square wave anodic stripping voltammetry. However, Dar et al. (2012) reported a very modest, reliable, and facile approach to fabricate a liquid ammonia chemical sensor using crystalline hexagonal-shaped ZnO nanopencils as an efficient electron facilitator. A low temperature facile hydrothermal method was used to synthesize ZnO nanopencils. From the results, they concluded that the sensitivity of ~26.58 $\mu A\ cm^{-2}\mu M^{-1}$ and detection limit

TABLE 8.7

Properties of Volatile Organic Compounds

VOC Name	Formula	Properties	Chemical Structure
Acetone	C_3H_6O	Colorless, volatile, flammable	
Ethanol	C_2H_6O	Colorless, volatile, flammable, slightly odor, low toxicity	
Trichloromethane	$CHCl_3$	Colorless, volatile, sweet smelling, dense liquid	
Methanol	CH_3OH	Colorless, volatile, highly toxic	
Formaldehyde	CH_2O	Colorless, volatile, pungent, strong odor	
Toluene	C_7H_8	Colorless, volatile, highly toxic, water insoluble liquid	
Benzene	C_6H_6	Colorless, volatile, flammable, high toxicity	

of ~5 nM with a correlation coefficient (R) of 0.9965 and a response time of less than 10 s were perceived for the fabricated liquid ammonia by I–V technique.

8.5.3   GAS SENSOR

Gas sensors technology has an important effect on many aspects in our daily life and has gained much progress driven by the advancement of nanotechnology. Generally, gas sensors are a device that operates on the principle of converting target gas concentration in its vicinity into a measurable signal. Typical applications of gas sensors include environmental pollution monitoring, indoor air quality, workplace health and safety, and homeland security (Liu et al. 2012, Tian et al. 2013, Zhang et al. 2016). Several requirements should be considered for good and efficient gas sensors: high sensitivity, excellent selectivity, fast response and recovery time, low operating temperature and temperature independence, stability in performances, low fabrication cost, and low analyst consumption (Wang and Yeow 2009, Wei et al. 2011, Liu et al. 2012). Metal oxide semiconductors (MOS) as a sensing element for gas sensors have been widely used and studied owing to several advantages such as low cost of fabrication, wide detection range, a strong response of target gases, long-term stability, and high compatibility with microelectronic processing (Fine et al. 2010, Choopun et al. 2012, Sun et al. 2012, Kim and Lee 2014). In recent years, one dimensional (1-D) metal oxide semiconductor materials have become intensively studied in gas sensor applications. Gas sensing tests can be carried out in different testing setups that suit the respective researchers. However, their overall principle remains the same. Figure 8.19 shows a general schematic of a MOS gas sensor device. The setup mainly consists of a target gas, a mass flow controller unit, a multi-component gas mixer, a testing chamber (may consist UV light and heater), a source meter, and a computer for data collection and controller setup. LabVIEW based software is mainly used to control all testing parameters and measurements during analysis (Kanan et al. 2009). The testing chamber consists of UV light, heater, and more importantly, a MOS sensor structure for target gas detection as illustrated in Figure 8.19.

Generally, the working principle of the MOS gas sensor is based on chemi-resistance change in electrical conductivity, or resistivity of MOS sensing elements upon exposure to a target gas. When the target gas adsorbs onto the MOS surface, a redox reaction of target gas occurs either by electron donor or acceptor that is determined by the nature of the target gas (either oxidative or reductive compare to oxygen molecule). Then the reaction between gas-solid interfaces alter the charge carrier concentration near the MOS surface; hence, modifying the conductivity of the MOS sensing element (Zhang et al. 2016). Generally, MOS materials can be divided into two groups: n-type (where electrons are the major carrier) and p-type (where holes are the major carrier). Therefore, different sensing behavior is observed when MOS with different types of conduction bands are exposed to the same target gas. Figure 8.20 shows the schematic diagram for the change of MOS gas sensor resistance upon exposure to the target gas (reducing and oxidizing gas) for n-type MOS gas sensor. In the atmosphere, oxygen plays a predominant role in the adsorption process due to its high electronegativity of 3.65 and lone pairs of electrons, making it easy to be adsorbed in the surface of MOS (Shankar and Rayappan 2015). Oxygen molecules gain electrons from the MOS and turn it into ionic forms of O_2^- leading to the increase of the electron depletion

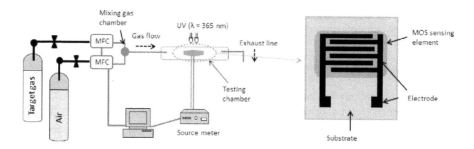

FIGURE 8.19   General set up for gas sensor measurement and fabricated MOS sensor structure.

layer and increasing the resistance. When the sensor is exposed to a reducing gas, the electrons trapped by oxygen adsorbate return into the MOS surface leading to a decrease in the electron depletion layer, as the carrier concentration increases and thus reduces the resistance of the MOS sensing element. For n-type MOS gas sensors, when exposed to oxidizing gas more electrons are extracted from the MOS surface by oxidizing gas as illustrated in Figure 8.20b. This increases the electron depletion layer and subsequently increases the resistance of the MOS sensing element. The decrease of resistance can be monitored by the electrical instruments.

On the other hand, for p-type MOS under normal air, oxygen molecules are adsorbed onto the MOS surface and further extract electrons from the electron/hole pair, leaving holes. As a result, the number of charge carriers increases, resulting in a decrease of resistance. When the p-type MOS sensor is exposed to reducing gas (target gas), electrons are injected back into the MOS surface and recombine with the holes. This reaction reduces holes and charge carrier availability, which lead to an increase in sensor resistance. In the case of oxidizing gas, the change of resistance is opposite to the above discussion. A schematic diagram for change of sensor resistance upon exposure to target gas

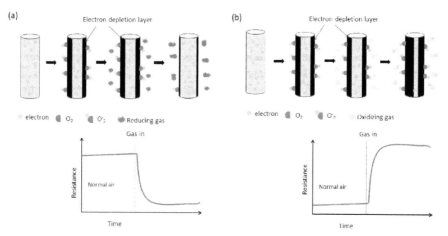

FIGURE 8.20   Schematic diagram of n-type MOS gas sensor resistance upon exposure to: (a) reducing gas and (b) oxidizing gas.

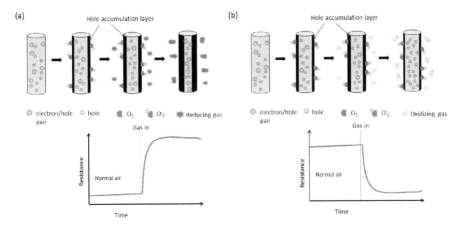

FIGURE 8.21 Schematic diagram of p-type MOS gas sensor resistance upon exposure to: (a) reducing gas and (b) oxidizing gas.

(reducing and oxidizing gas) is illustrated in Figure 8.21. In details, the sensing mechanism of MOS can be described into two different models: ionosorption model and oxygen-vacancy model (Zhang et al. 2017, Gurlo and Riedel 2007). The ionosorption model is described as the space-change effects or the changes of electric surface potential that result from the gas adsorption. Meanwhile the oxygen-vacancy model is based on the reaction between oxygen vacancies and gas molecules, and the variation of the amount of the sub surface/surface oxygen vacancies and their reduction-reoxidation mechanism (Zhang et al. 2017). The second model is considered the determining factor in the chemi-resistive behavior of the MOS gas sensor (Gurlo and Riedel 2007).

Target gas can be categorized into two groups: oxidizing and reducing gas (Huang and Wan 2009, Arafat et al. 2012, Shankar and Rayappan 2015). For oxidizing gas such as NO_2, O_2 and CO_2 have the tendency to accept electrons from the MOS surface, while reducing gas such as H_2, CO, NH_3, H_2S and C_2H_5OH act as electron donors when interacting with MOS surface (Shankar and Rayappan 2015). Therefore, depending on the type of MOS use and type of target gas to be detected, a different signal is obtained. Table 8.8 lists a summary of prominent target gas, the gas characteristic, and its area of detection.

8.5.3.1 Pristine 1-D MOS Based Gas Sensor

The first generation of MOS gas sensors were based on thin films as sensing platforms made of sintered materials. However, thin film structure-based gas sensors often result in low sensitivity and selectivity due to limited surface-to-volume ratio for the target gas to be adsorbed. Moreover, most thin film gas sensors operate at high temperature (200°C–600°C) to achieve enhanced chemical reactivity (Tseng 2009). Therefore, to enhance the gas sensor performance, immense research have been conducted to increase the surface-to-volume ratio in which MOS with 1-D nanostructures such as nanowire, nanorod, and nanotube were used as gas sensor platforms. Resmini et al. 2016 reported the influence of the surface area of ZnO nanorod on the NO_2 and H_2 gas response. They

TABLE 8.8

Summary of Prominent Target Gas, the Gas Characteristic, and Its Area of Detection

Target gas	Description	Application for Detection	Reference
Nitrogen dioxide, NO_2	By product of hydrocarbons Can cause acid rain Flammable but not explode Yellow-brown gas Pungent, acid odor Effect on lungs and respiratory tract. Breathing difficulties	Internal combustion engines (diesel engines) Thermal power station	Binions and Naik (2013)
Carbon monoxide, CO	By product of poorly combustion organic materials: petrol, oil or methane gas Colorless gas Flammable, able to explode Highly toxic	Faulty gas-powered boilers (home monitoring safety) Process monitoring in industrial plants Release gas from exhaust of car	Binions and Naik (2013)
Hydrogen, H_2	Mainly used as clean fuel in transportation, power fuel and industrial process Not toxic but highly flammable Colorless and odorless	Important to detect H_2 leaks from valve or tank into air, as the mixture can explode	Anand et al. (2014), Kanan et al. (2009)
Hydrogen sulphide, H_2S	By product of decomposition Highly flammable and easy explode Toxic especially to respiratory tract Colorless Rotten egg odor	Detection in crude petroleum, natural gas, volcanic gas, food processing, kraft paper mill, petroleum refineries	Kanan et al. (2009)

present three types of nanorod with different aspect ratio (8, 87, and 129) grown at different hydrothermal times (50, 210, and 20 min), respectively. NO_2 (operating temperature 150°C) and H_2 (200°C) gas response increased with increases of nanorods aspect ratio. This is due to the increase in number of adsorption sites available for gas chemisorption at a high aspect ratio of ZnO nanorod. Secondly, the small diameter of ZnO nanorods make it comparable with Deby length, which eventually allows the depletion layer to extend throughout the whole nanorod section (Resmini et al. 2016). Oros et al. 2016 presented comparable studies between SnO_2 thin films and SnO_2 nanorods and the effect of annealing treatment for NO_2 gas detection. The SnO_2 nanorod shows a higher and faster resistance change especially at low NO_2 gas concentrations compared to SnO_2 thin films. This is attributed to the high surface area available to provide more active sites for NO_2 gas adsorption than those thin films. Moreover, vertically growth nanorods allows gas penetration deep into the bottom of the structures, while dense thin film only allows the gas interaction on the very top surface (Oros et al. 2016). The effect of annealing treatment showed that when SnO_2 nanorods were annealed (T: 300°C and 400°C), the gas sensor response was enhanced compared to the non-annealed SnO_2 nanorods. When the SnO_2 nanorod was treated with annealing, the crystallinity of SnO_2 nanorods was enhanced, resulting in a more reactive surface of NO_2 chemisorption (Oros et al. 2016). A high operating temperature (>100°C) sets a huge limitation in MOS gas sensors for achieving a wide application due to the power consumption issues (Song et al. 2016b). Therefore, research nowadays is keen to develop sensitive room temperature gas detection. Hosseini and Mortezaali 2015 demonstrated a high response and selectivity of the ZnO nanorod to low concentration of H_2S gas at room temperature compared to 250°C. As the operating temperature increases to 250°C, it enhances the desorption ability of O_2 onto the ZnO nanorod surface. This reaction activated a large fraction of oxygen vacancy defects that then increased the electron concentration and electrical conductance. The increase in electron concentration obscures target gas (H_2S) to be detected as the surface reactions reduce. Generally, high operating temperatures of gas sensors influenced the adsorption and desorption rates of O_2 gas molecule, H_2S gas molecule and reaction products, surface decomposition rate of H_2S gas, and charge carrier concentration (Lee and Reedy 1999). Table 8.9 lists the performance of 1-D MOS gas sensor for different types of gas.

8.5.3.2 Hybrid and Decorated 1-D MOS Based Gas Sensor

There have been important advances to combine and integrate nanocomposites as an individual component to improve the performance of gas sensors. Several reports on MOS nanocomposites have been reported such as n-type MOS hybrid with n-type MOS (An et al. 2012, Lu et al. 2012, Nikan et al. 2013, Mondal et al. 2014, Park et al. 2014a), p-type MOS hybrid with n-type MOS (Na et al. 2011, Van Hieu et al. 2012, Park et al. 2013, Shao et al. 2013), and metallic nanoparticle hybrid on 1-D MOS (Huang et al. 2010, Hoa et al. 2014, Horprathum et al. 2014, Takács et al. 2016, Tian et al. 2018) for enhancing the gas response and selectivity of target gas detection. This is because nanocomposite structures have greater physical and chemical properties due to the modification of their electronic state (Zou et al. 2016). Mondal et al. (2014) reported the effect of ZnO content-doped SnO_2 nanofibers on response and selective detection of ethanol (C_2H_5OH) and carbon monoxide (CO) in the presence of methane

TABLE 8.9

Performance Comparison of Pristine 1-D MOS Based Gas Sensor at Different Types of Gas

Target Gas	MOS Sensor Structure	Average Dimension	Gas Concentration/ Sensitivity	Response Time/ Operating Temperature	References
NO$_2$	Branch ZnO NW	—	5 ppm/106%	12 sec/300°C	An et al. (2013)
	ZnO NW	D: 10–15 nm L: 0.5–1 μm	10 ppm/336	-/RT	Xia et al. (2016)
	ZnO NR	D: 50 nm, L: 500 nm	10 ppb/800	15 min/250°C	Oh et al. (2009)
	ZnO NR	D: 100 nm, L: 1.62 μm	1 ppm/1.316 (under halogen lamp)	1.3 min/200°C	Şahin et al. (2014)
	ZnO NB	Wi: 100–200 nm L:-	50 ppm/6.5 50 ppm/54	-/RT 135 sec/200°C	Kaur et al. (2017)
	WO$_3$ NT	D: 100–200 nm, L: 3–4 μm, WT: 20–30 nm	5 ppm/677%	30 sec/300°C	An et al. (2014)
	WO$_3$ NR	D: 100–150 nm	5 ppm/1.8	136 sec/RT	Ya-Qiao et al. (2014)
	SnO$_2$ NB	Wi: 100–150 nm	50 ppm/65	10 min/200°C	Suman et al. (2015)
	SnO$_2$ NR	D: 40–80 nm, L: 400–450 nm	5 ppm/~5310	2.5 min/150°C	Oros et al. (2016)
	SnO$_2$ NW	D: 20–40 nm, L: 1–10 μm	10 ppm/6.8	-/200°C	Hung et al. (2017)
	ZnO NB	T: 10 nm, Wi: 50 nm, L: ~1 μm	10 ppm/1.81	350°C	Sadek et al. (2005)
CO	TiO$_2$ NT	ID: 70.5 ± 9.5 nm, L: ~1.8 μm	200 ppm/68.3%	<1 min/350°C	Jang et al. (2014)
	SnO$_2$ NW	D: 70 nm, L:.7.7 μm	200 ppm/14.7%	-/260°C	Hernandez-Ramirez et al. (2007)
	ZnO NB+NR	D: 119 nm, Wi: 492	200 ppm/1.4 (under UV)	-/RT	Amin et al. (2012)
H$_2$	SnO$_2$ NW	-	1000 ppm/11.5	<2 min/100°C	El-Maghraby (2013)
	SnO$_2$ NW	ESA: ~35m²/g	1000 ppm/5.5	-/150°C	Shen et al. (2015)
	SnO$_2$ NR	ESA: ~25m²/g	1000 ppm/2.8	-/200°C	

(Continued)

TABLE 8.9 (*Continued*)
Performance Comparison of Pristine 1-D MOS Based Gas Sensor at Different Types of Gas

Target Gas	MOS Sensor Structure	Average Dimension	Gas Concentration/ Sensitivity	Response Time/ Operating Temperature	References
	ZnO NW (single unit)	D: 100 nm	100 ppm/34%	-/RT	Lupan et al. (2010)
	ZnO NW	D: 100–150 nm	100 ppm/~5.5	-/200°C	Khan et al. (2010)
	ZnO NB	T: 10 nm, Wi: 50 nm, L: ~1 μm	1%/15.32	-/385°C	Sadek et al. (2005)
	ZnO NW	D: 50–250 nm	1500 ppm/5	2 min/400°C	Huerta et al. (2013)
	WO$_3$ NR	D:~50 nm	0.05%/1.2	28 sec/150°C	Ahmad et al. (2013)
	WO$_3$ NR	D: 50 nm	0.05%/1.1	/170°C	Ahmad, Kang, et al. (2012)
H$_2$S	SnO$_2$ NW	D: 3.5 nm	50 ppm/38.4	2 sec/70°C	Song et al. (2016a)
	SnO$_2$ NR	D: 90 nm, L: 500 nm	50 ppm/130	-/350°C	Xu et al. (2009)
	SnO$_2$ NW	D: 4.2 nm	50 ppm/33	2 sec/RT	Song et al. (2016b)
	TiO$_2$ NT	ID: 110 nm WT: 16 nm L: 3.8 μm	50 ppm/26.2	22 sec/300°C	Tong et al. (2017)
	TiO$_2$ NT	ID: 100 nm, WT: 30 nm, L: 12 μm	38 ppm/144	146 sec/70°C	Perillo and Rodríguez (2016)
	ZnO NR	D: 300–500 nm, L: 7–9.5 μm	5 ppm/581	-/RT	Hosseini and Mortezaali (2015)

Note: NO$_2$: Nitrogen dioxide, CO: Carbon monoxide, H$_2$: Hydrogen, H$_2$S: Hydrogen sulphide, ZnO: Zinc oxide, SnO$_2$: Tin oxide, TiO$_2$: Titanium dioxide, WO$_3$: Tungsten oxide, NR: nanorod, NB: nanobelt, NT: nanotube, NW: nanowire.

(CH_4). ZnO doping concentration was varied from 0 to 5wt% with an optimum amount of 1wt% of ZnO doped significantly enhancing the gas response of CO and C_2H_5OH gas. Moreover, the operating temperature of 1wt% ZnO doped decreases by 50°C compared to undoped SnO_2. Lu et al. (2012) incorporated SnO_2 nanoparticles synthesis by hydrolyzing tin (II) chloride on ZnO nanorods as the sensing element for NO_2 gas detection at room temperature with and without UV light illumination. The maximum response can be achieved at a molar ratio of 1:1 for ZnO to SnO_2 compared to other ratios of ZnO to SnO_2. They also demonstrated that with the aid of UV light, the response time can be reduced from 12 min (without UV) to 7 min (with UV) and the recovery time was shortened from 14 min (without UV) to 8 min (with UV). This is due to the sufficiently large energy from UV light to cause the faster and higher adsorption and desorption of NO_2 gas onto SnO_2 NP/ZnO NR surface. Under UV light, ZnO acts as a photocatalyst which ensures an efficient charge separation (between holes and electrons) and increases the lifetime of charge carriers (Lu et al. 2012).

Another technique combining MOS is by core-shell nanostructure. Park et al. (2014a) reported the synthesis of ZnO core/WO_3-shell nanowire by thermal oxidation of ZnO followed by WO_3 sputter deposition on the enhancement of H_2 gas detection response. They demonstrated that by encapsulating ZnO nanowire with WO_3 led to a high response (1.1–2.7 folds) to 200–1000 ppb of H_2 compared to pristine ZnO nanowire. This is due to an increase of the depletion layer width and a potential barrier built up at the ZnO-WO_3 interface due to electron trapping into interface states. The interface acts as a lever that facilitates electron transfer resulting in sensing performance enhancement (Park et al. 2014a). Besides incorporating two materials of n-type MOS to enhance the gas sensitivity, extensive research on 1-D n-type MOS decorated with p-type MOS nanoparticles have been studied and reported. Particularly CuO nanoparticles decorated on 1-D SnO_2 have been widely studied in H_2S gas detection with excellent gas performance (Na et al. 2011, Van Hieu et al. 2012, Park et al. 2013). This phenomenon is due to the particular interaction between CuO and H_2S gas which transforms CuO into metallic CuS, which then changes a standard p-n junction to a metal-semiconductor interface (Shao et al. 2013). By incorporating metal nanoparticles (e.g. Pt, Au, Pd) with 1-D MOS, gas sensor performance would increase significantly due to the catalytic properties of metal nanoparticles. Kim et al. (2013) reported a hierarchy structure consisting of Pd nanoparticles decorated on branched ZnO nanowires growing on the SnO_2 nanowires. By incorporating Pd nanoparticles, much lower NO_2 gas concentration (1 ppm) was able to be detected compared to ZnO nanowires branched SnO_2 nanowires only (500 ppm). This is due to the increase of the exposed area on the branch structure towards target gas and the properties of Pd as a transition metal with high catalytic activity that enhances the response of the gas sensor. Huang et al. (2010) demonstrated an increase in H_2 gas sensor performance after Pt nanoparticles were decorated on SnO_2 nanorods. The improvement of the sensing response was due to the effect of "chemical sensitization mechanism" in which Pt nanoparticles promote the reaction of H_2 gases with O_2 species adsorbed on the surface of SnO_2 nanorod resulting in sensing response and also a decrease in operating temperature (Huang et al. 2010). Other materials employed for hybridization in the construction of gas sensor are presented in Table 8.10 along with their associated sensitivity, response time, and operating temperature.

TABLE 8.10

Performance Comparison of Hybrid/Decorated 1-D MOS Based Gas Sensor at Different Types of Gas

Target Gas	MOS Sensor Structure	Gas Concentration/Sensitivity	Response Time/Operating Temperature	References
NO_2	Pd NP/ZnO NW (branches)/SnO_2 (stem)	1 ppm/22.51	295 sec/250°C	Kim et al. (2013)
	Au NP/WO_3 NR	5 ppm/800	64.2 sec/250°C	Kabcum et al. (2017)
	Au NP/ZnO NR	50 ppm/4.14	-/300°C	Rai et al. (2012)
	Co_3O_4/ZnO NW	5 ppm/45.4	-/200°C	Van Hieu et al. (2012)
	Pt NP/ZnO NR	10 ppm/53	40 sec/150°C	Tian et al. (2018)
	QCM/ZnO NR	1000 ppm/0.85 Hz	-/RT	Van Quy et al. (2011)
	SnO_2 NP/ZnO NR	500 ppb/1000 (under UV)	420 sec (under UV)/RT	Lu et al. (2012)
	ZnO NR (core)/WO_3 film (shell)	5 ppm/281%	90 sec/300°C	An et al. (2012)
	CuO NS/ZnO NW	20 ppm/7100%	-/250°C	Fuadi et al. (2013)
CO	ZnO doped SnO_2 NW	300 ppm /90	70 sec/300 °C	Mondal et al. (2014)
	Au NP/SnO_2 NB	500 ppm/-	60 sec/400°C	Qian et al. (2006)
	Pd NP/SnO_2 NW	400 ppm/7	10 sec/400°C	Hoa et al. (2014)
	Pt NP/SnO_2 NR	1000 ppm/50	-/410°C	Huang et al. (2010)
	Au NP/ZnO NR	1000 ppm/12	-/150°C	Rai et al. (2012)
H_2	Pd doped SnO_2 NR	50 ppm/30	120 sec/300°C	Lee et al. (2008)
	Pd NP/ZnO NR	1000 ppm/91%	18 sec/RT	Rashid et al. (2013)
	Pd doped TiO_2	1000 ppm/5.8	-/150°C	Meng (2014)
	Pd NP/SnO_2 NW	10 ppm/1.5	-/150°C	Nguyen et al. (2017)
	PdO NP/WO_3 NR	40 ppm/23.5	5.3 sec/100°C	Geng et al. (2017)
	Pt NP/WO_3 NR	150 ppm/1530	-/200°C	Horprathum et al. (2014)
	Pt NP/SnO_2 NR	1000 ppm/250	-/400°C	Huang et al. (2010)
	ZnO NW(core)/WO_3 film (shell)	200 ppm/141 (under UV)	-/RT	Park et al. (2014a)

(Continued)

TABLE 8.10 (*Continued*)
Performance Comparison of Hybrid/Decorated 1-D MOS Based Gas Sensor at Different Types of Gas

Target Gas	MOS Sensor Structure	Gas Concentration/Sensitivity	Response Time/Operating Temperature	References
H_2S	Au/WO$_3$	10 ppm/~100	~30 sec/291°C	Vuong et al. (2015)
	Au NP/ZnO NR	6 ppm/1270	-/RT	Hosseini et al (2015)
	Au NP/ZnO NR	5 ppm/79.4	40 sec/RT	Ramgir et al. (2013)
	CoOx/SnO$_2$ NW	2 ppm/1000	-/250°C	Rumyantseva et al. (2017)
	Cu doped SnO$_2$ NW	50 ppm/10	420 sec/RT	Kumar et al. (2009)
	CuO NP/SnO$_2$ NW	20 ppm/~800	14 sec/300°C	Hwang et al. (2009)
	CuO NP/SnO$_2$ NW	5 ppm/7.4	150 sec/250°C	Shao et al. (2013)
	CuO NP/WO$_3$ NW	100 ppm/672.52	-/300°C	Park et al. (2014b)
	Pt NP/WO$_3$ NR	10 ppm/40	30–40 sec/270°C	Takács et al. (2016)
	Au NP/WO$_3$ NR	10 ppm/50	30–40 sec/270°C	Takács et al. (2016)

ACKNOWLEDGMENT

The authors appreciate the technical assistance of the School of Materials and Mineral Resources Engineering, Institute for Research in Molecular Medicine and NOR laboratory, USM. This research was supported by TRGS Grant 203/Pbahan/6763001 and RU Top Down 1001/Pbahan/870049.

REFERENCES

Ahmad, M., J. Kang, A. Sadek et al., 2012. Synthesis of WO$_3$ nanorod based thin films for ethanol and H$_2$ sensing. *Procedia Engineering*. 47: 358–361.

Ahmad, M.Z., J. Kang, A.S. Zoolfakar et al., 2013. Gas sensing studies of pulsed laser deposition deposited WO$_3$ nanorod based thin films. *Journal of Nanoscience and Nanotechnology*. 13: 8315–8319.

Ahmad, R., N. Tripathy, and Y.-B. Hahn, 2014. Highly stable urea sensor based on ZnO nanorods directly grown on Ag/glass electrodes. *Sensors and Actuators B: Chemical*. 194: 290–295.

Ahmad, R., N. Tripathy, J.H. Kim et al., 2012. Highly selective wide linear-range detecting glucose biosensors based on aspect-ratio controlled ZnO nanorods directly grown on electrodes. *Sensors and Actuators B: Chemical*. 174: 195–201.

Ahmad, R., N. Tripathy, S.H. Kim et al., 2014. High performance cholesterol sensor based on ZnO nanotubes grown on Si/Ag electrodes. *Electrochemistry Communications*. 38: 4–7.

Ali, A., A.A. Ansari, A. Kaushik et al., 2009. Nanostructured zinc oxide film for urea sensor. *Materials Letters*. 63: 2473–2475.

Ali, S.M.U., Z.H. Ibupoto, S. Salman et al., 2011. Selective determination of urea using urease immobilized on ZnO nanowires. *Sensors and Actuators B: Chemical*. 160: 637–643.

Amin, M., U. Manzoor, M. Islam et al., 2012. Synthesis of ZnO nanostructures for low temperature CO and UV sensing. *Sensors*. 12: 13842–13851.

An, S., S. Park, H. Ko et al., 2012. Enhanced NO$_2$ gas sensing properties of WO$_3$ nanorods encapsulated with ZnO. *Applied Physics A*. 108: 53–58.

An, S., S. Park, H. Ko et al., 2014. Fabrication of WO$_3$ nanotube sensors and their gas sensing properties. *Ceramics International*. 40: 1423–1429.

An, S., S. Park, H. Ko et al., 2013. Enhanced gas sensing properties of branched ZnO nanowires. *Thin Solid Films*. 547: 241–245.

Anand, K., O. Singh, M.P. Singh et al., 2014. Hydrogen sensor based on graphene/ZnO nanocomposite. *Sensors and Actuators B: Chemical*. 195: 409–415.

Ansari, A.A., A. Kaushik, P.R. Solanki et al., 2009. Electrochemical cholesterol sensor based on tin oxide-chitosan nanobiocomposite film. *Electroanalysis*. 21: 965–972.

Ansari, S.G., Z.A. Ansari, H.-K. Seo et al., 2008. Urea sensor based on tin oxide thin films prepared by modified plasma enhanced CVD. *Sensors and Actuators B: Chemical*. 132: 265–271.

Arafat, M., B. Dinan, S.A. Akbar et al., 2012. Gas sensors based on one dimensional nanostructured metal-oxides: A review. *Sensors*. 12: 7207–7258.

Arain, M., A. Nafady, Sirajuddin et al., 2016. Simpler and highly sensitive enzyme-free sensing of urea via NiO nanostructures modified electrode. *RSC Advances*. 6: 39001–39006.

Arduini, F., L. Micheli, D. Moscone et al., 2016. Electrochemical biosensors based on nanomodified screen-printed electrodes: Recent applications in clinical analysis. *TrAC Trends in Analytical Chemistry*. 79: 114–126.

Arya, S.K., M. Datta, and B.D. Malhotra, 2008. Recent advances in cholesterol biosensor. *Biosensors and Bioelectronics*. 23: 1083–1100.

Bankar, S.B., M.V. Bule, R.S. Singhal et al., 2009. Glucose oxidase–An overview. *Biotechnology Advances*. 27: 489–501.

Binions, R. and A. Naik, 2013. Metal oxide semiconductor gas sensors in environmental monitoring. *Semiconductor Gas Sensors*. 433–466.

Cao, S., L. Zhang, Y. Chai et al., 2013. An integrated sensing system for detection of cholesterol based on TiO$_2$–graphene–Pt–Pd hybridnanocomposites. *Biosensors and Bioelectronics*. 42: 532–538.

Charan, C. and V.K. Shahi, 2014. Nanostructured manganese oxide–chitosan-based cholesterol sensor. *Journal of Applied Electrochemistry*. 44: 953–962.

Chen, T., X. Li, C. Qiu et al., 2014. Electrochemical sensing of glucose by carbon cloth-supported Co$_3$O$_4$/PbO$_2$ core-shell nanorod arrays. *Biosensors and Bioelectronics*. 53: 200–206.

Chen, Y., W.H. Tse, L. Chen et al., 2015. Ag nanoparticles-decorated ZnO nanorod array on a mechanical flexible substrate with enhanced optical and antimicrobial properties. *Nanoscale Research Letters*. 10: 106.

Ching, K.-L., G. Li, Y.-L. Ho et al., 2016. The role of polarity and surface energy in the growth mechanism of ZnO from nanorods to nanotubes. *CrystEngComm*. 18: 779–786.

Choopun, S., H. Tabata, and T. Kawai, 2005. Self-assembly ZnO nanorods by pulsed laser deposition under argon atmosphere. *Journal of Crystal Growth*. 274: 167–172.

Choopun, S., N. Hongsith, and E. Wongrat 2012. Metal-oxide nanowires for gas sensors, In *Nanowires-Recent Advances*. (Ed.) X. Peng. InTechOpen. https://www.intechopen.com/books/nanowires-recent-advances/metal-oxide-nanowires-for-gas-sensors.

Clark, L.J. and C. Lyons, 1962. Electrode systems for continuous monitoring in cardiovascular surgery. *Annals of the New York Academy of Sciences*. 102: 29.

Dar, G., A. Umar, S. Zaidi et al., 2012. Ce-doped ZnO nanorods for the detection of hazardous chemical. *Sensors and Actuators B: Chemical*. 173: 72–78.

Dar, G., A. Umar, S.A. Zaidi et al., 2012. Ultra-high sensitive ammonia chemical sensor based on ZnO nanopencils. *Talanta*. 89: 155–161.

Das, R., S. Basak, and A. Maity, 2011. Sol-gel synthesized aligned ZnO nanorods growth: studies on structural and optoelectronic properties. In *Nanoscience, Technology and Societal Implications (NSTSI), 2011 International Conference*. IEEE.

Dedova, T., I. Gromyko, M. Krunks et al., 2015. Spray pyrolysis deposition and characterization of highly c-axis oriented hexagonal ZnS nanorod crystals. *Crystal Research and Technology*. 50: 85–92.

Deng, C., M. Li, Q. Xie et al., 2006. New glucose biosensor based on a poly(o phenylendiamine)/glucose oxidase-glutaraldehyde/Prussian blue/Au electrode with QCM monitoring of various electrode-surface modifications. *Analytica Chimica Acta*. 557: 85–94.

Devan, R.S., R.A. Patil, J.H. Lin et al., 2012. One-dimensional metal-oxide nanostructures: Recent developments in synthesis, characterization, and applications. *Advanced Functional Materials*. 22: 3326–3370.

Devi, P., M. Kaur, and R. Thakur. Novel carbon/manganese oxide nanocomposite for electrochemical detection of arsenic in water-a step towards portable real time sensor. In *Signal Processing, Computing and Control (ISPCC), 2015 International Conference*. 2015. IEEE.

Dhawan, G., G. Sumana, and B.D. Malhotra, 2009. Recent developments in urea biosensors. *Biochemical Engineering Journal*. 44: 42–52.

Dutta, D., S. Chandra, A.K. Swain et al., 2014. SnO$_2$ quantum dots-reduced graphene oxide composite for enzyme-free ultrasensitive electrochemical detection of urea. *Analytical Chemistry*. 86: 5914–5921.

Dwivedi, C. and V. Dutta, 2012. Vertically aligned ZnO nanorods via self-assembled spray pyrolyzed nanoparticles for dye-sensitized solar cells. *Advances in Natural Sciences: Nanoscience and Nanotechnology*. 3: 015011.

El-Maghraby, E.M., A. Qurashi, and T. Yamazaki, 2013. Synthesis of SnO$_2$ nanowires their structural and H$_2$ gas sensing properties. *Ceramics International*. 39: 8475–8480.

Faisal, M., S.B. Khan, M.M. Rahman et al., 2011. Synthesis, characterizations, photocatalytic and sensing studies of ZnO nanocapsules. *Applied Surface Science*. 258: 672–677.

Faisal, M., S. Bahadar Khan, M. M Rahman et al., 2013. Fabrication of highly sensitive chemi-sensor and efficient photocatalyst based on ZnO nanostructured material. *Micro and Nanosystems*. 5: 35–46.

Febrianti, Y., N. Putri, I. Sugihartono et al., 2017. Synthesis and characterization of Co-doped zinc oxide nanorods prepared by ultrasonic spray pyrolysis and hydrothermal methods. In *AIP Conference Proceedings*, 1862. AIP Publishing.

Fine, G.F., L.M. Cavanagh, A. Afonja et al., 2010. Metal oxide semi-conductor gas sensors in environmental monitoring. *Sensors*. 10: 5469–5502.

Fioravanti, A., A. Bonanno, M. Carotta et al., 2014. ZnO as functional material for sub-ppm acetone detection. In *SENSORS*, 2014. IEEE.

Fuadi, M., D. Yang, C. Park et al., 2013. Highly sensitive NO$_2$ gas sensor based on zinc oxide/copper oxide hybrid-nanostructures. In *Solid-State Sensors, Actuators and Microsystems (TRANSDUCERS & EUROSENSORS XXVII), 2013 Transducers & Eurosensors XXVII: The 17th International Conference*. IEEE.

Geng, X., Y. Luo, B. Zheng et al., 2017. Photon assisted room-temperature hydrogen sensors using PdO loaded WO$_3$ nanohybrids. *International Journal of Hydrogen Energy*. 42: 6425–6434.

Giri, A.K., C. Charan, S.C. Ghosh et al., 2016. Phase and composition selective superior cholesterol sensing performance of ZnO@ZnS nano-heterostructure and ZnS nanotubes. *Sensors and Actuators B: Chemical*. 229: 14–24.

Goode, J.A., J.V.H. Rushworth, and P.A. Millner, 2014. Biosensor regeneration: A review of common techniques and outcomes. *Langmuir*. 31: 6267–6276.

Grieshaber, D., R. MacKenzie, J. Voros et al., 2008. Electrochemical biosensors–Sensor principles and architectures. *Sensors*. 8: 1400.

Gurlo, A. and R. Riedel, 2007. In situ and operando spectroscopy for assessing mechanisms of gas sensing. *Angewandte Chemie International Edition*. 46: 3826–3848.

Haarindraprasad, R., U. Hashim, S.C.B. Gopinath et al., 2016. Fabrication of interdigitated high-performance zinc oxide nanowire modified electrodes for glucose sensing. *Analytica Chimica Acta*. 925: 70–81.

Hernandez-Ramirez, F., A. Tarancon, O. Casals et al., 2007. High response and stability in CO and humidity measures using a single SnO$_2$ nanowire. *Sensors and Actuators B: Chemical*. 121: 3–17.

Hoa, N.D., P. Van Tong, N. Van Duy et al., 2014. Effective decoration of Pd nanoparticles on the surface of SnO$_2$ nanowires for enhancement of CO gas-sensing performance. *Journal of Hazardous Materials*. 265: 124–132.

Horprathum, M., T. Srichaiyaperk, B. Samransuksamer et al., 2014. Ultrasensitive hydrogen sensor based on Pt-decorated WO$_3$ nanorods prepared by glancing-angle dc magnetron sputtering. *ACS Applied Materials & Interfaces*. 6: 22051–22060.

Hosseini, Z. and A. Mortezaali, 2015. Room temperature H$_2$S gas sensor based on rather aligned ZnO nanorods with flower-like structures. *Sensors and Actuators B: Chemical*. 207: 865–871.

Hosseini, Z., A. Mortezaali, and S. Fardindoost, 2015. Sensitive and selective room temperature H$_2$S gas sensor based on Au sensitized vertical ZnO nanorods with flower-like structures. *Journal of Alloys and Compounds*. 628: 222–229.

Huang, H., C. Ong, J. Guo et al., 2010. Pt surface modification of SnO$_2$ nanorod arrays for CO and H$_2$ sensors. *Nanoscale*. 2: 1203–1207.

Huang, J. and Q. Wan, 2009. Gas sensors based on semiconducting metal oxide one-dimensional nanostructures. *Sensors*. 9: 9903–9924.

Huang, K.-J., D.-J. Niu, X. Liu et al., 2011. Direct electrochemistry of catalase at amine-functionalized graphene/gold nanoparticles composite film for hydrogen peroxide sensor. *Electrochimica Acta*. 56: 2947–2953.

Huerta, A., G. Perez-Sanchez, F. Chavez et al. ZnO nanowires synthesized by CSS and their application as a hydrogen gas sensor. In *Electrical Engineering, Computing Science and Automatic Control (CCE), 2013 10th International Conference*. IEEE.

Hung, P.T., V.X. Hien, J.-H. Lee et al., 2017. Fabrication of SnO_2 nanowire networks on a spherical Sn surface by thermal oxidation. *Journal of Electronic Materials*. 1–8.

Hwang, I.-S., J.-K. Choi, S.-J. Kim et al., 2009. Enhanced H_2S sensing characteristics of SnO_2 nanowires functionalized with CuO. *Sensors and Actuators B: Chemical*. 142: 105–110.

Ikhmayies, S.J. and M.B. Zbib, 2017. Spray pyrolysis synthesis of ZnO micro/nanorods on glass substrate. *Journal of Electronic Materials*. 46: 5629–5634.

Jang, N.-S., M.S. Kim, S.-H. Kim et al., 2014. Direct growth of titania nanotubes on plastic substrates and their application to flexible gas sensors. *Sensors and Actuators B: Chemical*. 199: 361–368.

Kabcum, S., N. Kotchasak, D. Channei et al., 2017. Highly sensitive and selective NO_2 sensor based on Au-impregnated WO_3 nanorods. *Sensors and Actuators B: Chemical*. 252: 523–536.

Kafi, A.K.M., G. Wu, and A. Chen, 2008. A novel hydrogen peroxide biosensor based on the immobilization of horseradish peroxidase onto Au-modified titanium dioxide nanotube arrays. *Biosensors and Bioelectronics*. 24: 566–571.

Kanan, S.M., O.M. El-Kadri, I.A. Abu-Yousef et al., 2009. Semiconducting metal oxide based sensors for selective gas pollutant detection. *Sensors*. 9: 8158–8196.

Kärber, E., T. Raadik, T. Dedova et al., 2011. Photoluminescence of spray pyrolysis deposited ZnO nanorods. *Nanoscale Research Letters*. 6: 359.

Karnati, P., A. Haque, M. Taufique et al., 2018. A systematic study on the structural and optical properties of vertically aligned zinc oxide nanorods grown by high pressure assisted pulsed laser deposition technique. *Nanomaterials*. 8.

Kashif, M., U. Hashim, M. Ali et al., 2013. Morphological, structural, and electrical characterization of sol-gel-synthesized ZnO nanorods. *Journal of Nanomaterials*. 2013.

Kaur, M., S. Kailasaganapathi, N. Ramgir et al., 2017. Gas dependent sensing mechanism in ZnO nanobelt sensor. *Applied Surface Science*. 394: 258–266.

Kawakami, M., A.B. Hartanto, Y. Nakata et al., 2003. Synthesis of ZnO nanorods by nanoparticle assisted pulsed-laser deposition. *Japanese Journal of Applied Physics*. 42: L33.

Khan, R., H.-W. Ra, J. Kim et al., 2010. Nanojunction effects in multiple ZnO nanowire gas sensor. *Sensors and Actuators B: Chemical*. 150: 389–393.

Khun, K., Z.H. Ibupoto, X. Liu et al., 2015. The ethylene glycol template assisted hydrothermal synthesis of Co_3O_4 nanowires; structural characterization and their application as glucose non-enzymatic sensor. *Materials Science and Engineering: B*. 194: 94–100.

Kim, H.-J. and J.-H. Lee, 2014. Highly sensitive and selective gas sensors using p-type oxide semiconductors: Overview. *Sensors and Actuators B: Chemical*. 192: 607–627.

Kim, S.S., H.G. Na, S.-W. Choi et al., 2013. Decoration of Pd nanoparticles on ZnO-branched nanowires and their application to chemical sensors. *Microelectronic Engineering*. 105: 1–7.

Kissinger, P.T., 2005. Biosensors–A perspective. *Biosensors and Bioelectronics*. 20: 2512–2516.

Kumar, V., S. Sen, K. Muthe et al., 2009. Study of H_2S sensitivity of pure and Cu doped SnO_2 single nanowire sensors. In *AIP Conference Proceedings*. AIP.

Kung, C.-W., C.-Y. Lin, Y.-H. Lai et al., 2011. Cobalt oxide acicular nanorods with high sensitivity for the non-enzymatic detection of glucose. *Biosensors and Bioelectronics*. 27: 125–131.

Lee, A.P. and B.J. Reedy, 1999. Temperature modulation in semiconductor gas sensing. *Sensors and Actuators B: Chemical*. 60: 35–42.

Lee, S.H., J. Yang, Y.J. Han et al., 2015. Rapid and highly sensitive MnOx nanorods array platform for a glucose analysis. *Sensors and Actuators B: Chemical.* 218: 137–144.

Lee, Y.-M., C.-M. Huang, H.-W. Chen et al., 2013. Low temperature solution-processed ZnO nanorod arrays with application to liquid ethanol sensors. *Sensors and Actuators A: Physical.* 189: 307–312.

Lee, Y., H. Huang, O. Tan et al., 2008. Semiconductor gas sensor based on Pd-doped SnO$_2$ nanorod thin films. *Sensors and Actuators B: Chemical.* 132: 239–242.

Leonhardt, A., H. Liepack, K. Biedermann et al., 2005. Synthesis of SiC nanorods by chemical vapor deposition. *Fullerenes, Nanotubes, and Carbon Nanostructures.* 13: 91–97.

Li, M., H. Gou, I. Al-Ogaidi et al., 2013. Nanostructured sensors for detection of heavy metals: A review.

Liu, J., Y. Li, X. Huang et al., 2010. Tin oxide nanorod array-based electrochemical hydrogen peroxide biosensor. *Nanoscale Research Letters.* 5: 1177.

Liu, J., P. Yan, G. Yue et al., 2006. Synthesis of doped ZnS one-dimensional nanostructures via chemical vapor deposition. *Materials Letters.* 60: 3471–3476.

Liu, J., C. Guo, C.M. Li et al., 2009. Carbon-decorated ZnO nanowire array: A novel platform for direct electrochemistry of enzymes and biosensing applications. *Electrochemistry Communications.* 11: 202–205.

Liu, S. and A. Chen, 2005. Coadsorption of horseradish peroxidase with thionine on TiO$_2$ nanotubes for biosensing. *Langmuir.* 21: 8409–8413.

Liu, X., S. Cheng, H. Liu et al., 2012. A survey on gas sensing technology. *Sensors.* 12: 9635–9665.

Liu, X., J. Zhang, S. Liu et al., 2013. Gold nanoparticle encapsulated-tubular TIO$_2$ nanocluster as a scaffold for development of thiolated enzyme biosensors. *Analytical Chemistry.* 85: 4350–4356.

Liu, Y. and M. Liu, 2007. Ordered ZnO nanorods synthesized by combustion chemical vapor deposition. *Journal of Nanoscience and Nanotechnology.* 7: 4529–4533.

Lu, G., J. Xu, J. Sun et al., 2012. UV-enhanced room temperature NO$_2$ sensor using ZnO nanorods modified with SnO$_2$ nanoparticles. *Sensors and Actuators B: Chemical.* 162: 82–88.

Lupan, O., V. Ursaki, G. Chai et al., 2010. Selective hydrogen gas nanosensor using individual ZnO nanowire with fast response at room temperature. *Sensors and Actuators B: Chemical.* 144: 56–66.

Lv, H., D. Sang, H. Li et al., 2010. Thermal evaporation synthesis and properties of ZnO nano/microstructures using carbon group elements as the reducing agents. *Nanoscale Research Letters.* 5: 620.

Ma, Q. and K. Nakazato, 2014. Low-temperature fabrication of ZnO nanorods/ferrocenyl-alkanethiol bilayer electrode and its application for enzymatic glucose detection. *Biosensors and Bioelectronics.* 51: 362–365.

Mahajan, A.P., S.B. Kondawar, R.P. Mahore et al., 2015. Polyaniline/MnO$_2$ nanocomposites based stainless steel electrode modified enzymatic urease biosensor. *Procedia Materials Science.* 10: 699–705.

Meng, D., T. Yamazaki, and T. Kikuta, 2014. Preparation and gas sensing properties of undoped and Pd-doped TiO$_2$ nanowires. *Sensors and Actuators B: Chemical.* 190: 838–843.

Miao, F., X. Lu, B. Tao et al., 2016. Glucose oxidase immobilization platform based on ZnO nanowires supported by silicon nanowires for glucose biosensing. *Microelectronic Engineering.* 149: 153–158.

Mirzaei, A., S. Leonardi, and G. Neri, 2016. Detection of hazardous volatile organic compounds (VOCs) by metal oxide nanostructures-based gas sensors: a review. *Ceramics International.* 42: 15119–15141.

Mondal, B., B. Basumatari, J. Das et al., 2014. ZnO–SnO$_2$ based composite type gas sensor for selective hydrogen sensing. *Sensors and Actuators B: Chemical.* 194: 389–396.

Monošík, R., M. Streďanský, and E. Šturdík, 2012. Biosensors–Classification, characterization and new trends. In *Acta Chimica Slovaca*. p. 109.

Mustafa, M., Y. Iqbal, U. Majeed et al., 2017. Effect of precursor's concentration on structure and morphology of ZnO nanorods synthesized through hydrothermal method on gold surface. In *AIP Conference Proceedings*. AIP Publishing.

Na, C.W., H.-S. Woo, I.-D. Kim et al., 2011. Selective detection of NO_2 and C_2H_5OH using a Co_3O_4-decorated ZnO nanowire network sensor. *Chemical Communications*. 47: 5148–5150.

Nandi, R., S.K. Appani, and S. Major, 2016. Vertically aligned ZnO nanorods of high crystalline and optical quality grown by dc reactive sputtering. *Materials Research Express*. 3: 095009.

Nguyen, K., C.M. Hung, T.M. Ngoc et al., 2017. Low-temperature prototype hydrogen sensors using Pd-decorated SnO_2 nanowires for exhaled breath applications. *Sensors and Actuators B: Chemical*. 253: 156–163.

Nikan, E., A.A. Khodadadi, and Y. Mortazavi, 2013. Highly enhanced response and selectivity of electrospun ZnO-doped SnO_2 sensors to ethanol and CO in presence of CH_4. *Sensors and Actuators B: Chemical*. 184: 196–204.

Niu, X., H. Wei, K. Tang et al., 2015. Solvothermal synthesis of 1D nanostructured Mn_2O_3: effect of Ni^{2+} and Co^{2+} substitution on the catalytic activity of nanowires. *RSC Advances*. 5: 66271–66277.

Oh, E., H.-Y. Choi, S.-H. Jung et al., 2009. High-performance NO_2 gas sensor based on ZnO nanorod grown by ultrasonic irradiation. *Sensors and Actuators B: Chemical*. 141: 239–243.

Oros, C., M. Horprathum, A. Wisitsoraat et al., 2016. Ultra-sensitive NO_2 sensor based on vertically aligned SnO_2 nanorods deposited by DC reactive magnetron sputtering with glancing angle deposition technique. *Sensors and Actuators B: Chemical*. 223: 936–945.

Orvatinia, M. and R. Imani, 2012. Temperature effect on structural and electronic properties of zinc oxide nanowires synthesized by carbothermal evaporation method. *International Journal of Nanoscience*. 11: 1250032.

Ou, S.-L., F.-P. Yu, and D.-S. Wuu, 2017. Transformation from film to nanorod via a sacrificial layer: Pulsed laser deposition of ZnO for enhancing photodetector performance. *Scientific Reports*. 7: 14251.

Ozdemir, C., O. Akca, E.I. Medine et al., 2012. Biosensing applications of modified core–shell magnetic nanoparticles. *Food Analytical Methods*. 5: 731–736.

Palanisamy, S., C. Karuppiah, S.-M. Chen et al., 2014. Direct electrochemistry of myoglobin at silver nanoparticles/myoglobin biocomposite: Application for hydrogen peroxide sensing. *Sensors and Actuators B: Chemical*. 202: 177–184.

Palomera, N., Balaguera M, Arya SK et al., 2011. Zinc oxide nanorods modified indium tin oxide surface for amperometric urea biosensor. *Journal of Nanoscience and Nanotechnology*. 11: 6683–6689.

Pandey, C.M., I. Tiwari, V.N. Singh et al., 2017. Highly sensitive electrochemical immunosensor based on graphene-wrapped copper oxide-cysteine hierarchical structure for detection of pathogenic bacteria. *Sensors and Actuators B: Chemical*. 238: 1060–1069.

Paraguay-Delgado, F., W. Antúnez-Flores, M. Miki-Yoshida et al., 2005. Structural analysis and growing mechanisms for long SnO_2 nanorods synthesized by spray pyrolysis. *Nanotechnology*. 16: 688.

Park, S., T. Hong, J. Jung et al., 2014a. Room temperature hydrogen sensing of multiple networked ZnO/WO_3 core–shell nanowire sensors under UV illumination. *Current Applied Physics*. 14: 1171–1175.

Park, S., S. Park, J. Jung et al., 2014b. H_2S gas sensing properties of CuO-functionalized WO_3 nanowires. *Ceramics International*. 40: 11051–11056.

Park, W.J., M.H. Kim, B.H. Koo et al., 2013. Alternatively driven dual nanowire arrays by ZnO and CuO for selective sensing of gases. *Sensors and Actuators B: Chemical*. 185: 10–16.

Patel, P.N., V. Mishra, and A.S. Mandloi, 2010. Optical biosensors: Fundamentals & trends. *Journal of Engineering Research and Studies*. 1: 15–34.

Perednis, D. and L.J. Gauckler, 2005. Thin film deposition using spray pyrolysis. *Journal of Electroceramics*. 14: 103–111.

Perillo, P.M. and D.F. Rodríguez, 2016. TiO$_2$ nanotubes membrane flexible sensor for low-temperature H$_2$S detection. *Chemosensors*. 4: 15.

Polsongkram, D., P. Chamninok, S. Pukird et al., 2008. Effect of synthesis conditions on the growth of ZnO nanorods via hydrothermal method. *Physica B: Condensed Matter*. 403: 3713–3717.

Pujol, L., D. Evrard, K. Groenen-Serrano et al., 2014. Electrochemical sensors and devices for heavy metals assay in water: The French groups' contribution. *Frontiers in Chemistry*. 2: 19.

Pushkariov, V., A. Nikolaev, and E. Kaidashev, 2014. Synthesis and characterization of ZnO nanorods obtained by catalyst-free thermal technique. In *Journal of Physics: Conference Series*. IOP Publishing.

Putzbach, W., and N.J. Ronkainen, 2013. Immobilization techniques in the fabrication of nanomaterial-based electrochemical biosensors: A review. *Sensors*. 13: 4811–4840.

Qadir Israr, M., J. Rana Sadaf, M. Asif et al., 2010. Potentiometric cholesterol biosensor based on ZnO nanorods chemically grown on Ag wire. *Thin Solid Films*. 519: 1106–1109.

Qian, L., K. Wang, Y. Li et al., 2006. CO sensor based on Au-decorated SnO$_2$ nanobelt. *Materials Chemistry and Physics*. 100: 82–84.

Rai, P., Y.-S. Kim, H.-M. Song et al., 2012. The role of gold catalyst on the sensing behavior of ZnO nanorods for CO and NO$_2$ gases. *Sensors and Actuators B: Chemical*. 165: 133–142.

Ramgir, N.S., P.K. Sharma, N. Datta et al., 2013. Room temperature H$_2$S sensor based on Au modified ZnO nanowires. *Sensors and Actuators B: Chemical*. 186: 718–726.

Rashid, T.-R., D.-T. Phan, and G.-S. Chung, 2013. A flexible hydrogen sensor based on Pd nanoparticles decorated ZnO nanorods grown on polyimide tape. *Sensors and Actuators B: Chemical*. 185: 777–784.

Resmini, A., U. Anselmi-Tamburini, S. Emamjomeh et al., 2016. The influence of the absolute surface area on the NO$_2$ and H$_2$ gas responses of ZnO nanorods prepared by hydrothermal growth. *Thin Solid Films*. 618: 246–252.

Rizal, U., S. Das, D. Kumar et al., 2016. Synthesis and characterization of TiO$_2$ nanostructure thin films grown by thermal CVD. In *AIP Conference Proceedings*. AIP Publishing.

Rogers, K.R., 2000. Principles of affinity-based biosensors. *Molecular Biotechnology*. 14: 109–129.

Rumyantseva, M., S. Vladimirova, N. Vorobyeva et al., 2017. p-CoO x/n-SnO$_2$ nanostructures: New highly selective materials for H$_2$S detection. *Sensors and Actuators B: Chemical*. 255: 564–571.

Saber-Tehrani, M., A. Pourhabib, S.W. Husain et al., 2013. A simple and efficient electrochemical sensor for nitrite determination in food samples based on Pt nanoparticles distributed poly(2-aminothiophenol) modified electrode. *Food Analytical Methods*. 6: 1300–1307.

Sadek, A., W. Wlodarski, K. Kalantar-Zadeh et al. 2005. ZnO nanobelt based conductometric H$_2$ and NO$_2$ gas sensors. in *Sensors*. IEEE.

Şahin, Y., S. Öztürk, N. Kılınç et al., 2014. Electrical conduction and NO$_2$ gas sensing properties of ZnO nanorods. *Applied Surface Science*. 303: 90–96.

Shang, Z., Z. Liu, P. Shang et al., 2012. Synthesis of single-crystal TiO$_2$ nanowire using titanium monoxide powder by thermal evaporation. *Journal of Materials Science & Technology*. 28: 385–390.

Shankar, P. and J.B.B. Rayappan, 2015. Gas sensing mechanism of metal oxides: The role of ambient atmosphere, type of semiconductor and gases-A review. *Science Letters Journal.* 4: 126.

Shao, F., M.W. Hoffmann, J.D. Prades et al., 2013. Heterostructured p-CuO (nanoparticle)/n-SnO$_2$ (nanowire) devices for selective H$_2$S detection. *Sensors and Actuators B: Chemical.* 181: 130–135.

Shen, Y., W. Wang, A. Fan et al., 2015. Highly sensitive hydrogen sensors based on SnO$_2$ nanomaterials with different morphologies. *International Journal of Hydrogen Energy.* 40: 15773–15779.

Shukla, M., Pramila, T. Dixit et al., 2017. Influence of aspect ratio and surface defect density on hydrothermally grown ZnO nanorods towards amperometric glucose biosensing applications. *Applied Surface Science.* 422: 798–808.

Singh, J., P. Kalita, M.K. Singh, and et al., 2011. Nanostructured nickel oxide-chitosan film for application to cholesterol sensor. *Applied Physics Letters.* 98: 123702.

Solanki, P.R., A. Kaushik, V.V. Agrawal et al., 2011. Nanostructured metal oxide-based biosensors. *NPG Asia Materials.* 3: 17–24.

Solanki, P.R., A. Kaushik, A.A. Ansari et al., 2008. Zinc oxide-chitosan nanobiocomposite for urea sensor. *Applied Physics Letters.* 93: 163903.

Song, J., L. Xu, R. Xing et al., 2013. Ag nanoparticles coated NiO nanowires hierarchical nanocomposites electrode for nonenzymatic glucose biosensing. *Sensors and Actuators B: Chemical.* 182: 675–681.

Song, Z., S. Xu, M. Li et al., 2016a. Solution-processed SnO$_2$ nanowires for sensitive and fast-response H$_2$S detection. *Thin Solid Films.* 618: 232–237.

Song, Z., Z. Wei, B. Wang et al., 2016b. Sensitive room-temperature H$_2$S gas sensors employing SnO$_2$ quantum wire/reduced graphene oxide nanocomposites. *Chemistry of Materials.* 28: 1205–1212.

Stradiotto, N.R., H. Yamanaka, and M.V.B. Zanoni, 2003. Electrochemical sensors: A powerful tool in analytical chemistry. *Journal of the Brazilian Chemical Society.* 14: 159–173.

Suman, P., A. Felix, H. Tuller et al., 2015. Comparative gas sensor response of SnO2, SnO and Sn$_3$O$_4$ nanobelts to NO$_2$ and potential interferents. *Sensors and Actuators B: Chemical.* 208: 122–127.

Sun, Y.-F., S.-B. Liu, F.-L. Meng et al., 2012. Metal oxide nanostructures and their gas sensing properties: A review. *Sensors.* 12: 2610–2631.

Sun, Y., G.M. Fuge, and M.N. Ashfold, 2006. Growth mechanisms for ZnO nanorods formed by pulsed laser deposition. *Superlattices and Microstructures.* 39: 33–40.

Takács, M., D. Zámbó, A. Deák et al., 2016. WO$_3$ nano-rods sensitized with noble metal nano-particles for H$_2$S sensing in the ppb range. *Materials Research Bulletin.* 84: 480–485.

Tao, W., D. Pan, Y. Liu et al., 2005. An amperometric hydrogen peroxide sensor based on immobilization of hemoglobin in poly(o-aminophenol) film at iron–cobalt hexacyanoferrate-modified gold electrode. *Analytical Biochemistry.* 338: 332–340.

Tian, H., H. Fan, J. Ma et al., 2018. Pt-decorated zinc oxide nanorod arrays with graphitic carbon nitride nanosheets for highly efficient dual-functional gas sensing. *Journal of Hazardous Materials.* 341: 102–111.

Tian, W.-C., Y.-H. Ho, C.-H. Chen et al., 2013. Sensing performance of precisely ordered TiO$_2$ nanowire gas sensors fabricated by electron-beam lithography. *Sensors.* 13: 865–874.

Tiwari, J.N., R.N. Tiwari, and K.S. Kim, 2012. Zero-dimensional, one-dimensional, two-dimensional and three-dimensional nanostructured materials for advanced electrochemical energy devices. *Progress in Materials Science.* 57: 724–803.

Tong, X., W. Shen, X. Chen et al., 2017. A fast response and recovery H$_2$S gas sensor based on free-standing TiO$_2$ nanotube array films prepared by one-step anodization method. *Ceramics International.* 43: 14200–14209.

Tseng, T.-Y. 2009. *Handbook of Nanoceramics and Their Based Nanodevices: Sensors, Fuel Cells, and Biomedical Application.* Stevenson Ranch, CA: American Scientific Publishers.

Turner, A.P.F., 2013. Biosensors: Sense and sensibility. *Chemical Society Reviews.* 42: 3184–3196.

Tyagi, M., M. Tomar, and V. Gupta, 2013. NiO nanoparticle-based urea biosensor. *Biosensors and Bioelectronics.* 41: 110–115.

Tyagi, M., M. Tomar, and V. Gupta, 2014. Glad assisted synthesis of NiO nanorods for realization of enzymatic reagentless urea biosensor. *Biosensors and Bioelectronics.* 52: 196–201.

Umar, A., M.M. Rahman, A. Al-Hajry et al., 2009. Highly-sensitive cholesterol biosensor based on well-crystallized flower-shaped ZnO nanostructures. *Talanta.* 78: 284–289.

Van Hieu, N., P. Thi Hong Van, L. Tien Nhan et al., 2012. Giant enhancement of H_2S gas response by decorating n-type SnO_2 nanowires with p-type NiO nanoparticles. *Applied Physics Letters.* 101: 253106.

Van Quy, N., V.A. Minh, N. Van Luan et al., 2011. Gas sensing properties at room temperature of a quartz crystal microbalance coated with ZnO nanorods. *Sensors and Actuators B: Chemical.* 153: 188–193.

Vuong, N.M., D. Kim, and H. Kim, 2015. Porous Au-embedded WO_3 nanowire structure for efficient detection of CH_4 and H_2S. *Scientific Reports.* 5: 11040.

Wang, C., X. Tan, S. Chen et al., 2012. Highly-sensitive cholesterol biosensor based on platinum–gold hybrid functionalized ZnO nanorods. *Talanta.* 94: 263–270.

Wang, X., C.J. Summers, and Z.L. Wang, 2004. Mesoporous single-crystal ZnO nanowires epitaxially sheathed with Zn_2SiO_4. *Advanced Materials.* 16: 1215–1218.

Wang, Y. and J.T. Yeow, 2009. A review of carbon nanotubes-based gas sensors. *Journal of Sensors.* 2009.

Wang, Y., M. Tang, X. Lin et al., 2012. Sensor for hydrogen peroxide using a hemoglobin-modified glassy carbon electrode prepared by enhanced loading of silver nanoparticle onto carbon nanospheres via spontaneous polymerization of dopamine. *Microchimica Acta.* 176: 405–410.

Wei, A., L. Pan, and W. Huang, 2011. Recent progress in the ZnO nanostructure-based sensors. *Materials Science and Engineering: B.* 176: 1409–1421.

Wei, N., X. Xin, J. Du et al., 2011. A novel hydrogen peroxide biosensor based on the immobilization of hemoglobin on three-dimensionally ordered macroporous (3DOM) gold-nanoparticle-doped titanium dioxide (GTD) film. *Biosensors and Bioelectronics.* 26: 3602–3607.

Wei, Y., Y. Li, X. Liu et al., 2010. ZnO nanorods/Au hybrid nanocomposites for glucose biosensor. *Biosensors and Bioelectronics.* 26: 275–278.

Xia, Y., J. Wang, X. Li et al., 2016. Nanoseed-assisted rapid formation of ultrathin ZnO nanorods for efficient room temperature NO_2 detection. *Ceramics International.* 42: 15876–15880.

Xu, J., D. Wang, L. Qin et al., 2009. SnO_2 nanorods and hollow spheres: Controlled synthesis and gas sensing properties. *Sensors and Actuators B: Chemical.* 137: 490–495.

Xu, S., H. Qi, S. Zhou et al., 2014. Mediatorless amperometric bienzyme glucose biosensor based on horseradish peroxidase and glucose oxidase cross-linked to multiwall carbon nanotubes. *Microchimica Acta.* 181: 535–541.

Ya-Qiao, W., H. Ming, and W. Xiao-Ying, 2014. A study of transition from n-to p-type based on hexagonal WO_3 nanorods sensor. *Chinese Physics B.* 23: 040704.

Yang, J., M. Cho, and Y. Lee, 2016. Synthesis of hierarchical $NiCo_2O_4$ hollow nanorods via sacrificial-template accelerate hydrolysis for electrochemical glucose oxidation. *Biosensors and Bioelectronics.* 75: 15–22.

Yang, Y., L. Xiao, Y. Zhao et al., 2008. Hydrothermal synthesis and electrochemical characterization of a-MnO$_2$ nanorods as cathode material for lithium batteries. *International
Journal of Electrochemical Science*. 3: 67–74.

Yang, Y., T. Yu, C. Jin et al., 2013. Effect of growth conditions on morphology and photoluminescence properties of ZnO nanowires fabricated by thermal evaporation without a
metal catalyst. In *Advanced Materials Research*. Trans Tech Publications.

Zhang, J., X. Liu, G. Neri et al., 2016. Nanostructured materials for room-temperature gas
sensors. *Advanced Materials*. 28: 795–831.

Zhang, J., Z. Qin, D. Zeng et al., 2017. Metal-oxide-semiconductor based gas sensors:
Screening, preparation, and integration. *Physical Chemistry Chemical Physics*. 19:
6313–6329.

Zhang, Y., M.K. Ram, E.K. Stefanakos et al., 2012. Synthesis, characterization, and applications of ZnO nanowires. *Journal of Nanomaterials*. 2012: 20.

Zheng, W., Y.F. Zheng, K.W. Jin et al., 2008. Direct electrochemistry and electrocatalysis of
hemoglobin immobilized in TiO$_2$ nanotube films. *Talanta*. 74: 1414–1419.

Zhou, F., W. Jing, Q. Wu et al., 2016. Effects of the surface morphologies of ZnO nanotube arrays on the performance of amperometric glucose sensors. *Materials Science in
Semiconductor Processing*. 56: 137–144.

Zou, C., J. Wang, and W. Xie, 2016. Synthesis and enhanced NO$_2$ gas sensing properties of
ZnO nanorods/TiO$_2$ nanoparticles heterojunction composites. *Journal of Colloid and
Interface Science*. 478: 22–28.

Zou, X., P.-P. Wang, C. Li et al., 2014. One-pot cation exchange synthesis of 1D porous
CdS/ZnO heterostructures for visible-light-driven H$_2$ evolution. *Journal of Materials
Chemistry A*. 2: 4682–4689.

9 Titanium Dioxide Nanotube Arrays for Solar Harvesting Applications to Address Environmental Issues

Srimala Sreekantan, Khairul Arifah Saharudin,
Nyein Nyein, and Zainovia Lockman

CONTENTS

9.1 INTRODUCTION

Global warming is an important issue faced by humankind, which causes an unprecedented onslaught of deadly and costly weather disasters, such as severe storms, droughts, heat waves, rising seas, and floods throughout the world. The combustion of fossil fuel from human activities, such as transportation and industry, massively increase the energy consumption thus contributing to global warming by releasing billions of tons of greenhouse gases into the atmosphere such as carbon dioxide (CO_2), methane (CH_4), nitrous oxide (N_2O), and fluorinated gases. Statistical data of fossil fuel consumption and CO_2 emission reported by British Petroleum (BP) shows that fossil fuel consumption increases continuously year by year and the overall consumptions are estimated at approximately eight times higher than alternative sources, for example nuclear- and hydro-energy (Maggio and Cacciola 2012).

High consumption of fossil fuel also leads to an energy crisis which strongly impacts the economic globalization by rising energy prices. And also, the emissions of effluents, gaseous, or wasted water from human activities that contribute deleterious effects on the Earth ecosystems (Akpan and Hameed 2009). Pursuant to aforementioned issues, it is indispensable to explore sustainable solutions to save Earth and mankind. Over past decades, abundant resources, including geothermal, solar, and wind have drawn considerable attention as alternative resources for inexhaustible and non-polluting "green" energy. However, they can only subsidize approximately 15% of overall electricity generated. Solar energy, radiant light and heat from the sun, is the most abundant clean energy source available. It is a fact that the solar energy striking the Earth in one hour is relatively higher than the energy consumed by humans for an entire year (Lewis 2007). Hence, extensive research and development of materials that can efficiently harvest solar irradiation and convert it to clean and renewable energy are essential.

Photocatalysis which utilizes solar energy to activate the chemical reactions via oxidation and reduction is a sustainable technology to provide solutions for energy crisis and environmental issues (Kudo and Miseki 2009, Vinu and Madras 2012). Photoelectrochemical (PEC) water splitting (Fujishima and Honda 1972) and dye-sensitized solar cells (DSSC) (O'regan and Grätzel 1991) appear as potential means to convert solar energy to hydrogen (H_2) fuel and electricity, respectively. Several types of photocatalysts, such as TiO_2, ZnO, Fe_2O_3, ZrO_2, V_2O_5, Nb_2O_5, and WO_3, have been used in PECs and DSSCs. Among them, TiO_2 is one of the most promising candidates because of its superior properties, e.g., light absorption capabilities, chemical inertness, and stability (Ni et al. 2007, Chen et al. 2007) with PECs and DSSCs. Furthermore, its strong oxidation nature offers a great ability in detoxifying hazardous waste and air contaminant (Fujishima and Honda 1972).

9.2 ONE-DIMENSIONAL NANOSTRUCTURED MATERIALS

One-dimension (1-D) nanostructured material has been discovered (Iijima 1991), and it has trigged enormous efforts in physics, chemistry, and materials science. TiO_2, with a well-aligned nanotubular structure, provides unique electronic properties, such as high e^- mobility, low quantum confinement effects, high specific surface area, excellent ability to harvest photon energy (hv), and high mechanical strength (Mohapatra et al. 2007a, Baker and Kamat 2009, Roy et al. 2011b). Furthermore, vectorial charge transport facilitates the photoelectrochemical properties and photocatalytic efficiency (Baker and Kamat 2009, Mohamed and Rohani 2011). These advantages render TiO_2 nanotube arrays as a promising candidate for various applications, such as pollutant detoxification (Chen et al. 2009, Sreekantan et al. 2010, Sangpour et al. 2010), PECs (Park et al. 2006, Mohapatra et al. 2007a, Palmas et al. 2010) and DSSCs (Jennings et al. 2008, Lei et al. 2010, Premalal et al. 2012, Chang et al. 2012). The performance of PECs and DSSCs are greatly determined by structural morphologies of nanotube arrays. Hence, related literatures on the development of TiO_2 nanotube arrays for solar harvesting applications are elaborated in this chapter.

9.3   SOLAR ENERGY HARVESTING TECHNOLOGIES

Innovations in materials science related to photoelectrochemical technology play a key role in the paradigm shift from fossil fuel to clean and renewable sources (Varghese et al. 2009). Photoelectrochemcials (Fujishima and Honda 1972) and DSSCs (O'regan and Grätzel 1991) have been recognized as potential technologies to aforementioned transition by generating clean and renewable energy from solar energy. Photoelectrochemical solar energy conversion devices involve photosensitizers that absorb light and engage in electron-transfer reaction (Kalyanasundaram and Grätzel 1998). Figure 9.1 illustrates the main steps involved in the photocatalytic water splitting process under illumination.

Numerous n-type semiconductors, such as TiO_2, ZnO, Fe_2O_3, ZrO_2, V_2O_5, Nb_2O_5, and WO_3 have been integrated for PECs and DSSCs as photoanodes (Kalyanasundaram and Grätzel 1998, Ghicov and Schmuki 2009). However, relative positions of band edges play a key role in determining their chemical applications. For PEC, the bottom level of the conduction band (C_B) edge needs to be more negative than the redox potential of H^+/H_2 (0 V vs. normal hydrogen electrode: NHE), while the top level of the valence band (V_B) be more positive than the redox potential of H_2O/O_2 (c.a., 1.23 eV) (see Figure 9.2). Therefore, the theoretical minimum band gap

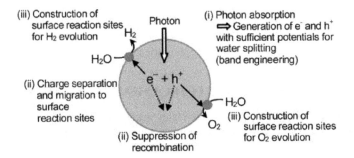

FIGURE 9.1   Main processes in photocatalytic water splitting. (Reprinted with permission from Kudo, A. and Miseki, Y., *Chem. Soc. Rev.*, 38, 253–278, 2009. Copyright 2008, Royal Society of Chemistry.)

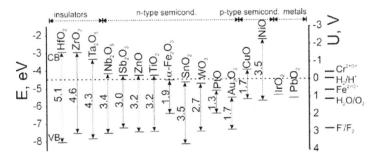

FIGURE 9.2   Electronic structures of different metal oxides and relative position of their band edges vs. some important redox potential. (Reprinted with permission from Ghicov, A. and Schmuki, P., *Chem. Commun.*, 2791–2808, 2009. Copyright 2009, The Royal Society of Chemistry.)

energy for water splitting is 1.23 eV, that corresponds to the wavelength (λ) of light at approximately 1,100 nm (Ghicov and Schmuki 2009, Kudo and Miseki 2009).

Among qualified candidates, TiO_2 is the most promising candidate to be used as a photoelectrode because of its superior properties, for example strong oxidation, high stability, and against photocorrosion (Chen et al. 2007, Varghese et al. 2009, Lei et al. 2010). TiO_2 exists in three different structures (anatase, rutile, and brookite). The anatase and rutile phases have been widely used for solar energy harvesting applications (Mohapatra et al. 2007a, Kang et al. 2009, Premalal et al. 2012). The band gap energies of anatase and rutile are 3.2 and 3.0 eV, corresponding to absorption thresholds at 390 and 415 nm, respectively (Rao et al. 1980). This infers that rutile phase can absorb a more extensive range of light. The V_B positions of these two phases is almost the same (*c.a.* 3.0 V), but the C_B positions are slightly different. The C_B potential of rutile is slightly different from the NHE potential, whereas anatase is shifted cathodically by 0.2 V. This indicates a relatively higher driving force of anatase phase for the reduction of water as compared to the rutile phase (Leung et al. 2010).

Remarkable advantages of anatase TiO_2 render them as efficient photoanodes in DSSCs (O'regan and Grätzel 1991, Jennings et al. 2008, Wang and Lin 2009, Lei et al. 2010). However, the photoconversion efficiency of DSSCs is strongly dependent on the dye semiconductor and a redox mediator (Kalyanasundaram and Grätzel 1998, Adachi et al. 2007). Therefore, it is dispensable to understand the operation principle of PEC and DSSC in order to utilize TiO_2 nanotube arrays as efficient photoanodes for solar energy harvesting applications.

9.4 PRINCIPLE OF PHOTOELECTROCHEMICAL WATER SPLITTING

Hydrogen is an ideal, renewable, and clean energy carrier and one of the most promising alternates for fossil fuel in the future; thus, environmental issues related to global warming could be addressed. The utilization photocatalyst for solar energy harvesting using PEC is essential for sustainable H_2 production (Zhang et al. 2010, Gong et al. 2010). The PEC system consists of a semiconductor working electrode (i.e., nanocrystalline TiO_2 film) and platinum (Pt) counter electrode, both immersed in an aqueous electrolyte (Fujishima and Honda 1972). Upon illumination with sufficient light energy ($h\nu$), electrons (e^-) in the V_B are excited into the C_B, creating electron-hole pairs in the semiconductor. The e^- are transported from the C_B via an external wire to the Pt cathode where the H_2 evolution reaction occurs through the reduction of water (H_2O). On the other hand, the holes (h^+) are transported to the photoanode surface where they oxidize H_2O to produce oxygen gas (O_2). By this means, H_2 and O_2 are produced at different electrodes and can be collected in separate storage volumes. The electrochemical reactions at the cathode and photoanode are:

$$TiO_2 + h\nu \rightarrow e^- + h^+ \qquad \text{(Reaction 9.1)}$$

Anode: $$2H_2O + 4h^+ \rightarrow O_2 + 4H^+ \qquad \text{(Reaction 9.2)}$$

Cathode: $$2H^+ + 2\,e^- \rightarrow H_2 \qquad \text{(Reaction 9.3)}$$

Overall: $$2H_2O + 4h\nu \rightarrow 2H_2 + O_2 \qquad \text{(Reaction 9.4)}$$

In addition, the flat band potential (V_{fb}) at the photoanode/electrolyte interface is also essential to determine the efficiency of PEC. The band bending at the photoanode/electrolyte interface is a result of solid-electrolyte interface phenomena. The Fermi level (E_F) in a semiconductor and electrochemical reactions in the electrolyte (E_{redox}) are equal at the equilibrium state. The space charge region is formed at the interface. This space charge region provides a strong electric field that is indispensable for an effective separation of photoexcited e^- from h^+. On the other hand, if light is absorbed in the bulk of the photoanode, the photoexcited e^- and h^+ are created, but there is a high possibility for recombination to occur during water photolysis (Radecka et al. 2008).

9.5 PRINCIPLE OF DYE-SENSITIZED SOLAR CELLS

The dye-sensitized solar cell has been recognized as a viable competitor to the well-developed silicon solar cell which is relatively expensive (Varghese et al. 2009, Lei et al. 2010). A typical DSSC is assembled with a nanocrystalline TiO_2 film on fluorine-doped tin oxide (FTO) glass that is covered with a monolayer of dye molecules, redox mediator, and Pt-coated FTO glass (O'regan and Grätzel 1991, Wang and Lin 2009). Several types of ruthenium bipyridyl dyes (i.e., black dye, N3, and N719) are often used as the dye, while iodide/tri-iodide (I^-/I_3^-) redox electrolyte is used as the redox mediator (Adachi et al. 2007, Lei et al. 2010, Zhang et al. 2011). Considerable efforts have been devoted to develop TiO_2 nanotube arrays on FTO glass and used as a photoanode (TiO_2/FTO) for front-side illumination (see Figure 9.3a) (Varghese et al. 2009, Lei et al. 2010). However, the fabrication of TiO_2/FTO involves the sputtering of Ti-films on FTO glass, thus leading to high fabrication costs (Lei et al. 2010).

TiO_2 nanotube arrays on Ti substrate (TiO_2/Ti) have later been developed to overcome the aforementioned problem (Kuang et al. 2008). TiO_2/Ti-based DSSCs require back-side illumination (Figure 9.3b), which limits the enhancement of photoelectrochemical performance because the platinized-counter electrode partially reflects light, and also induces high resistance at the metal/oxide interface (Varghese et al. 2009, Wang and Lin 2009).

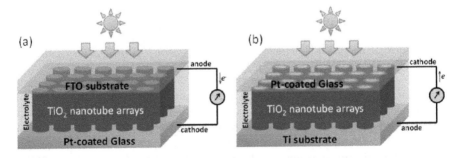

FIGURE 9.3 Schematic diagrams of (a) front-side and (b) back-side illumination modes DSSCs using TiO_2 nanotube arrays. (Reprinted with permission from Lei, B.-X. et al., *J. Phys. Chem. C*, 114, 15228–15233, 2010. Copyright 2010, American Chemical Society.)

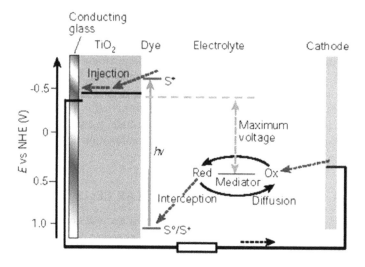

FIGURE 9.4 Schematic of operation of the DSSC. (With kind permission from Springer Science+Business Media: *Nature*, Photoelectrochemical cells, 414, 338–344, 2001, Grätzel, M. Copyright 2001, Springer Nature.)

Figure 9.4 illustrates the reaction in DSSC under illumination. Optical excitation of the dye with $h\upsilon$ leads to the transition of e^- in the dye from equilibrium state to an excited state (Reaction 9.5). Photoexcited e^- rapidly moves from dye into the C_B of TiO_2, thereby remaining a positively charged state (Reaction 9.6). The dye accepts e^-, then regenerate I_3^- from $3I^-$ in the electrolyte (Reaction 9.7). The I_3^- formed in the dye regeneration process diffuse through the liquid phase to the cathode, where they are reduced back to $3I^-$ to complete the cycle by receiving an e^- from redox mediator (see Reaction 9.8) (Kalyanasundaram and Grätzel 1998, Adachi et al. 2007, Jennings et al. 2008).

$$Dye + h\upsilon \rightarrow Dye* \qquad \text{(Reaction 9.5)}$$
$$Dye* + TiO_2 \rightarrow TiO_2^{\cdot} + Dye^+ \qquad \text{(Reaction 9.6)}$$
$$Dye^+ + 3I^- \rightarrow Dye + I_3^- \qquad \text{(Reaction 9.7)}$$
$$I_3^- + CE^{-\cdot} \rightarrow 3I^- + CE \qquad \text{(Reaction 9.8)}$$

In either model, the e^- may be localized near the surface or in the bulk, and during their transit the e^- may partially loss by transfer across the solid/liquid interface to I_3^-. The efficiency of collecting the photoexcited e^-, which is determined by competition between electron transport to the anode and electron transfer to I_3^- in the electrolyte, is critical to enhance the device performance (Jennings et al. 2008).

9.6   ANODIC GROWTH OF SELF-ORGANIZED TiO$_2$ NANOTUBE ARRAYS

TiO$_2$ nanotubes have been fabricated through various methods, such as sol-gel (Maiyalagan et al. 2006, Kang et al. 2009), hydrothermal (Yu et al. 2008, Sreekantan and Wei 2010), chemical vapor deposition (Hsieh et al. 2010) and electrochemical anodization (Gong et al. 2001, Macak and Schmuki 2006, Rohani 2009). However, electrochemical anodization has been recognized as an efficient and facile approach to produce integrative, vertically oriented, highly ordered nanotube arrays with controllable structural morphologies without any additional process (Roy et al. 2011a, Mohamed and Rohani 2011).

Electrochemical anodization can be defined as well-desired electrochemical growth of an oxide film on a metal substrate by polarizing the metal anodically in an electrochemical cell. Anodization can be conducted by (1) applying a constant potential difference between anode and cathode: potentiostatic mode, (2) by imposing constant current: galvanostatic mode, or (3) by sweeping the anode potential at given rate: potentiodynamic mode (Pasquale et al. 2002, Kaneco et al. 2007, Miraghaei et al. 2011, Vanhumbeeck and Proost 2009). Potentiostatic anodization has widely been used to grow oxide films due to its extreme simplicity (Zwilling et al. 1999, Mohapatra et al. 2007a, Baker and Kamat 2009, Roy et al. 2011a). Zwilling et al. (1999) first demonstrated the formation of self-organized porous TiO$_2$ via potentiostatic anodization of Ti in chromic acid (H_2CrO_4) containing hydrofluoric acid (HF). The obtained tube-like structure was not highly organized, and it showed considerable sidewall inhomogeneity. From this origin, the presence of fluoride ions (F^-) in electrolyte has been recognized as an essential in anodic growth of self-organized oxide structures. The anodic growth of self-organized structures mainly involves electrochemical oxidation and chemical dissolution. A compact oxide layer initially forms on the metal surface; M involving metal ion formation; M^{z+} (Reaction 9.9), and consequently reacted with oxide ion (O^{2-}) and/or hydroxyl ions (OH^-) via field-assisted oxidation (see Reactions 9.10 through 9.12) (Shankar et al. 2007, Roy et al. 2011a). The oxide formation mechanism on a metal is illustrated in Figure 9.5a.

$$M \rightarrow M^{z+} + ze^- \qquad \text{(Reaction 9.9)}$$

$$M + \tfrac{z}{2}H_2O \rightarrow MO_{z/2} + zH^+ + ze^- \qquad \text{(Reaction 9.10)}$$

$$M^{z+} + zH_2O \rightarrow M(OH)_z + zH^+ \qquad \text{(Reaction 9.11)}$$

$$M(OH)_z \rightarrow MO_{z/2} + \tfrac{z}{2}H_2O \qquad \text{(Reaction 9.12)}$$

In principle, four different morphologies, including compact structure, random porous structure, oriented porous structure, and oriented tubular structure can be obtained by anodization (Figure 9.5b). Chemical dissolution plays an important role to determine the structural morphologies of anodic oxide. After the formation of an

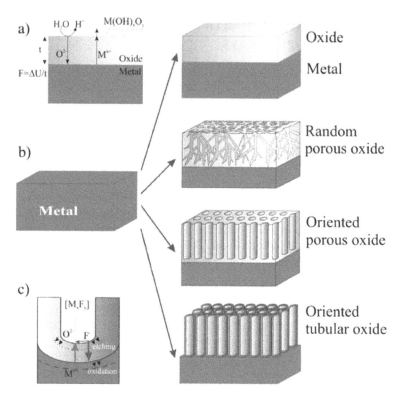

FIGURE 9.5 (a) Oxide formation mechanism on a metal, (b) morphologies which can be obtained by anodization of Ti metal—a compact oxide film, a disordered nanoporous layer, a self-ordered nanoporous or a self-ordered nanotube layer, (c) nanotube formation mechanism. (Reprinted with permission from Ghicov, A. and Schmuki, P., *Chem. Commun.*, 2791–2808, 2009. Copyright 2009, The Royal Society of Chemistry.)

initial oxide layer, the O^{2-} and/or OH^- ions migrate through the oxide layer reaching the metal/oxide interface and then react with the metal. A high electric field at the initial oxide layer dominates the polarization of the $M^{z+}-O$ bond (Mor et al. 2006, Zhang and Han 2010). The M^{z+} consequently migrates from the metal at the metal/oxide interface and moves outwards the oxide/electrolyte interface.

Besides, the presence of F^- in electrolyte strongly affects the anodization process, as F^- subsequently forms soluble complex $[MF_z]^{2-}$ species (Figure 9.5c). On the other hand, complexation occurs with M^{z+} ion migrated from the oxide/electrolyte interface (Reaction 9.13), and consequently attaches to the formed oxide (see Reaction 9.14) (Ghicov and Schmuki 2009, Roy et al. 2011a).

$$M^{z+} + zF^- \rightarrow [MF_z]^{2-} \qquad \text{(Reaction 9.13)}$$
$$MO_z + zF^- + zH^+ \rightarrow [MF_z]^{2-} + H_2O \qquad \text{(Reaction 9.14)}$$

Gong et al. (2001) investigated the significance of electrochemical dissolution and chemical dissolution in aqueous electrolyte on the anodic growth of TiO_2 oxide by varying applied potential over the range of 5–20 V, and HF concentration of 0.5–3.5 wt%. Porous and particulate structure was formed in aqueous electrolyte containing 0.5 wt% HF under applied potential lower than 10 V. Self-organized tubular structure was successfully formed at 10–20 V. It was noticeable that an appropriately applied potential for the formation of TiO_2 nanotube arrays is disproportional to the HF concentration. These reveal the essentials of anodization parameters on the formation and structural morphologies of TiO_2 nanotube arrays.

9.7 CRYSTALLIZATION OF TiO_2 NANOTUBE ARRAYS

As-anodized TiO_2 nanotube arrays are amorphous in nature, which exhibit poor electrical, optical, and mechanical properties (Alivov et al. 2009, Oh et al. 2011). The crystallinity and the isomorph present at the desired operating conditions greatly determine their potential applications. Anatase phase is preferred in PEC (Raja et al. 2006, Mohapatra et al. 2007b), DSSC (Kang et al. 2009, Premalal et al. 2012) and catalyst (Sreekantan et al. 2010), while rutile is mostly used in the area of dielectrics (Kim et al. 2005) and high-temperature oxygen gas sensors (Lu et al. 2008). Hence, the comprehension on influence of heat treatment temperature and atmosphere on the crystal structure, and photoelectrochemical and photocatalytic efficiency of TiO_2 nanotube arrays are essential for achieving high-efficiency PEC and DSSC devices.

Varghese et al. (2003) investigated the significance of heat treatment conditions on the crystallization and structural morphologies of TiO_2 nanotube arrays formed in aqueous solution containing 0.5 wt% HF by annealing at temperatures of 230°C–880°C in dry oxygen (O_2), and argon (Ar) atmospheres. Under O_2 atmosphere, TiO_2 nanotube arrays crystallized in the anatase phase at ~280°C. The rutile phase emerged at ~430°C and completely crystallized at temperatures of 620°C–680°C. No discernible changes in structural morphologies were observed at temperature lower than 580°C. Small protrusions emerged through the tubular structure at 550°C–580°C, and dominated uniformly at 680°C. Above this temperature, the tubular structure collapsed and formed dense rutile crystallites at the oxide/metal interface. Similar behavior was reported for TiO_2 nanotube arrays formed in organic-based electrolytes (i.e., ethylene glycol (EG) and glycerol) containing NH_4F (Yang et al. 2008, Yang et al. 2011, Sun et al. 2011). However, the anatase-to-rutile transformation temperature in TiO_2 nanotube arrays formed in organic-based electrolytes was observed at ~550°C, which is relatively higher than that formed in aqueous solution (Varghese et al. 2003). Heat treatment temperature also greatly affects the diffusion of adsorbed carbonate species from EG electrolyte into the structure (Yang et al. 2011), and thus determines the amount of H_2 production in the PEC cell (Sun et al. 2011). The evolution rate of H_2 was proportional with annealing temperature, and reached a maximum of 122 µmol h^{-1} cm^{-2} at 450°C. The formation of the rutile phase at a higher temperature resulted in the reduction of the H_2 evolution due to a greater recombination of photogenerated e^-/h^+ as compared to anatase phase (Paulose et al. 2006, Gong et al. 2010).

Annealing atmosphere also plays an essential role in determining the photoelectrochemical properties of TiO_2 nanotube arrays by controlling crystallization and defect structure. Inert atmosphere has also been used to crystallize TiO_2 nanotube arrays (Ghicov et al. 2006, Salari et al. 2011, Saharudin et al. 2014). However, annealing in inert atmosphere such as Ar caused structural shrinkage at 480°C and started to collapse at 580°C onwards. Moreover, Salari et al. (2011) explained that an annealing in a neutral oxygen-free environment (Ar, N_2, and H_2) induced partial reduction of Ti^{4+} to lower valence Ti^{3+}. At the same time, $V_O^{\bullet\bullet}$ can be generated in the TiO_2 structure due to the partial loss of oxygen. The presence of $V_O^{\bullet\bullet}$ accelerates the rearrangement of Ti–O bond in anatase phase to form the rutile phase (Salari et al. 2011, Hardcastle et al. 2011). The presence of $V_O^{\bullet\bullet}$ was also found to retard the photoelectrochemical properties due to high recombination of photogenerated e^-/h^+ (Ghicov et al. 2006).

High-temperature crystallization of the anatase structure hinders the integration of TiO_2 nanostructures with polymeric substrates or temperature-sensitive devices (Allam et al. 2008). Matsuda and co-workers (2000) have suggested a facile method to crystallize various types of oxide, for example SiO_2-TiO_2 (Matsuda et al. 2000, 2006, Kotani et al. 2001) and ZrO_2 nanoparticles (Prastomo et al. 2010, Sakamoto et al. 2011), and TiO_2 thin films by immersion of these oxides in hot water. The resulting products showed superior characteristics, such as improved photocatalytic activity, optical properties, and wettability. Several studies have suggested that the adsorbed-OH species on the photocatalyst surfaces could stimulate the photocatalytic efficiency (Schindler and Gamsjäger 1972, Nagaveni et al. 2004, Du et al. 2008). There are also works done to increase the –OH species by anodizing TiO_2 in a rather alkaline electrolyte (Taib et al. 2017).

9.8   PHOTOELECTROCHEMICAL PERFORMANCE

The overall splitting of water using solar energy has drawn considerable attention in view of the current importance of hydrogen as an efficient renewable resource for inexhaustible as well as non-polluting "green" energy. To bring hydrogen to the point of commercial readiness and viability in terms of performance and cost, substantial research on the development of n-type semiconductor have become a major topic of current research in photoelectrochemical cells (PEC) (Grätzel 2001, Paulose et al. 2006). TiO_2 nanotubes are one of the most promising candidates in this n-type semiconductor category. Figure 9.6 shows the schematic diagram of the water-splitting system using TiO_2 nanotubes and Figure 9.7 clearly shows the linear increase in the volume of evolved H_2 with increasing irradiation time. The samples with tubular structures (Figure 9.7b–f) exhibited a significantly higher H_2 evolution rate than those with a mixture of porous-like and cylindrical structures (Figure 9.7a). This result is due to the greater capability of the tubular structure to harvest the photon excitation and produce larger charge carrier density. The photogenerated h^+ subsequently reacted with the OH species at the photoelectrode/electrolyte interface through the reaction; $2H_2O + 2h^+ \rightarrow O_2 + 4H^+$, thus inducing H_2 evolution at the Pt electrode via the reaction; $2H^+ + 2e^- \rightarrow H_2$. Photoelectrodes with high

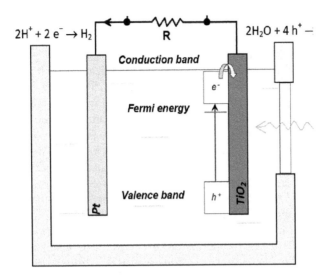

FIGURE 9.6   Circuit of PEC cell for water splitting.

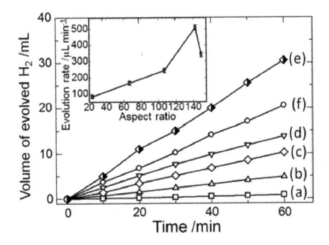

FIGURE 9.7 H_2 evolution of the TiO_2 nanotubes with aspect ratio of (a) 27.8, (b) 23.6, (c) 64.1, (d) 109.4, (e) 142.5, and (f) 193.2 under illumination. The inset shows a plot of the evolution rate of the samples with different aspect ratios.

aspect ratios, thick walls, and large pore sizes contributed to better photocatalytic efficiency. The results show that the capability of the photoelectrode for H_2 evolution is proportional to the aspect ratio (Krengvirat et al. 2012). A maximum evolution rate of 508.3 mL min^{-1}cm^{-2} was achieved from the photoelectrode with an aspect ratio of 142.5 (inset in Figure 9.7). However, the photoelectrode with an ultra-high aspect ratio (193) showed a significant decrease in H_2 evolution rate. This result is ascribed to the high recombination of the charge carriers in photoelectrodes with ultra-high aspect ratios. Therefore, photoelectrodes with nanotubular structures with an aspect ratio of 142.5 exhibit superior photoelectrochemical properties and possess high H_2 generation capability.

Park et al. (2006) also have examined the significance of structural morphology, i.e., TiO_2 nanotube and nanoporous, as well as P25 nanoparticulate film on the photoelectrochemical efficiency under simulated solar irradiation. The photocurrent density (j_p) of TiO_2 nanotube arrays with length of 2 μm was more than 60% higher than that of TiO_2 nanoporous with similar thickness. Furthermore, TiO_2 nanotube arrays exhibited the j_p more than 10 times higher than a 15 μm nanoparticulate film. In addition, Palmas et al. (2010) found that tubular structure exhibited j_p approximately 10 times higher, while charge carrier density was more than 40 times greater than compact oxide film. These reveal the significance of the tubular structure on the hv harvesting and charge transport.

Structural characteristics, including pore size, wall thickness, and nanotube length, also greatly affect the performance of PEC. The nanotube arrays with a larger pore size allow better light penetration, thus enabling the diffusive transport of photogenerated h^+ to react with OH^- in the electrolyte (Ruan et al. 2005, Wan et al. 2009, Sun et al. 2010). In addition, the nanotube arrays with thick walls allow the formation of a space charge. The potential barriers are created near both the inner and outer surfaces of the nanotubes by the bending of C_B and V_B. These barriers inhibit the e^- transfer into the redox couples in the electrolyte and thus contribute to better photoelectrochemical properties by minimizing e^- loss (Sun et al. 2010). In addition, long nanotubes provide a large reaction site between the photocatalyst and the electrolyte (Sun et al. 2011), resulting in the enhancement of photoconversion efficiency (η). However, the nanotube arrays with ultra-long nanotube lengths and very high aspect ratios did not fully absorb hv, and thus induced the recombination of the charge carriers (Paulose et al. 2006, 2007).

Shankar et al. (2007) demonstrated the influence of structural characteristics of TiO_2 nanotubes on the photoelectrochemical properties under UV irradiation. Different anodization voltages resulted in a variation of nanotube array length and tube outer diameter: 15 V resulted in a nanotube array with 8.2 μm long, and 80 nm outer diameter; 20 V resulted in a nanotube array with 14.4 μm long, and 94 nm outer diameter; 25 V resulting in a nanotube array 16 μm long, and 140 nm outer diameter. The 20 V nanotubes achieved a photoconversion efficiency of 14.42%. A tradeoff between nanotube length and internal surface area was found to play an important role in PEC performance. The light absorption of the 15 V nanotubes with a large internal surface area is limited due to short nanotubes. Similarly, the 25 V with long nanotubes also have limited light absorption due to less surface area. These indicate the critical role of aspect ratio and surface area on the light absorption and electrochemical properties. A maximum η was obtained for nanotubes anodically formed at 20 V (Shankar et al. 2007).

The photoelectrochemical performance of water-treated TiO_2 nanotube arrays were investigated in comparison to amorphous, thermally and hydrothermally annealed TiO_2 nanotube arrays under AM 1.5 solar irradiation (Liu et al. 2012). Thermally annealed TiO_2 nanotube arrays yielded the highest photocurrent, followed by the hydrothermal treatment, water-treated, and as-formed (amorphous) nanotube arrays.

However, it was found that water-treated nanotube arrays have comparable amounts of dye absorption to thermally annealed nanotube arrays, which is approximately three times higher than as-formed nanotube arrays. This provides an insight that the incorporation between water treatment and thermal annealing would provide significant improvement in the photoelectrochemical performance (Liu et al. 2012).

9.9 DYE-SENSITIZED SOLAR CELLS PERFORMANCE

The illumination modes of DSSCs, i.e., front-side and back-side illuminations, significantly affect their efficiency (Varghese et al. 2009, Wang and Lin 2009). Lei and co-workers (2010) investigated the significance of illumination modes on the performance of DSSCs. They found that back-side illumination DSSC with 20.8 μm long exhibited $V_{oc} = 0.820$ mA cm^{-2}, $J_{sc} = 13.18$ mA cm^{-2}, $FF = 0.675$ and $\eta = 7.29\%$, which is relatively lower than that of front-side DSSC; $V_{oc} = 0.814$ mA cm^{-2}, $J_{sc} = 15.46$ mA cm^{-2}, $FF = 0.641$ and $\eta = 8.07\%$ (Figure 9.8a).

Structural morphologies of TiO$_2$ nanotube arrays, including nanotube length and pore size, play significant roles on the η of DSSCs. Long nanotube arrays with large pore sizes could absorb larger amounts of dye (Sun et al. 2010), thus increasing performance, i.e., J_{sc} and η (Shankar et al. 2007, Kuang et al. 2008, Chen et al. 2009). Higher concentrations of adsorbed-dye resulted in higher incident-to-photon conversion efficiency; IPCE (Macak et al. 2005). The increase in the internal resistance of the cell with thick oxide however resulted in the decrease of V_{oc} and FF (Kalyanasundaram and Grätzel 1998, Adachi et al. 2007). Furthermore, the increase in surface area provided additional charge-recombination sites and thus lowered the V_{oc}. Lei et al. (2010) investigated the influence of nanotube length on the performance of front-side illumination DSSCs under AM 1.5 solar irradiation (Figure 9.8b). The highest η of 8.07% was achieved for a 20.8 μm length TiO$_2$ nanotube arrays, with $V_{oc} = 0.814$ mA cm^{-2}, $J_{sc} = 15.46$ mA cm^{-2} and $FF = 0.64$. Shankar et al. (2007)

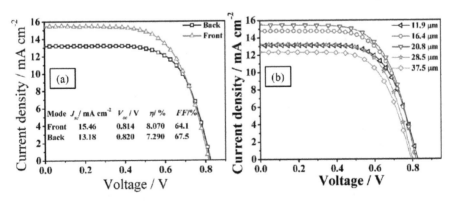

FIGURE 9.8 (a) comparison of I–V curves of TiO$_2$ nanotube arrays-based DSSCs in front-side and back-side illumination modes and (b) I–V curves of front-side irradiation DSSCs constructed using TiO$_2$ nanotube arrays with various lengths. (Reprinted with permission from Lei, B.-X. et al., *J. Phys. Chem. C*, 114, 15228–15233, 2010. Copyright 2010, American Chemical Society.)

evaluated photoelectrochemical performance of 30 back-side illumination DSSCs using TiO_2 nanotube array electrodes with different nanotube length, and found that nanotube arrays with 20 µm in length exhibited highest performance with $V_{oc} = 0.817$ mA cm^{-2}, $J_{sc} = 12.72$ mA cm^{-2}, $FF = 0.663$ and $\eta = 6.89\%$.

Lin et al. (2012) proved that water treatment for 1–3 days together with thermal annealing greatly contributed to the performance of DSSCs. The specific surface area of TiO_2 nanotube arrays increased from 20.2 m^2 g^{-1} to 39.9 and 42.7 m^2 g^{-1} after water treatment for 2 and 3 days. The rough surface of nanotubes covered with nanoparticles could absorb large amounts of dye, thus enhancing the performance (Figure 9.9) (Lin et al. 2012).

Alternatively, Liao et al. (2011) have suggested hot water treatment as an effective approach to crystallize TiO_2 nanotube arrays. Hot water treatment at 92°C induced the rapid formation of anatase crystallites. The structural transformation from nanotubes to nanotube-nanoparticle hybrid structure in hot water-treated nanotube arrays significantly increases specific surface area approximately two-fold as compared to that of thermal annealing. Furthermore, hot water treatment induces the OH adsorption on the nanotube surface, and thus contributes to the photocatalytic efficiency by instantaneous reaction at the surface of nanotubes (Schindler and Gamsjäger 1972, Nagaveni et al. 2004). These observations underline the significance of structural morphologies of nanotube arrays, heat treatment, and illumination mode on the performances.

Fabrication of TiO_2 nanotubes decorated with nanoparticles (NPs) of noble metals like silver (Ag) or gold (Au) has been thought to be beneficial to further improving photoelectric conversion efficiency of DSSCs due to the localized surface plasmon resonance of the NPs (Nyein et al. 2017). Decoration of Ag NPs into TiO_2 nanotube (TNT) arrays can be done by irradiating TiO_2 nanotube arrays immersed in Ag-precursor solution for a period of time. Generally, Ag NP-deposited TNT arrays

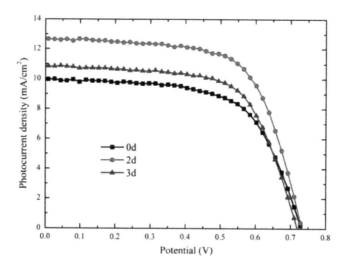

FIGURE 9.9 I–V curves from DSSCs constructed using TiO_2 nanotube layers of a length of 15 µm after water treatment for 0 day, 2 day, and 3 day. (Reprinted with permission from Lin, J. et al., *Nanoscale*, 4, 5148–5153, 2012. Copyright 2012, The Royal Society of Chemistry.)

should exhibit higher J_{sc} than TNTs without Ag NPs. However, there are many parameters that need to be considered, for example the size and distribution of the nanoparticles, interfaces properties, and purity of the nanoparticles added (Nyein et al. 2017). Moreover, the process for Ag NPs deposition need to be optimized as certain processes may affect the adherence of the TNTs to the substrate, thus hindering the flow of electrons to the outer circuit.

9.10 CONCLUSIONS

In summary, we have elaborated on electrochemical anodization for formation of TiO_2 nanotube arrays and the relationship within structural properties of TiO_2 nanotube arrays and their performance in solar energy harvesting applications such as PECs and DSSCs to address environmental issues. The key factors to increase the efficiency of TiO_2 photoanodes are (1) increase the surface area by controlling structural morphologies, (2) minimize the recombination of photogenerated e^-/h^+ by rapid charge separation, and (3) to extend the light absorption to the visible-light region by modifying band gap energy. This chapter provided a comprehensive review on the formation of TiO_2 nanotube arrays via anodization in aqueous electrolytes and the influences of heat treatment processes on the crystal structure as well as the performance of PECs and DSSCs.

ACKNOWLEDGMENT

The authors are thankful to the Ministry of Education (MOE) Malaysia for funding this work under Transdisciplinary Research Grant Scheme (TRGS) grant no. 6769002. The authors are very much grateful to Universiti Sains Malaysia (USM) for providing the necessary facilities to carry out the research work and financial support under Research University (RU) grant no. 814281.

REFERENCES

Adachi, M., J. Jiu, and S. Isoda. 2007. "Synthesis of morphology-controlled titania nanocrystals and application for dye-sensitized solar cells." *Current Nanoscience* 3 (4):285–295.
Akpan, U. G., and B. H. Hameed. 2009. "Parameters affecting the photocatalytic degradation of dyes using TiO_2-based photocatalysts: A review." *Journal of Hazardous Materials* 170 (2):520–529.
Alivov, Y., M. Pandikunta, S. Nikishin, and Z. Y. Fan. 2009. "The anodization voltage influence on the properties of TiO_2 nanotubes grown by electrochemical oxidation." *Nanotechnology* 20 (22):225602. doi:10.1088/0957-4484/20/22/225602.
Allam, N. K., K. Shankar, and C. A. Grimes. 2008. "A general method for the anodic formation of crystalline metal oxide nanotube arrays without the use of thermal annealing." *Advanced Materials* 20 (20):3942–3946.
Baker, D. R., and P. V. Kamat. 2009. "Photosensitization of TiO_2 nanostructures with CdS quantum dots: Particulate versus tubular support architectures." *Advanced Functional Materials* 19 (5):805–811.
Chang, S., Q. Li, X. Xiao, K. Y. Wong, and T. Chen. 2012. "Enhancement of low energy sunlight harvesting in dye-sensitized solar cells using plasmonic gold nanorods." *Energy & Environmental Science* 5 (11):9444–9448.

Chen, C. C., W. D. Jehng, L. L. Li, and E. W.-G. Diau. 2009. "Enhanced efficiency of dye-sensitized solar cells using anodic titanium oxide nanotube arrays." *Journal of the Electrochemical Society* 156 (9):C304–C312.

Chen, D., Z. Jiang, J. Geng, Q. Wang, and D. Yang. 2007. "Carbon and nitrogen co-doped TiO_2 with enhanced visible-light photocatalytic activity." *Industrial & Engineering Chemistry Research* 46 (9):2741–2746.

Du, P., A. Bueno-Lopez, M. Verbaas, A. R. Almeida, M. Makkee, J. A. Moulijn, and G. Mul. 2008. "The effect of surface OH-population on the photocatalytic activity of rare earth-doped P25-TiO_2 in methylene blue degradation." *Journal of Catalysis* 260 (1):75–80.

Fujishima, A., and K. Honda. 1972. "Electrochemical photolysis of water at a semiconductor electrode." *Nature* 238 (5358):37–38.

Ghicov, A., and P. Schmuki. 2009. "Self-ordering electrochemistry: A review on growth and functionality of TiO_2 nanotubes and other self-aligned MOx structures." *Chemical Communications* (20):2791–2808.

Ghicov, A., H. Tsuchiya, J. M. Macak, and P. Schmuki. 2006. "Annealing effects on the photoresponse of TiO_2 nanotubes." *Physica Status Solidi (a)* 203 (4):R28–R30.

Gong, D., C. A. Grimes, O. K. Varghese, W. C. Hu, R. S. Singh, Z. Chen, and E. C. Dickey. 2001. "Titanium oxide nanotube arrays prepared by anodic oxidation." *Journal of Materials Research* 16 (12):3331–3334.

Gong, J., Y. Lai, and C. Lin. 2010. "Electrochemically multi-anodized TiO_2 nanotube arrays for enhancing hydrogen generation by photoelectrocatalytic water splitting." *Electrochimica Acta* 55 (16):4776–4782.

Grätzel, M. 2001. "Photoelectrochemical cells." *Nature* 414 (6861):338–344.

Hardcastle, F. D., H. Ishihara, R. Sharma, and A. S. Biris. 2011. "Photoelectroactivity and Raman spectroscopy of anodized titania (TiO_2) photoactive water-splitting catalysts as a function of oxygen-annealing temperature." *Journal of Materials Chemistry* 21 (17):6337–6345.

Hsieh, C.-T., M.-H. Lai, and C. Pan. 2010. "Synthesis and visible-light-derived photocatalysis of titania nanosphere stacking layers prepared by chemical vapor deposition." *Journal of Chemical Technology and Biotechnology* 85 (8):1168–1174.

Iijima, S. 1991. "Helical microtubules of graphitic carbon." *Nature* 354 (6348):56.

Jennings, J. R., A. Ghicov, L. M. Peter, P. Schmuki, and A. B. Walker. 2008. "Dye-sensitized solar cells based on oriented TiO_2 nanotube arrays: Transport, trapping, and transfer of electrons." *Journal of the American Chemical Society* 130 (40):13364–13372.

Kalyanasundaram, K., and M. Grätzel. 1998. "Applications of functionalized transition metal complexes in photonic and optoelectronic devices." *Coordination Chemistry Reviews* 177 (1):347–414.

Kaneco, S., Y. Chen, P. Westerhoff, and J. C. Crittenden. 2007. "Fabrication of uniform size titanium oxide nanotubes: Impact of current density and solution conditions." *Scripta Materialia* 56 (5):373–376.

Kang, T.-S., A. P. Smith, B. E. Taylor, and M. F. Durstock. 2009. "Fabrication of highly-ordered TiO_2 nanotube arrays and their use in dye-sensitized solar cells." *Nano Letters* 9 (2):601–606.

Kim, J. Y., D.-W. Kim, H. S. Jung, and K. S. Hong. 2005. "Influence of anatase–rutile phase transformation on dielectric properties of sol–gel derived TiO_2 thin films." *Japanese Journal of Applied Physics* 44 (8R):6148.

Kotani, Y., T. Matoda, A. Matsuda, T. Kogure, M. Tatsumisago, and T. Minami. 2001. "Anatase nanocrystal-dispersed thin films via sol–gel process with hot water treatment: Effects of poly (ethylene glycol) addition on photocatalytic activities of the films." *Journal of Materials Chemistry* 11 (8):2045–2048.

Krengvirat, W., S. Sreekantan, A. F. Mohd Noor, N. Negishi, S. Yul Oh, G. Kawamura, H. Muto, and A. Matsuda. 2012. "Carbon-incorporated TiO_2 photoelectrodes prepared via rapid-anodic oxidation for efficient visible-light hydrogen generation." *International Journal of Hydrogen Energy* 37 (13):10046–10056.

Kuang, D., J. Brillet, P. Chen, M. Takata, S. Uchida, H. Miura, K. Sumioka, S. M. Zakeeruddin, and M. Grätzel. 2008. "Application of highly ordered TiO_2 nanotube arrays in flexible dye-sensitized solar cells." *ACS Nano* 2 (6):1113–1116.

Kudo, A., and Y. Miseki. 2009. "Heterogeneous photocatalyst materials for water splitting." *Chemical Society Reviews* 38 (1):253–278.

Lei, B.-X., J.-Y. Liao, R. Zhang, J. Wang, C.-Y. Su, and D.-B. Kuang. 2010. "Ordered crystalline TiO_2 nanotube arrays on transparent FTO glass for efficient dye-sensitized solar cells." *The Journal of Physical Chemistry C* 114 (35):15228–15233.

Leung, D. Y. C., X. Fu, C. Wang, M. Ni, M. K. H. Leung, X. Wang, and X. Fu. 2010. "Hydrogen production over titania-based photocatalysts." *ChemSusChem* 3 (6):681–694.

Lewis, N. S. 2007. "Toward cost-effective solar energy use." *Science* 315 (5813):798–801.

Lin, J., X. Liu, M. Guo, W. Lu, G. Zhang, L. Zhou, X. Chen, and H. Huang. 2012. "A facile route to fabricate an anodic TiO_2 nanotube–nanoparticle hybrid structure for high efficiency dye-sensitized solar cells." *Nanoscale* 4 (16):5148–5153.

Liu, N., S. P. Albu, K. Lee, S. So, and P. Schmuki. 2012. "Water annealing and other low temperature treatments of anodic TiO_2 nanotubes: A comparison of properties and efficiencies in dye sensitized solar cells and for water splitting." *Electrochimica Acta* 82:98–102.

Lu, H. F., F. Li, G. Liu, Z.-G. Chen, D.-W. Wang, H.-T. Fang, G. Q. Lu, Z. H. Jiang, and H.-M. Cheng. 2008. "Amorphous TiO_2 nanotube arrays for low-temperature oxygen sensors." *Nanotechnology* 19 (40):405504.

Macak, J. M., P. J. Barczuk, H. Tsuchiya, M. Z. Nowakowska, A. Ghicov, M. Chojak, S. Bauer, S. Virtanen, P. J. Kulesza, and P. Schmuki. 2005. "Self-organized nanotubular TiO_2 matrix as support for dispersed Pt/Ru nanoparticles: Enhancement of the electrocatalytic oxidation of methanol." *Electrochemistry Communications* 7 (12):1417–1422.

Macak, J. M., and P. Schmuki. 2006. "Anodic growth of self-organized anodic TiO_2 nanotubes in viscous electrolytes." *Electrochimica Acta* 52 (3):1258–1264.

Maggio, G., and G. Cacciola. 2012. "When will oil, natural gas, and coal peak?" *Fuel* 98:111–123.

Maiyalagan, T., B. Viswanathan, and U. V. Varadaraju. 2006. "Fabrication and characterization of uniform TiO_2 nanotube arrays by sol-gel template method." *Bulletin of Materials Science* 29 (7).

Matsuda, A., Y. Higashi, K. Tadanaga, and M. Tatsumisago. 2006. "Hot-water treatment of sol–gel derived SiO_2–TiO_2 microparticles and application to electrophoretic deposition for thick films." *Journal of Materials Science* 41 (24):8101–8108.

Matsuda, A., Y. Kotani, T. K., M. Tatsumisago, and T. Minami. 2000. "Transparent anatase nanocomposite films by the sol–gel process at low temperatures." *Journal of the American Ceramic Society* 83 (1):229–231.

Miraghaei, S., F. Ashrafizadeh, K. Raeissi, M. Santamaria, and F. D. Quarto. 2011. "An electrochemical investigation on the adhesion of as-formed Anodic TiO_2 nanotubes grown in organic solvents." *Electrochemical and Solid-State Letters* 14 (1):K8–K11.

Mohamed, A. E. R., and S. Rohani. 2011. "Modified TiO_2 nanotube arrays (TNTAs): Progressive strategies towards visible light responsive photoanode, a review." *Energy & Environmental Science* 4 (4):1065–1086.

Mohapatra, S. K., M. Misra, V. K. Mahajan, and K. S. Raja. 2007a. "A novel method for the synthesis of titania nanotubes using sonoelectrochemical method and its application for photoelectrochemical splitting of water." *Journal of Catalysis* 246 (2):362–369.

Mohapatra, S. K., M. Misra, V. K. Mahajan, and K. S. Raja. 2007b. "Design of a highly efficient photoelectrolytic cell for hydrogen generation by water splitting: Application of TiO(2-)xC(x) nanotubes as a photoanode and Pt/TiO$_2$ nanotubes as a cathode." *Journal of Physical Chemistry C* 111 (24):8677–8685.

Mor, G. K., O. K. Varghese, M. Paulose, K. Shankar, and C. A. Grimes. 2006. "A review on highly ordered, vertically oriented TiO$_2$ nanotube arrays: Fabrication, material properties, and solar energy applications." *Solar Energy Materials and Solar Cells* 90 (14):2011–2075.

Nagaveni, K., M. S. Hegde, N. Ravishankar, G. N. Subbanna, and G. Madras. 2004. "Synthesis and structure of nanocrystalline TiO$_2$ with lower band gap showing high photocatalytic activity." *Langmuir* 20 (7):2900–2907.

Nyein N., W. K. Tan, G. Kawamura, A. Matsuda, and Z. Lockman. 2017. "TiO$_2$ nanotube arrays formation in fluoride/ethylene glycol electrolyte containing LiOH or KOH as photoanode for dye-sensitized solar cell." *Journal of Photochemistry and Photobiology A: Chemistry* 343: 33–39.

Nyein N., M. A. Zulkifli, W. K. Tan, A. Matsuda, and Z. Lockman. 2017 "Effect of NaOH concentration on the formation of TiO$_2$ nanotube arrays by anodic oxidation process for photoelectrochemical cell." *Solid State Phenomena* 264: 152–155.

Ni, M., M. K. H. Leung, D. Y. C. Leung, and K. Sumathy. 2007. "A review and recent developments in photocatalytic water-splitting using TiO$_2$ for hydrogen production." *Renewable and Sustainable Energy Reviews* 11 (3):401–425.

O'regan, B., and M. Grätzel. 1991. "A low-cost, high-efficiency solar cell based on dye-sensitized colloidal TiO$_2$ films." *Nature* 353 (6346):737.

Oh, H.-J., S. Lee, B. Lee, Y. Jeong, and C.-S. Chi. 2011. "Surface characteristics and phase transformation of highly ordered TiO$_2$ nanotubes." *Metals and Materials International* 17 (4):613–616.

Palmas, S., A. M. Polcaro, J. R. Ruiz, A. Da Pozzo, M. Mascia, and A. Vacca. 2010. "TiO$_2$ photoanodes for electrically enhanced water splitting." *International Journal of Hydrogen Energy* 35 (13):6561–6570.

Park, J. H., S. Kim, and A. J. Bard. 2006. "Novel carbon-doped TiO$_2$ nanotube arrays with high aspect ratios for efficient solar water splitting." *Nano Letters* 6 (1):24–28.

Pasquale, M. A., S. L. Marchiano, and A. J. Arvia. 2002. "Transitions in the growth mode of branched silver electrodeposits under isothermal and non-isothermal ionic mass transfer kinetics." *Journal of Electroanalytical Chemistry* 532 (1):255–268.

Paulose, M., H. E. Prakasam, O. K. Varghese, L. Peng, K. C. Popat, G. K. Mor, T. A. Desai, and C. A. Grimes. 2007. "TiO$_2$ nanotube arrays of 1000 μm length by anodization of titanium foil: Phenol red diffusion." *The Journal of Physical Chemistry C* 111 (41):14992–14997.

Paulose, M., K. Shankar, S. Yoriya, H. E. Prakasam, O. K. Varghese, G. K. Mor, T. A. Latempa, A. Fitzgerald, and C. A. Grimes. 2006. "Anodic growth of highly ordered TiO$_2$ nanotube arrays to 134 μm in length." *Journal of Physical Chemistry B* 110 (33):16179–16184.

Prastomo, N., H. Muto, M. Sakai, and A. Matsuda. 2010. "Formation and stabilization of tetragonal phase in sol–gel derived ZrO$_2$ treated with base-hot-water." *Materials Science and Engineering: B* 173 (1):99–104.

Premalal, E. V. A., N. Dematage, G. R. A. Kumara, R. M. G. Rajapakse, K. Murakami, and A. Konno. 2012. "Shorter nanotubes and finer nanoparticles of TiO$_2$ for increased performance in dye-sensitized solar cells." *Electrochimica Acta* 63:375–380.

Radecka, M., M. Rekas, A. Trenczek-Zajac, and K. Zakrzewska. 2008. "Importance of the band gap energy and flat band potential for application of modified TiO$_2$ photoanodes in water photolysis." *Journal of Power Sources* 181 (1):46–55.

Raja, K. S., M. Misra, V. K. Mahajan, T. Gandhi, P. Pillai, and S. K. Mohapatra. 2006. "Photo-electrochemical hydrogen generation using band-gap modified nanotubular titanium oxide in solar light." *Journal of Power Sources* 161 (2):1450–1457.

Rao, M. V., K. Rajeshwar, V. R. Pai Verneker, and J. DuBow. 1980. "Photosynthetic production of hydrogen and hydrogen peroxide on semiconducting oxide grains in aqueous solutions." *The Journal of Physical Chemistry* 84 (15):1987–1991.

Rohani, S. 2009. Synthesis of titania nanotube arrays by anodization. Paper read at *AIDIC Conf. Ser.*

Roy, P., S. Berger, and P. Schmuki. 2011a. "TiO$_2$ nanotubes: Synthesis and applications." *Angewandte Chemie International Edition in English* 50 (13):2904–2939.

Roy, P., S. Berger, and P. Schmuki. 2011b. "TiO$_2$ nanotubes: Synthesis and applications." *Angewandte Chemie International Edition* 50 (13):2904–2939.

Ruan, C., M. Paulose, O. K. Varghese, G. K. Mor, and C. A. Grimes. 2005. "Fabrication of highly ordered TiO$_2$ nanotube arrays using an organic electrolyte." *The Journal of Physical Chemistry B* 109 (33):15754–15759.

Saharudin, K. A., S. Sreekantan, and C. W. Lai. 2014. "Fabrication and photocatalysis of nanotubular C-doped TiO$_2$ arrays: Impact of annealing atmosphere on the degradation efficiency of methyl orange." *Materials Science in Semiconductor Processing* 20:1–6.

Sakamoto, H., M. A. M. Nor, N. H. B. Zakaria, G. Kawamura, H. Muto, and A. Matsuda. 2011. "Low temperature fabrication of titanium oxide composite films by hot-water treatment and application for dye-sensitized solar cells." *Electrochemistry* 79 (10):817–820.

Salari, M., K. Konstantinov, and H. Kun Liu. 2011. "Enhancement of the capacitance in TiO$_2$ nanotubes through controlled introduction of oxygen vacancies." *Journal of Materials Chemistry* 21 (13):5128–5133.

Sangpour, P., F. Hashemi, and A. Z. Moshfegh. 2010. "Photoenhanced degradation of methylene blue on cosputtered M: TiO$_2$ (M= Au, Ag, Cu) nanocomposite systems: a comparative study." *The Journal of Physical Chemistry C* 114 (33):13955–13961.

Schindler, P. W., and H. Gamsjäger. 1972. "Acid—base reactions of the TiO$_2$ (Anatase)—water interface and the point of zero charge of TiO$_2$ suspensions." *Colloid & Polymer Science* 250 (7):759–763.

Shankar, K., G. K. Mor, H. E. Prakasam, S. Yoriya, M. Paulose, O. K. Varghese, and C. A. Grimes. 2007. "Highly-ordered TiO$_2$ nanotube arrays up to 220 μm in length: Use in water photoelectrolysis and dye-sensitized solar cells." *Nanotechnology* 18 (6):065707.

Sreekantan, S., K. A. Saharudin, Z. Lockman, and T. W. Tzu. 2010. "Fast-rate formation of TiO$_2$ nanotube arrays in an organic bath and their applications in photocatalysis." *Nanotechnology* 21 (36):365603.

Sreekantan, S., and L. C. Wei. 2010. "Study on the formation and photocatalytic activity of titanate nanotubes synthesized via hydrothermal method." *Journal of Alloys and Compounds* 490 (1–2):436–442.

Taib, M. A. A., K. A. Razak, M. Jaafar, and Z. Lockman. 2017. "Initial growth study of TiO$_2$ nanotube arrays anodised in KOH/fluoride/ethylene glycol electrolyte." *Materials and Design* 128:195–205.

Sun, L., S. Zhang, X. Sun, and X. He. 2010. "Effect of the geometry of the anodized titania nanotube array on the performance of dye-sensitized solar cells." *Journal of Nanoscience and Nanotechnology* 10 (7):4551–4561.

Sun, Y., K. Yan, G. Wang, W. Guo, and T. Ma. 2011. "Effect of annealing temperature on the hydrogen production of TiO$_2$ nanotube arrays in a two-compartment photoelectrochemical cell." *The Journal of Physical Chemistry C* 115 (26):12844–12849.

Vanhumbeeck, J.-F., and J. Proost. 2009. "Current understanding of Ti anodisation: Functional, morphological, chemical and mechanical aspects." *Corrosion Reviews* 27 (3):117–204.

Varghese, O. K., D. W. Gong, M. Paulose, C. A. Grimes, and E. C. Dickey. 2003. "Crystallization and high-temperature structural stability of titanium oxide nanotube arrays." *Journal of Materials Research* 18 (1):156–165.

Varghese, O. K., M. Paulose, T. J. LaTempa, and C. A. Grimes. 2009. "High-rate solar photocatalytic conversion of CO_2 and water vapor to hydrocarbon fuels." *Nano Letters* 9 (2):731–737.

Vinu, R., and G. Madras. 2012. "Environmental remediation by photocatalysis." *Journal of the Indian Institute of Science* 90 (2):189–230.

Wan, J., X. Yan, J. Ding, M. Wang, and K. Hu. 2009. "Self-organized highly ordered TiO_2 nanotubes in organic aqueous system." *Materials Characterization* 60 (12):1534–1540.

Wang, J., and Z. Lin. 2009. "Dye-sensitized TiO_2 nanotube solar cells with markedly enhanced performance via rational surface engineering." *Chemistry of Materials* 22 (2):579–584.

Yang, B., C. K. Ng, M. K. Fung, C. C. Ling, A. B. Djurišić, and S. Fung. 2011. "Annealing study of titanium oxide nanotube arrays." *Materials Chemistry and Physics* 130 (3):1227–1231.

Yang, X., C. Cao, L. Erickson, K. Hohn, R. Maghirang, and K. Klabunde. 2008. "Synthesis of visible-light-active TiO(2)-based photocatalysts by carbon and nitrogen doping." *Journal of Catalysis* 260 (1):128–133.

Yu, K.-P., W.-Y. Yu, M.-C. Kuo, Y.-C. Liou, and S.-H. Chien. 2008. "Pt/titania-nanotube: A potential catalyst for CO_2 adsorption and hydrogenation." *Applied Catalysis B: Environmental* 84 (1):112–118.

Zhang, L., and Y. Han. 2010. "Effect of nanostructured titanium on anodization growth of self-organized TiO_2 nanotubes." *Nanotechnology* 21 (5):055602.

Zhang, S., F. Peng, H. Wang, H. Yu, S. Zhang, J. Yang, and H. Zhao. 2011. "Electrodeposition preparation of Ag loaded N-doped TiO_2 nanotube arrays with enhanced visible light photocatalytic performance." *Catalysis Communications* 12 (8):689–693.

Zhang, X.-Y., H.-P. Li, X.-L. Cui, and Y. Lin. 2010. "Graphene/TiO_2 nanocomposites: Synthesis, characterization and application in hydrogen evolution from water photocatalytic splitting." *Journal of Materials Chemistry* 20 (14):2801–2806.

Zwilling, V., M. Aucouturier, and E. Darque-Ceretti. 1999. "Anodic oxidation of titanium and TA6V alloy in chromic media. An electrochemical approach." *Electrochimica Acta* 45 (6):921–929.

10 Carbon Nanotube-Metal Oxide Hybrid Nanocomposites Synthesis and Applications

Zaid Aws Ali Ghaleb and Mariatti Jaafar

CONTENTS

10.1 INTRODUCTION

The discovery of carbon nanotubes (CNTs) can be traced back to the chemistry of fullerenes (buckyball, C_{60}) developed by Smalley and co-workers at Rice University in 1985 (Kroto et al., 1985). Fullerenes have provided an exciting new insight into carbon nanostructures built from geometric cage-like architectures of sp^2 carbon atoms that are composed of hexagonal and pentagonal faces. A few years later, in 1991, Sumio Iijima (1991) a Japanese researcher working in NEC, Japan, published his paper in *Nature* on the discovery of CNTs that are one dimensional carbon materials with an aspect ratio of 1000 or more. CNTs are long slender fullerenes where the walls of the tubes are hexagonal carbon and usually has at least one capped at each end. Recently, attaching CNTs with desired functional components caused new chances for enhancing the physical and chemical properties of CNT-based nanocomposites, and thus leading to a broad range of practical applications. A variety of functional components including metals such as platinum nanoparticles (Pt) and gold nanoparticles (Au) (Ershadifar et al., 2017, Wu et al., 2017), semiconductors such as amine functionalized cadmium selenide (CdSe), cadmium telluride nanocrystals (CdTe), (Chinta et al., 2017), metal oxide semiconductors such as zinc oxide nanorods, nanoparticles (ZnO), and titanium dioxide (TiO_2) (Ahmad et al., 2014, Oweis et al., 2014) were successfully assembled with CNTs by covalent or non-covalent interactions. Over past decades, a combination of the carbon nanotube with metal oxide is an effective way to build hybrid carbon architectures with fascinating new properties. This chapter summarizes the recent advances on the principle and techniques of preparation and functionalization of the CNTs. The effects of combining CNTs with metal oxides such as aluminum dioxide (Al_2O_3), titanium dioxide, zinc oxide, iron oxide (Fe_2O_3), etc., and how they are applied for devices is discussed. Development of nano-hybrids with multifunctional properties for a broad range of applications such as sensor, supercapacitors, absorbent, photocatalytic, and photovoltaic is covered in this chapter as well.

10.2 TYPES OF CARBON NANOTUBES

There are two main types of CNTs: (1) single-wall carbon nanotubes (SWCNTs), which are one-atom thick sheets of graphite, referred to as graphene as shown in Figure 10.1A, rolled up in cylinders and (2) multi-wall carbon nanotubes (MWCNTs), which consist of multiple layer of graphite as shown in Figure 10.1B, rolled on themselves to form a tube shape. Single-walled carbon nanotubes are often capped with half of a fullerene dome consisting of five- and six-member rings at both ends, with a diameter of close to 1 nm and a length ranging from 10 to 100 μm or up to a few centimeters. The van der Waals forces keep the individual SWCNTs together, forming a triangular lattice with a lattice constant of 0.34 nm. Multi-walled CNTs are composed of individual SWCNTs concentrically with different diameters and separated by an interlayer distance of about 0.34 nm. The diameters and length of the inner tubes are similar to SWCNTs; the outer diameters, depending on the number of individual nanotubes, and can reach values up to 50 nm (Li et al., 2010).

According to the rolling angle of the graphene sheet, the atomic structure of CNTs can be categorized into three chirality groups, namely armchair, zigzag, and chiral.

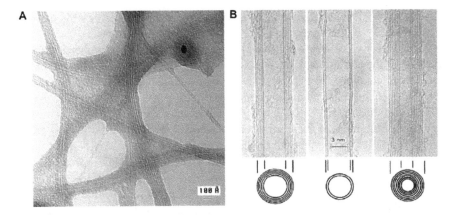

FIGURE 10.1 TEM images of different CNTs (A: SWCNTs; B: MWCNTs with different layers of 5, 2, and 7). (Reprinted from *Nature*, 354, Iijima, S., Helical microtubules of graphitic carbon, 56–58, 1991. Copyright 2008, with permission from Elsevier; Reprinted from *Nature*, 363, Bethune, D. et al., Cobalt-catalysed growth of carbon nanotubes with single-atomic-layer walls, 605–607, 1993. Copyright 2008, with permission from Elsevier.)

The CNTs chirality is defined by chiral angle, θ and the chiral vector, $C_h = na_1 + ma_2$. As shown in Figure 10.2, we can visualize cutting the graphite sheet along the dotted lines and rolling the tube so that the tip of the chiral vector touches its tail. Where the integers (n, m) are the number of steps along the unit vectors (a_1 and a_2) of the hexagonal lattice (Thostenson et al., 2001). The chiral angle determines the amount of twist in the carbon nanotubes. The two limiting cases exist where the chiral angle is as 0°, referred to as zig-zag, and 30°, referred to as armchair, based on the orientation of the carbon atoms around the nanotube circumference. The difference in zig-zag and armchair nanotube structures is shown in Figure 10.2. Using the chiral vector, the zig-zag nanotubes if m = 0 and the armchair nanotubes if n = m. Otherwise, they are

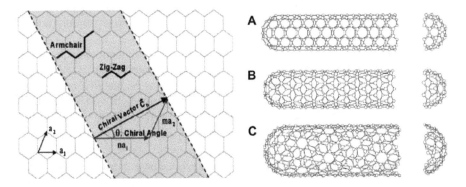

FIGURE 10.2 Schematic diagram showing how a hexagonal sheet of graphene is rolled to form a CNTs with different chiralities (A: armchair; B: zigzag; C: chiral). (Reprinted with permission from Thostenson, E.T. et al., *Compos. Sci. Technol.*, 61, 1899–1912, 2001; Dresselhaus, M.S. et al., *Carbon*, 33, 883–891, 1995. Copyright 2018.)

called "chiral." The chiral vector of the carbon nanotube also defines the nanotube diameter since the inter-atomic spacing of the carbon atom is known.

The chirality of nanotubes has significant impact on material properties. In particular, the tube chirality is known to have a strong impact on the electronic properties. Carbon nanotubes can be either metallic or semiconducting, depending on tube chirality. For a given (n, m) carbon nanotube, if (2n + m) is a multiple of 3, then the CNTs is metallic, otherwise the CNTs is a semiconductor. Each MWCNTs contains a multi-layer of graphene and each layer can have different chiralities, so the prediction of its physical properties is more complicated than that of SWCNTs (Ma et al., 2010). Investigations on the influence of chirality on the mechanical properties was evaluated by researchers. An interesting molecular structural mechanics model for the mechanical properties of defect-free CNTs was proposed by Xiao et al. (2005). The calculated Young's modulus tended to approach graphite (1.13 TPa), a tensile strength for armchair (126.2 GPa) and zig-zag (94.5 GPa) structures were predicted respectively around 23.1% and 15.6% to 17.5% of strain (Rossi and Meo, 2009).

10.3 PROPERTIES OF CARBON NANOTUBES

The fullerenes chemistry forms of CNTs have been shown to exhibit exceptional material properties that are a consequence of their symmetric structure and the chemical bonding of CNTs which is composed fully of sp^2 carbon-carbon bonds. The sp^2 bond is stronger than the sp^3 bond found in diamonds, for example, and provides CNTs with extremely high mechanical properties. It is well know that CNTs possess many unique and remarkable properties relative to the numerous available fillers (Meunier et al., 2016). Although there is no consensus in the literature on the exact properties of CNTs, theoretical and experimental results have shown unusual mechanical properties of CNTs. For example, in term of tensile strength SWCNTs and MWCNTs exhibit tensile strength of 100 and 200 GPa, respectively, compared to 1.2 GPa for steel. CNTs have also a very high Young's modulus, 1000–1200 GPa for SWCNTs and 270–950 GPa for MWCNTs (Peng et al., 2008). These make CNTs the strongest and stiffest materials known. In addition to the exceptional mechanical properties, they also possess other useful physical properties summarized in Table 10.1 (Koo, 2006, Ma et al., 2010).

Carbon nanotubes possess superior electrical and thermal conducting properties varying from metallic to moderate band-gap semi-conductive behaviors depending on their chirality, size, and purity. The SWCNTs may be metallic or semiconducting depending on the chirality with an electrical conductivity of approximately 100 S/m, while MWCNTs are metallic with an electrical conductivity of approximately 1850 S/cm (Sulong et al., 2009). Carbon nanotubes are thermally stable up to 2800°C in vacuum with the ability of thermal conductivity at room temperature to be as high as 3500 W/m K (Martin-Gallego et al., 2011). These properties offer CNTs great potential for wide applications in field emission, conducting plastics, thermal conductors, energy storage, conductive adhesives, thermal interface materials, structural materials, biological applications, air, and water filtration, etc.

TABLE 10.1
Physical Properties of Different Carbon Materials

Property	Graphite	Diamond	Fullerene	SWCNT	MWCNT
Specific gravity (g/cm³)	1.9–2.3	3.5	1.7	0.8	1.8
Electrical conductivity (S/cm)	4000^p 3.3^c	10^{-2}–10^{-15}	10^{-5}	10^2–10^6	10^3–10^5
Electron mobility (Cm²/Vs)	2×10^4	1800	0.5^{-6}	$\sim 10^5$	10^4–10^5
Thermal conductivity (W/mK)	298^p 2.2^c	900–2320	0.4	6000	2000
Coefficient of thermal expansion (K⁻¹)	-1×10^{-6p} 2.9×10^{-5c}	$1 \sim 3 \times 10^{-6}$	6.2×10^{-5}	Negligible	Negligible
Thermal stability in air (°C)	450–650	<600	~600	>600	>600

Source: Ma, P.-C. et al., *Compos. Part A Appl. Sci. Manuf.*, 41, 1345–1367, 2010; Koo, J.H., *Polymer Nanocomposites*, McGraw-Hill Professional, New York, 2006.

Note: p: in plane; c: c-axis.

10.4 SYNTHETIC METHODS FOR CARBON NANOTUBES

Since the discovery of CNTs, different methods and techniques have been employed to synthesize and produce CNTs which mainly involve gas phase processes. The most common synthetic methods are carbon arc-discharge, laser vaporization or laser ablation, and the chemical vapor deposition (CVD) methods from various carbon precursors. Carbon nanotubes have many structures differing in thickness, length, spiral types, and number of layers, although they are formed from essentially the same graphite sheet.

Arc-discharge techniques uses higher temperatures (above 1700°C) for CNT synthesis which typically causes the expansion of CNTs with fewer structural defects in comparison with other methods. In this method, a direct current is passed through two graphite electrodes in an inert atmosphere at a pressure of 100–1000 torr. A low voltage (about 12–25 V) and high current (50–120 A) power supply may be used while argon (Ar) or helium (He) gas is used to create the inert atmosphere. An arc is produced across a 1 mm gap between the two graphite electrodes. The anode is consumed and a deposit which contains MWCNTs, polyhedral particles, and amorphous carbon is formed on the cathode (Szabó et al., 2010, Karimi et al., 2015). The SWCNT may also be produced by carbon arc discharge method, but a mixed metal catalyst such as iron:cobalt (Fe:Co) or nickel:yttrium (Ni:Y) is required and inserted into the anode via the doping process. After arcing, SWCNTs are found in the chamber as a fluffy web like material. In general, the nanotubes produced by carbon arc discharge method require extensive purification before use (Bystrzejewski et al., 2008).

The first MWCNTs were produced via the carbon-arc discharge method by Iijima (1991) where nanotubes were formed on the cathode electrode along with soot and fullerenes. As for SWCNTs, Iijima and Ichihashi (1993) and Bethune et al. (1993) were the first to report on the production of SWCNTs. To produce SWCNTs, Iijima's group used an iron:carbon (Fe:C) anode in a methane:argon (CH_4:Ar) environment while Bethune's group used a Co:C anode in a He environment. To date, several variations and modifications were made to this method to tailor for different needs. Most growth is carried out in an Ar:He gas, and by adjusting the Ar:He gas ratio the diameter of SWCNTs can be controlled (Chiang and Sankaran, 2008). Synthesis of MWCNTs in a magnetic field produced defect free and high purity (>95%) nanotubes while synthesis of MWCNTs using a plasma rotating electrode system helps to produce MWCNTs economically (Eatemadi et al., 2014).

The laser vaporization method uses a 1.2% of cobalt/nickel (Co/Ni) with 98.8% of graphite composite target that is placed in a 1200°C quartz tube furnace with an inert atmosphere Ar or He and vaporized with a laser pulse at a pressure of 500 torr. Metal catalyst particles are formed in the plume of vaporized graphite and the particles help to catalyze SWCNTs growth in the plasma plume. Bundles of nanotubes and by-products are collected via condensation on a cold finger downstream from the target (Murty et al., 2013). Studies have shown the diameter of the nanotubes depends upon the laser power. When the laser pulse power is increased, the diameter of the tubes became thinner (José-Yacamán et al., 1993). This method has a potential for production of SWCNTs with high purity and high quality. The laser vaporization method, sometimes referred as laser ablation method, has some similarities as the carbon-arc discharge method. Both methods use the condensation of carbon atoms generated from the vaporization of graphite targets. Both methods also used a metal-catalyzed graphite target (or anode) to produce SWCNTs and pure graphite to produce MWCNTs (Ismail et al., 2008). But in the laser vaporization method, the needed energy is provided by a laser which hits a pure graphite pellet holding catalyst materials (frequently Co or Ni).

Both arc discharge and laser vaporization methods were first used to synthesize CNTs but due to high temperature preparation techniques, these methods have been substituted by low temperature chemical vapor deposition methods (<800°C), since the nanotube length, diameter, alignment, purity, density, and orientation of CNTs can be accurately controlled in the low temperature chemical vapor deposition methods. Chemical vapor deposition is a common method for the commercial production of CNTs. This method is a two-step process, consisting of catalyst preparation step and nanotubes synthesis step. The catalyst is prepared by sputtering transition metal such as Ni, Fe, Co, or a combination which is sputtered onto a substrate. The substrate is thermally annealed to induce the nucleation of catalyst particles. Thermal annealing results in metal cluster formation on the substrate, from which the nanotubes grow (Govindaraj and Rao, 2006). The nanotubes synthesis step entails the use of a carbon source in the gas phase (such as methane, ethylene, ethanol, carbon monoxide, and acetylene) and a plasma or a resistively heated coil, to transfer the energy to the gaseous carbon molecule. The energy source cracks the molecule into atomic carbon and the carbon then diffuses toward the catalyst coated substrate. The carbon is transported to the edges of catalyst particles where nanotubes can be produced. Studies have shown the conventionally accepted models are base growth and tip growth (Tempel et al., 2010). The synthesis or CNTs via CVD (chemical vapor deposition) is generally carried out in the temperature range of 650°C–900°C. A variety of CVD processes have been developed to synthesize CNTs, which include catalytic CVD (CCVD) either thermal, plasma enhanced CVD (PECVD) oxygen-assisted, water-assisted CVD, microwave plasma (MPECVD), thermal chemical CVD, aerogel-supported CVD, radio frequency CVD (RF-CVD), hot filament (HFCVD) and laser-assisted CVD (Govindaraj and Rao, 2006, Caglar, 2010). But catalytic chemical vapor deposition (CCVD) is currently the standard technique for the synthesis of carbon nanotubes. The advantage and disadvantage of three common CNTs synthesis methods (arc-discharge, laser vaporization, and the chemical vapor deposition) are summarized in Table 10.2. Compaing CVD with arc-discharge and laser vaporization, CVD is an economically practical method for large-scale and quite pure CNTs production and so the important advantage of CVDs are high purity obtained material and easy control of the reaction course.

TABLE 10.2

Comparison on the Three Different Synthesis Method for the CNTs

Method	Yield Rate	SWCNT MWCNT	Advantage	Disadvantage
Arc discharge	>75%	Both	Simple, inexpensive large quality nanotubes	High temperature (above 1700°C), purification required, tangled nanotubes
Laser vaporization	>75%	Both	Relatively high purity Room temperature synthesis	Method limited to the lab scale, crude product purification required
Chemical vapor deposition CVD	>75%	Both	Simple, high purity, low temperature, large-scale production, aligned growth possible	synthesized CNTs are usually MWNTs, defects

Source: Eatemadi, A. et al., *Nanoscale Res. Lett.*, 9, 393, 2014.

10.5 FUNCTIONALIZATION OF CARBON NANOTUBES

Carbon nanotubes in nature are very difficult to disperse or dissolve in water and in organic media. In addition, they are very resistant to wetting (Geckeler and Premkumar, 2011). Carbon nanotubes tend to agglomerate and form highly entangled network structures due to the strong and attractive inter-tube van der Waals forces (Dettlaff et al., 2017). This structure and the difficulty to disperse the nanotubes eventually affects the properties of the polymer/CNTs composite as most properties of composites required good dispersion of nanotubes and good interfacial adhesion between the nanotubes and polymer matrix. Functionalization is a process to introduce chemical functional groups onto the surface of the CNTs; this represents a strategy for overcoming the barriers to produce composites with improved properties. A chemically functionalized nanotube might have mechanical, electrical, or thermal properties that are different from those of the non-functionalized nanotubes. Thus, functionalization may also be utilized for fine-tuning the chemistry and physics of CNTs (Ma et al., 2010). There are various ways to introduce functional groups onto the CNTs surface, for example, acid treatment, amine functionalization, and amino functionalization.

10.5.1 Acid Treatment

Acid treatment (also known as oxidative treatment) is a type of functionalization aimed at creating defects and introducing some oxygen-bearing functional groups such as carboxyl ($-COOH$), carboxamide ($-CONH_2$), aldehyde ($-CHO$), and hydroxyl ($-OH$) onto the nanotubes. This is achieved by an oxidation process with nitric acid (HNO_3), sulfuric acid (H_2SO_4), or mixture of them, or with oxidants such as potassium permanganate ($KMnO_4$), and ozone (Byl et al., 2005, Khani and Moradi, 2013). Possible functionalized CNT is shown in Figure 10.3. The oxidation process with strong acids is damaging because structural defects are introduced along the wall and at the caps of the tube, which may severely degrade the original mechanical properties of CNTs (Saifuddin et al., 2012). By using this acid treatment, the end cap of CNTs tip will be opened and the lengths of treated CNTs will be shorter than that of untreated CNTs (Zhu et al., 2003, Gabriel et al., 2006). The oxygen bearing functional groups are introduced at the end of the tubes and at the defect sites. Amr et al. (2011) had successfully introduced $-COOH$ onto MWCNT by refluxing the nanotubes with concentrated nitric acid at 130°C for 48 hours with continuous stirring. In a separate work, Yuen et al. (2007) modified MWCNTs by stirring the nanotubes with a mixture of nitric acid and sulfuric acid at 50°C for 24 hours to obtain $-COOH$ functional group.

10.5.2 Amine Functionalization

Among all the organic groups' methods, amidation is a most effective way to chemically functionalize CNTs, and is more widely employed than other methods (Sun et al., 2002, Gabriel et al., 2006, Abjameh et al., 2014). This chemical functionalization of CNTs is based on the amidation of carboxylic acid groups. According to Gabriel et al. (2006), amine functionalization is based on the amidation of

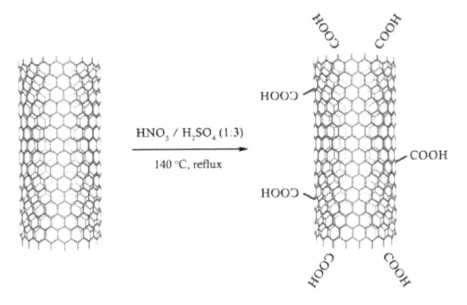

FIGURE 10.3 Functionalization reaction of CNTs. (Reprinted with permission from Bikiaris, D. et al., *Polym. Degrad. Stab.*, 93, 952–967, 2008. Copyright 2018.)

carboxylic acid groups and the strategies of amidation can be done either by thermal treatment alone (Scheme I in Figure 10.4) or thermal and acid treatment (Scheme II in Figure 10.4). For both strategies, there are two different routes to introduce the amine group onto CNTs. The first route involves converting the carboxylic acid groups (CNT-COOH) into the acyl chloride intermediate (CNT-COCL) by treatment with thionyl chloride ($SOCL_2$) prior to interaction with the amines (Scheme I(a) and II(a)) while the second route involves the interaction of amines directly with CNT-COOH (Scheme I(b) and II(b)). Results by Gabriel et al. (2006) indicated that chemical activation (with nitric acid and thionyl chloride) is necessary in order to functionalize CNTs and thus method I(b) is not efficient. Methods I(a), II(a), and II(b) are equivalently efficient.

10.5.3 Amino Functionalization

Amino-functionalized CNTs can be prepared after the following step: carboxylation, acylation and amidation (Shen et al., 2007). Shen and co-workers claimed that the amino-functionalized CNTs offered a pathway to a wide spectrum of nanotube derivatives suitable for numerous applications, for example, in polymer composites and coating. They had successfully modified MWCNTs with four different amines where they reported that different amino groups on the surface of the nanotubes have a great effect on their dispersibility in different solvent. On the other hand, Chen et al. (2008) successfully introduced high reactivity amino groups onto MWCNTs after the addition of carboxylalkyl radicals, acylation, and amidation. The detailed procedure used in their study is shown in Figure 10.5. In another separate work,

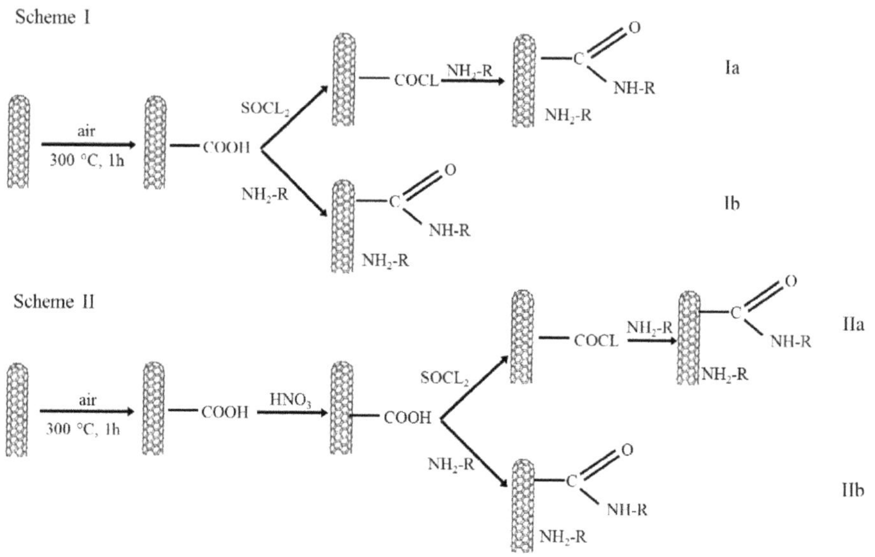

Scheme I

Scheme II

FIGURE 10.4   Schemes illustrating the different strategies used to functionalize CNTs, Scheme I: amidation by thermal treatment alone; Scheme II amidation by thermal and acid treatment.

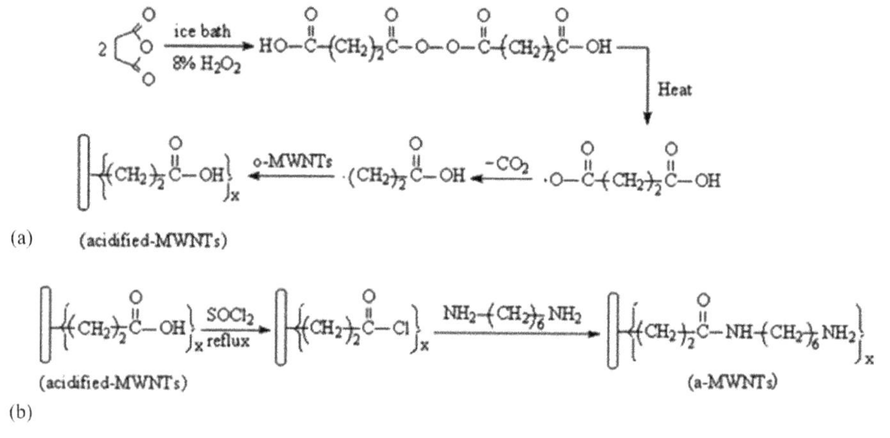

(a)   (acidified-MWNTs)

(b)

FIGURE 10.5   The detailed procedure illustrating amino functionalization of MWCNTs: (a) addition of carboxylalkyl radicals, (b) acylation, and (c) amidation. (Reprinted with permission from Chen, X. et al., *Mater. Sci. Eng. A*, 492, 236–242, 2008. Copyright 2018.)

Yang et al. (2009) produced amino-functionalized MWCNTs by treating the nanotubes with a mixture of concentrated sulfuric acid and nitric acid (H_2SO_4/HNO_3) first to obtain MWCNT-COOH. Then, the MWCNT-COOH was reacted with thionyl chloride ($SOCL_2$) to produce MWCNT-COCL. The MWCNT-COCL was further reacted with triethylenetetramine (TETA) to produce MWCNT-TETA. Figure 10.6

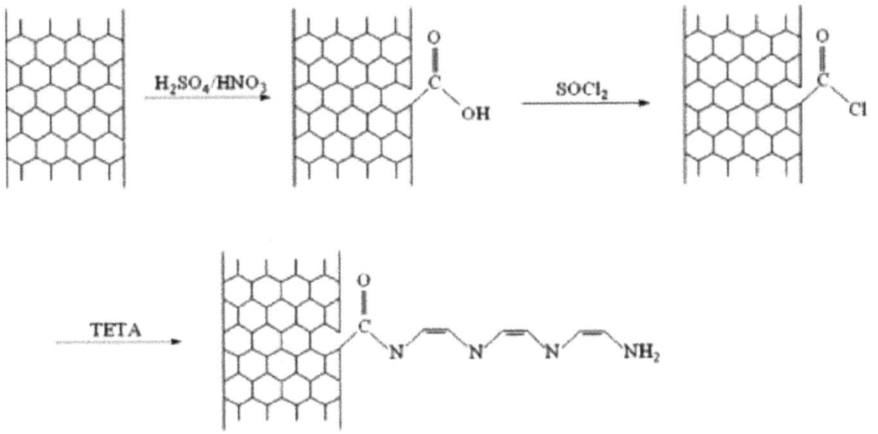

FIGURE 10.6 Scheme for the process of grafting TETA onto the MWCNTs surface. (Reprinted with permission from Yang, K. et al., *Carbon*, 47, 1723–1737, 2009. Copyright 2018.)

illustrates the process of grafting TETA onto the MWCNT surface as describe by Yang et al. (2009).

10.6 SYNTHESIS OF CARBON NANOTUBE/METAL OXIDE

Various methods have been defined to decorate metal or metal oxide nanoparticles onto the surface of CNTs. Figure 10.7 show the classification of the CNTs/metal oxide synthesis methods. These methods can be grouped roughly into two main approaches: (1) in situ synthesis, where the growths of CNTs are achieved directly within the same process, and (2) ex situ synthesis, where the decoration is performed in a separate step following the synthesis of CNTs. This section summarizes the synthesis methods of CNTs/metal oxide hybrid nanocomposites.

10.6.1 IN SITU SYNTHESIS

In this category, the metal oxides can be formed directly on the surface of CNTs or modified CNTs, sometimes both CNTs and metal oxides are synthesized simultaneously. The main benefit of this route is that the metal oxide can be deposited as a continuous amorphous or single-crystalline film with controlled thickness, or as separate units in the shape of nanoparticles, nanobeads, or nanorods. As a result, the CNTs may act as a support to stabilize uncommon or even novel crystal phases or prevent crystal growth during crystallization and phase-transformation processes, and finally can create interesting hybrids with new properties. A variety of chemical and physical synthesis techniques can be applied. The deposition process can be carried out in two phases: (1) the solution phase via electrochemical reduction of metal salts, electro- or electroless deposition, sol-gel processing,

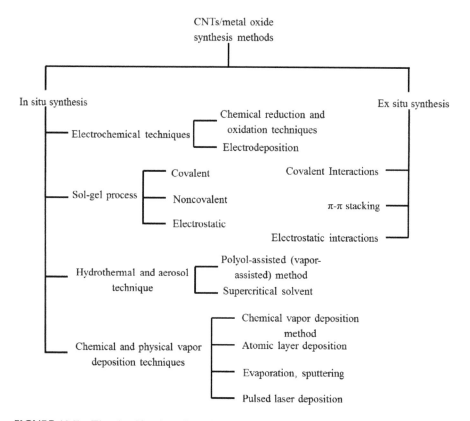

FIGURE 10.7   The classification of the CNTs/metal oxide synthesis methods.

and hydrothermal treatment with supercritical solvents, and (2) the gas phase via chemical vapor deposition (CVD) or physical deposition (laser ablation, electron beam deposition, thermal evaporation, or sputtering) (Yi et al., 2015, Mallakpour and Khadem, 2016).

10.6.1.1   Electrochemical Techniques

Electrochemical technique is an efficient and powerful technique for the deposition of various metal oxide nanoparticles because it enables effective control over nucleation and growth. The most research has been conducted on the deposition of metal oxides such as TiO_2 and iron (II) oxide Fe_2O as they are the metal oxides of choice for applications like heterogeneous electrocatalyst, sensors, and supercapacitors (Antonietti et al., 2014). The size of the metal oxide nanoparticles and their coverage on the sidewalls of CNTs can be controlled by the concentration of the metal salt and various electrochemical deposition parameters, including nucleation potential and deposition time (Hu and Guo, 2011). In general, metal oxide nanoparticles are obtained via reduction of metal complexes by chemical agents (chemical reduction and oxidation), or by electrons (electrodeposition).

TABLE 10.3

Advantages and Disadvantages of Electrochemical Techniques for CNTs/metal Oxide Hybrids Nanocomposites

Process Techniques	Advantages	Disadvantages
Chemical reduction and oxidation	Easy to produce bulk quantities	Longer time consuming Lower purity Coverage on the sidewalls of CNTs only
Electrodeposition	Shorter time consuming Higher purity Improved adhesion between CNTs and metal oxides Coverage both sidewalls and tips of CNTs	Difficult to produce bulk quantities

Table 10.3 summarizes the advantages and disadvantages of the chemical reduction and oxidation, and electrodeposition techniques.

Chemical reduction and oxidation techniques involve reactions, in which the reduction of the metal complexes is carried out with liquid or gaseous reducing agents with the aid of heat, light, ultrasound, microwave, or supercritical CO_2. Common reduction agents includes sodium borohydride ($NABH_4$), ethylene glycol, sodium citreate, and formic acid (Hu and Guo, 2011). For example, Sun et al. (2004) prepared CNTs/ZnO hybrid nanocomposites using an aqueous solution of sodium dodecylbenzene sulfonate (NaD-DBS), a mixture of Triton-X and cyclohexane, which resulted in very small water droplets on the CNT surface. Upon adding zinc acetate as the metal complex, the Zn^{2+} ions concentrated in the aqueous phase and then reacted with ammonia (NH_3) or lithium hydroxide (LiOH) to form spherical, hollow ZnOH nanoparticles. Subsequent calcinations oxidized them to create small and dense ZnO nanoparticles (Sun et al., 2004).

Electrodeposition by electron technique involves using electric current to reduce dissolved metal cations so that they form a coherent metal coating on an electrode. This technique exhibits a shorter time consumption and higher purity than chemical reduction and oxidation techniques. Adhesion of the CNTs and the metal oxide nanoparticles is improved due to van der Waals interactions (Hu and Guo, 2011). Another advantage of this technique is the electrodeposition occurs at the same extent on both the sidewalls and tips of CNTs. Therefore, the presence of carboxyl or hydroxyl groups as activators is not required (Georgakilas et al., 2007). As an example, Frank et al. (2007) studied the electrochemical behavior of CNT/TiO_2 hybrid nanocomposites. They used titanium (III) chloride ($TiCl_3$) as a precursor and electrolyte, kept at pH 2.5 with HCl/Na_2CO_3. The deposition was then carried out via galvanostatic oxidation with 1 mA/cm^2 and resulted in a rather irregular and partial coating of a mixture of anatase and TiO_2-B (Frank et al., 2007). The major drawback of this technique is that it is difficult to produce the CNTs/metal oxide in bulk quantities.

10.6.1.2   Sol-Gel Process

The sol-gel process is a solution based process that offers a flexible approach to producing various classes of materials, including ceramic, polymer, and glass materials in the form of nanoparticles, thin-film coatings, and nanoporous solids such as aluminosilicate zeolites (Owens et al., 2016). Sol-gel process involves two distinct phases: the liquid, colloidal "sol" phase and solid "gel" phase. The sol-gel process offers a number of advantages including low cost, ease of fabrication, and a low temperature technique that allows fine control of chemical composition and the introduction of the lowest concentrations of finely dispersed dopants. The major drawback is that the product typically consists of an amorphous phase rather than defined crystals and, thus, requires crystallization and post annealing steps (Hu and Guo, 2011).

The sol-gel process has been used extensively for mechanical mixing of CNTs within metal oxide nanoparticles such as TiO_2 (Yan et al., 2006, Chen et al., 2009, Yu et al., 2011). Yan et al. (2006) synthesized $MWCNT/TiO_2$ and they observed that the dispersion of pristine MWCNTs was more homogeneous and stabilized in TiO_2 sol via in situ sol-gel process. Generally, the thickness of the CNTs/metal oxide coating can be controlled by several parameters, such as (1) the reaction time, (2) the reaction composition, and (3) the precursor choice. Precursor modification in the liquid has been recognized as an important aspect in sol-gel chemistry (Hu and Guo, 2011, Brinker and Scherer, 2013). The use of titanium tetraisopropoxide (TTIP) produced irregular coatings of metal oxide on CNTs (Yu et al., 2005), on the other hand, use of tetraethoxy orthotitanate (TEOT) (Jitianu et al., 2004) or tetrabutoxy orthotitanate (TBOT) enabled a more uniform deposition (Eder and Windle, 2008). The sol-gel process was sometimes carried out under reflux (Rani et al., 2008), ultrasonication (Paulusse et al., 2007), or microwave (Huang et al., 2011), in order to enable faster and simultaneous nucleation resulting in a more homogeneous coating.

Covalent attractions used pristine CNTs, which is hydrophobic in nature providing little attractive interaction with the metal oxide and thus limiting the quality of the coating. Therefore, the common method to change the chemistry of CNTs surface is via chemical treatment in strong oxidizing acids such as H_2SO_4/HNO_3. The advantage of oxidizing acids introduces a variety of organic groups, such as carboxyl, carbonyl or hydroxyl groups, but with limited control on their number, type, and location on the surface of CNTs. However, the main disadvantage of this process is surface etching and the shortening of carbon nanotubes. Consequently, the metal oxide coatings on acid-treated CNTs were often non-uniform, although they provided better interaction in comparison with pristine CNTs (Hu and Guo, 2011). Therefore, to achieve a uniform coating on CNTs, non-covalent attractions by means of van der Waals, hydrogen bonding, π-π interactions, or electrostatic interaction provided a more facile and efficient alternative to homogenously coating the metal oxide on CNTs. Recently, a variety of CNT-surface modifiers have been developed and utilized to non-covalently functionalize CNTs to create versatile CNT-based metal oxides hybrid targeted to certain applications. Some amphiphilic molecules (surfactants) including ionic surfactants and aromatic compounds have been widely utilized to non-covalently functionalize CNTs surfaces. For the ionic

small molecular CNT-surface-modifiers, anionic sodium dodecylsulfate (SDS) surfactant has received the most enormous studies. For example Cao et al. (2006) modified CNTs with SDS, their hydrophobic aliphatic chain tends to be adsorbed onto
CNT surfaces by means of diverse hydrophobic interaction, while the hydrophilic
end attracted and interacted with metal ions of the ruthenium (III) chloride ($RuCl_3$)
precursor, which then reacted to form ruthenium (IV) oxide (RuO_2) (Govindaraj
and Rao, 2006).

Another example involves the use of benzene alcohol to non-covalent functionalized CNT, developed by Li (2011). They found that the benzyl alcohol adsorbs onto the
CNTs' surface via π-π interactions with the alcohol's benzene ring. Simultaneously, the
hydrophilic hydroxyl groups for the hydrolysis of the titanium precursor yields CNT/
titania hybrid nanostructures with uniform titania coatings. After removal of CNT
cores from CNT/titania nanohybrids via calcination treatment, anatase and rutile titania nanotubes can be obtained. This work further showed that benzyl alcohol strongly
affected the phase transition from anatase to rutile, providing very small and uniform
rutile nanocrystals with very high specific surface areas (60–100 m^2/g) without too
great a hindrance of the anatase to rutile transformation (Hu and Guo, 2011). The use
of electrostatic interactions for the in situ sol-gel route has been demonstrated for a few
metal oxides. For example, Hernadi et al. (2003) used CNTs that had been pretreated
with SDS, dried, and redispersed in 2-propanol. Using metal halides as precursors,
the authors could successfully produce coatings of aluminium oxide (Al_2O_3) and TiO_2
(Hernadi et al., 2003). Recently, Diaz Silva et al. (2017) prepared carbon nanohybrid
materials (CNF/Fe_2O_3) via a modified sol-gel technique consisting of two steps: functionalization of carbon nanofibers (CNF) in H_2SO_4/HNO_3 followed by synthesis using
$Fe(NO_3)_3$ $9H_2O$. As a result, the iron content of the CNF/Fe_2O_3 was increased by more
than twice from about 40% to about 87% mass, compared to the pristine CNF and
oxidized CNF specimens. The Fe_2O_3 nanoparticles are distributed very close to each
other due to the absence of reactive species used during the functionalization process
which can cause electrostatic interactions (Díaz Silva et al., 2017). Table 10.4 lists the
comprehensive collection regarding the surface modification of CNTs. The table summarizes the types of interaction and different surface modifiers.

TABLE 10.4
Sol-Gel Process with Different Types of Interaction by Using CNTs
Different Surface Modifiers

CNTs/Metal Oxide Types	Interaction Type	CNTs Surface Modifiers	References
CNTs/RuO_2	Hydrophobic interaction	SDS	Cao et al. (2006)
CNTs/TiO_2	π-π interactions	Benzene alcohol	Li (2011)
CNTs/Al_2O_3	Electrostatic interactions	SDS and isopropyl alcohol	Hernadi et al. (2003)
CNTs/TiO_2	Electrostatic interactions	SDS and isopropyl alcohol	Hernadi et al. (2003)
CNF/Fe_2O_3	Electrostatic interactions	H_2SO_4/HNO_3	Diaz Silva et al. (2017)

10.6.1.3   Hydrothermal and Aerosol Techniques

The hydrothermal technique has been realized as an excellent and cost-effective method for the processing of organic-inorganic hybrids. It typically enables the formation of crystalline particles or films without the postannealing and calcinations process. Moreover, the forced crystallization enables the metal oxide nanowires and nanorods formation (Yoshimura and Byrappa, 2008, Viet Quang and Hoai Chau, 2013). Polyol-assisted (vapor-assisted) method is the simplest case of the hydrothermal technique; it is the synthesis of metal-containing compounds using poly(ethylene glycol)s as the reaction medium, that plays a role of solvent, reducing agent, and complexing agent all at the same time with pristine or acid-treated CNTs added and treated in an autoclave at temperatures between 100°C and 240°C. This process was used to synthesize a wide range of highly crystalline metal oxide nanoparticles such as ZnO, TiO_2, indium tin oxide (ITO), gadolinium (III) oxide (Gd_2O_3), copper (I) oxide (Cu_2O), or Fe_2O_3 onto CNTs (Hu and Guo, 2011, Dhand et al., 2015). Zhang et al. (2006) used magnetron sputtering to precoated CNTs with a thin, amorphous layer of ZnO. Then the saturated solution of $Zn(OH)_4^{2-}$ was provided by using fine ZnO powder dissolved in NaOH at a pH of 10–12. The precoated CNT was then placed top-down in an autoclave at a temperature of 100°C for several hours, floating on the precursor solution. Finally, the Zn precursor nucleated on the CNT-ZnO film to grow ZnO nanowires perpendicular to the CNTs, with a thickness of 30–70 nm and lengths of up to 0.5 µm.

Supercritical solvent is a very important hydrothermal method utilized to produce metal oxide nanomaterials. It's shown that supercritical carbon dioxide (CO_2) could be used as a solvent or an anti-solvent or even as a solute itself within these processes to reduce the solvent strength of ethanol, resulting in the precipitation of the oxide due to high saturation. This method has been applied to deposit metal oxide, such as cerium (III) oxide (Ce_2O_3) and/or cerium (IV) oxide (CeO_2), Al_2O_3, lanthanum oxide (La_2O_3), Fe_2O_3, on the outer surfaces of CNTs through the decomposition of metal nitrate precursors in supercritical CO_2 modified with ethanol (Sun et al., 2007, Bozbag et al., 2012). Supercritical solvent is characterized with many unique properties such as low viscosity, high diffusivity, absence of surface tension, strong solvent power, readily delivers a high density of reactants, high uniformity and conformal coverage of complex surfaces, and poorly wettable substrates (Sun et al., 2007). Lin et al. (2005) showed that metal nanoparticles could be homogeneously decorated on the CNTs through the hydrogen reduction of organometallic precursors in supercritical carbon dioxide. Sun et al. (2006) used supercritical ethylenediamine as a solvent to produce thin coatings of RuO_2. They also observed various morphologies and structures of cerium oxide by simply changing the reaction temperature (Sun et al., 2007).

10.6.1.4   Chemical and Physical Vapor Deposition Techniques

Chemical vapor deposition (CVD) and physical vapor deposition (PVD) techniques are among the most common methods to produce metal oxide nanomaterials. These two techniques are used to create uniform and very thin layers of the oxide materials into carbon substrate with excellent control over the size and shape. The main difference between CVD and PVD is the processes they employ. In CVD, the oxide material is not pure as it is mixed with a volatile precursor as a carrier. The mixture is injected into the chamber that contains the carbon substrate and deposited into it. When the mixture is already adhered to the carbon substrate, the precursor eventually decomposes and leaves the desired layer of the source material in the carbon substrate. The by-product is then removed from the chamber via gas flow. The process of decomposition can be assisted or accelerated via the use of heat, plasma, or other processes. In PVD, a pure oxide material is gasified via evaporation, the application of high power electricity, laser ablation, and a few other techniques. The gasified material will then condense on the substrate material to create the desired layer. There are no chemical reactions that take place in the entire process. This subsection provides examples of the synthesis of various CNT/metal oxide hybrids using chemical techniques, including CVD and atomic layer deposition (ALD), and physical techniques such as evaporation, sputtering, and pulsed laser deposition (PLD).

Chemical vapor deposition (CVD) is a versatile technique that involves the chemical reaction of gaseous reactants on or near the vicinity of a heated substrate surface followed by the formation of a stable solid material. The CVD technique typically operates at medium temperatures (600°C–800°C) and at slightly reduced atmospheric pressures. CVD utilizes reagents of very high purity and can provide highly pure materials with structural control at atomic or nanometer level. Other advantages of CVD include production of single layer, multilayer, composite, nanostructured, and functionally graded coating materials with a high deposition rate, high degree of controlled dimension, unique structure at low processing temperature, and easy scalability. However, due to the fast deposition it is difficult to achieve uniform and defect-free coatings when scaling down to a few nanometers. The versatility of CVD had led to rapid growth and it has become one of the main processing methods for the semiconductor industry to produce thin films and coating (Carlsson and Martin, 2010, Jeong et al., 2017).

Many attempts have used CVD to synthesize CNT-metal oxide hybrids with Al_2O_3, SnO_2, Fe_2O_3, and RuO_2 (Šljukić et al., 2006, Zhang et al., 2009, Hu and Guo, 2011). For example Kuang et al. (2006) deposited acid-treated MWCNTs on a silicon (Si) wafer and heated them in a SnH_4 up to 550°C using N_2 atmosphere. The precursor decomposed at this temperature, attached to the functional groups of the CNTs and reacted with the oxygen impurities to produce SnO_2. On the other hand, Keshri et al. (2010) reported of the in situ growth of CNTs; they used a plasma-assisted CVD process to produce CNTs with a Al_2O_3 coating. Chemical vapor deposition was used to achieve a homogeneous dispersion of CNTs on Al_2O_3 powder. This powder was plasma sprayed onto a steel substrate to produce a 96% dense Al_2O_3 coating with CNT reinforcement. The addition of 1.5 wt% CNTs showed a 24% increase in the relative fracture toughness of the composite coating. The improvement in the fracture toughness is attributed to uniform dispersion of CNTs and toughening mechanism such as CNT bridging, crack deflection, and strong interaction between CNT/ Al_2O_3 interfaces.

Atomic layer deposition (ALD) is a chemical gas phase thin film deposition method based on an alternate saturated surface reaction. In contrast to chemical vapor deposition techniques, in ALD the source vapors are pulsed into the reactor alternately, one at a time, separated by purging or evacuation periods. Each precursor exposure step saturates the surface with a monomolecular layer of that precursor.

This results in a unique self-limiting growth mechanism that facilitates the growth of conformal thin films with simple and accurate thickness on large areas (Ritala and Leskelä, 2002, Su, 2011). For example, Boukhalfa et al. (2012) used nanostructured vanadium oxide coatings on the surface of MWCNT electrodes, thus offering a novel route for the formation of binder-free flexible composite electrode fabric for super capacitor applications with large thickness, controlled porosity, greatly improved electrical conductivity, and cycle stability. Min et al. (2003) demonstrated coatings of ruthenium (Ru) thin film at 300°C by ALD using $Ru(od)_3$/n-butylacetate solution and oxygen gas. This deposition temperature, below the temperature window for ALD, was chosen to avoid the oxidation of CNTs during the ALD process. Atomic layer deposition has main advantages over other thin film deposition methods, as it can be operated at low temperatures and allows exact control over the coating thickness. However, because of the sequential exposure of the precursors, the method has the lowest deposition rate compared with CVD (Hu and Guo, 2011).

Physical vapor deposition is a family of processes that is used to deposit layers of material from the vapor phase onto a solid substrate in a vacuum chamber. Two types of processes that commonly have been used are evaporation and sputtering. Evaporation is the process in which a target material evaporates in a crucible using either resistive heating (thermal evaporation) or bombardment with an electron beam (electron beam evaporation) under high vacuum that causes atoms from the target material to evaporate into the gaseous phase. These atoms then precipitate into solid form, coating everything in the vacuum chamber. A clear advantage of this process is that it permits direct transfer of energy to source during heating and very efficient in depositing pure evaporated material to carbon substrate. Also, deposition rate in this process can be as slow as 1 nm per minute to as high as few micrometers per minute. The deposition of metal oxides via thermal evaporation has been studied by Kim and Sigmund (2002); they mixed CNTs with Zn powder in a ratio of 1:12 (Kim and Sigmund, 2002). Based on the reaction temperature, the Zn particles reacted with oxygen impurities in argon to form a coating on the CNTs consisting either of spherical particles (450°C), nanowires (800°C), or short nanorods (900°C). In contrast, sputtering process (magnetron and radio frequency) relies on plasma (typically argon) and involves ejecting a target material that is a source onto a "substrate" (such as a silicon wafer) in a vacuum chamber. This effect is caused by the bombardment of the target by ionized gas, which often is an inert gas such as argon. Reactive sputtering involves a small amount of oxygen, which reacts with the sputtered material to deposit oxides. Radio frequency magnetron sputtering has been used to deposit nickel oxide (NiO) (Susantyoko et al., 2014), and ZnO (Mamat et al., 2016). In another example, Jina et al. (2007) have co-sputtered barium (Ba) and strontium (Sr) in an oxygen atmosphere to obtain a BaO/SrO coating on CNTs (Jina et al., 2007). In most of these works, the coating around the CNTs was generally conformal, although in the case of vertically aligned CNTs ("carpet"), the metal oxide material was deposited predominantly along the top of the carpet. The important advantage of sputtering is that even materials with very high melting points are easily sputtered while evaporation of these materials in a resistance evaporator is difficult and problematic. In addition, sputtered films typically have a better adhesion to the substrate than evaporated films and a composition closer to that of the source material.

Pulsed laser deposition (PLD) is a physical vapor deposition (PVD) technique where a high-power pulsed laser beam is focused inside a vacuum chamber to strike a target of the material that is to be deposited. This material is vaporized from the target and creates a plasma plume containing various energetic species, such as atoms, molecules, electrons, ions, clusters, particulates, and molten globules, which deposits as a thin film on a substrate (such as a silicon wafer facing the target). This process can occur in ultra-high vacuum or in the presence of a background gas, such as oxygen, which is commonly used when depositing oxides to fully oxygenate the deposited films (Salam Hamdy, 2010). Fejes et al. (2015) presented super growth of vertically aligned CNTs onto Al_2O_3 support and an Fe–Co catalyst layer system by PLD onto silicon wafer pieces. The effect of heat treatment at 750°C in nitrogen and in hydrogen of these PLD layers was compared. High-resolution electron microscopic images showed that treatment of catalyst layers in H_2 resulted in finer and denser catalytic particles. The main advantages of PLD include a simple concept as a laser beam vaporizes a target surface, producing a film with the same composition as the target, versatile as many materials can be deposited in a wide variety of gases over a broad range of gas pressures, cost-effective as one laser can serve many vacuum systems, fast, high quality samples can be grown reliably in 10 to 15 minutes, in addition to scalable complex oxides moving toward volume production (Vanalakar et al., 2015). Table 10.5 shows the difference between CVD, ALD and PVD, and PLD techniques.

10.6.2 Ex Situ Synthesis

Under ex situ synthesis, two hybrid components are first produced separately with their desired dimensions and morphology. Then either the metal oxide nanoparticles, or the CNT, or both of them are modified with functional groups or linker molecules, and finally are combined by the linking agents that utilize covalent, π-π stacking, or electrostatic interactions (Mallakpour and Khadem, 2016). The type of functionalization and, thus, the strength of interaction determine the distribution and concentration of the metal oxide nanoparticles on the CNT surface. The main advantage of this method is that control of the size, shape, structure, and morphology of the

TABLE 10.5
Comparison on the Different In Situ Gas-Phase Deposition Synthesis Methods

Properties	CVD	ALD	PVD	PLD
Vacuum	High/medium	Medium	High vacuum	Ultrahigh/high
Temperature range	Low	Wide	Low	Wide
Adhesion	Medium	Medium	Medium	Excellent
Porosity control	Poor	Poor	Poor	Good
Cost	High	High	High	Low
Scalabity to smaller geometries	Good	Good	Good	Good

nanoparticle is easy and, therefore, a good structure property relationship (Shirai et al., 2013). However the drawback of this synthesis method is the need for chemically modified CNTs or metal oxide. This chemical functionalization is often work-intensive, and alters the CNTs surface chemistry.

10.6.2.1 Covalent Interactions

Covalent interactions are used to attach metal oxide with various function group terminals to CNTs treated by acid. Amine-terminated or mercapto-terminated metal oxide nanoparticles are often used to attach carboxyl groups on the surface of CNTs treated by acid using amide bonds. The amide bonds can be achieved either by directly linking amine-terminated or mercapto-terminated metal oxide with the carboxyl groups of the CNTs, or by modifying these carboxyl groups into thiol groups, which then anchor to colloidal metal oxide (Paul and Mitra, 2016). In contrast to this, metal oxides can be attached to the carboxyl groups without any linking agent due to their hydrophilic nature, as recently demonstrated for manganese dioxide (MnO_2), magnesium oxide (MgO), TiO_2, and zirconium sulfate $Zr(SO_4)_2$. However, the authors observed relatively weak interactions between the oxides and the acid-terminated CNTs, resulting in rather non-uniform distributions of the nanoparticles. Better adhesion was observed when capping agents were used. For example, Sainsbury and Fitzmaurice produced capped TiO_2 and SiO_2 nanoparticles via a standard sol-gel process using titanium tetraisopropoxide (TTIP) as precursors with cetyltrimethylammonium bromide (CTAB) as the capping agent. MWCNTs were modified with 2-amino-ethylphosphoric acid and then mixed with the TiO_2 and SiO_2. The authors showed that the phosphonic acid groups on the CNTs were well-distributed and provided an excellent driving force for the attachment of TiO_2 and SiO_2 nanoparticles (Pandurangappa and Raghu, 2011).

10.6.2.2 π-π Stacking

This non covalent approach utilized the moderately strong interactions between delocalized π-electrons of the CNTs and those in aromatic organic compounds such as pyrene derivatives, phthalocyanines, porphyrins, benzyl alcohol, or triphenylphosphine. These molecules are modified with long alkyl chains that are terminated with groups such as thiol, amine, or acid groups which can connect to metal oxide nanoparticles and enable their attachment to pristine CNTs via π-π stacking (Hu and Guo, 2011). For example, Chen et al. (2001) reported that small molecules with rich conjugated π electrons such as pyrene derivatives could be bound onto the surface of carbon nanotubes through noncovalent π-π interactions. Another example, Georgakilas et al. (2005) used carboxylic derivative of pyrene as inter linker to bind capped magnetic nanoparticles on the carbon nanotubes. Strongly pyrene compound interaction on the CNT surface, even after processing steps, is a major advantage of this method and thus enhanced the solubility and continuous dispersing of the modified CNTs in various organic solvents and aqueous.

10.6.2.3 Electrostatic Interactions

This method used electrostatic interactions between modified CNTs and metal oxide nanoparticles. The most common route is the deposition of ionic polyelectrolytes to attract charged nanoparticles. Polyelectrolytes typically covalent bond to the CNTs functional groups. In contrast to the deposition of cationic polyelectrolyte such as polyethyleneimine (PEI) which is an amino-rich cationic polyelectrolyte and interacts with CNTs via physisorption (Georgakilas et al., 2007). For example Sun et al. (2006) deposited Al_2O_3, ZrO^2, and TiO_2 nanoparticles on charged CNTs in functionalization attachment of metal oxides to carbon nanotube surfaces. First to induce a positive surface charge, the CNTs were pretreated in NH_3 at 600°C. Then the addition of PEI increased the positive charges even further and enabled a better dispersion. Commercially available α-Al_2O_3 and 3Y-TZP were then dispersed in poly(acrylic acid) (PAA), which provided a negative surface potential over a wide range of pH values. Upon mixing, the Al_2O_3 and ZrO_2 nanoparticles formed strong electrostatic attractive interactions and covered the CNT surface completely (Sun and Gao, 2006).

In addition to Au nanoparticles, CNTs have also been combined with other types of nanoparticles. Indeed, because of their high specific surface area, SWCNTs have been used as a building scaffold for the adsorption of SPIONs and Au nanoparticles via electrostatic interactions (Wang et al., 2014). In another study, poly(diallyldimethylammonium chloride) (PDDA) was coated on the surface of acid-treated MWCNTs by electrostatic interactions (Liu et al., 2014). By using this polymer route, very dense, uniform distribution of either negative or positive charges over the entire CNTs surface can be obtained. This enables very dense assemblies of metal oxide nanoparticles over the entire CNTs (Hu and Guo, 2011).

10.7   APPLICATIONS OF CARBON NANOTUBES/METAL OXIDE

The interesting properties and improved performances of CNTs/metal oxide hybrid composites is due to combining the properties of CNTs and metal oxide creating new properties caused by the interaction between them. Besides that, combined CNTs and metal oxides can overcome some drawbacks. For example, the dispersion of the metal oxide on the surface of the CNT prevents metal oxide agglomeration caused by their dangling bond. Examples of application fields for the CNTs/metals oxides hybrids composites are such as absorbent, catalysis, chemical sensor, supercapacitors, photovaltic, field emission device and etc. In the following section, some applications of catalysis, chemical sensors, and supercapacitors are highlighted and summarized.

10.7.1   Catalysis

The area of catalysis is one of the most significant applications of CNTs/metal oxide hybrid composites. As a catalyst, two properties have the potential to revolutionize the field of catalysis: (a) the nanometer-scale dimensions of metal oxide nanoparticles facilitate and enhance diffusion rates; (b) fast electron transfer kinetics (Vairavapandian et al., 2008). CNTs have been widely used to support and enhance the catalytic activity of metal oxides. Therefore, various research groups have done extensive studies on the effect of using CNTs with metal oxide as a catalyst (Gupta and Saleh, 2011). For example, CNTs/tungsten trioxide (WO_3) nanocomposite have been used as catalysts for C6 olefin skeletal isomerization in different activation conditions. The CNTs/WO$_3$ nanocomposite showed higher skeletal isomerization selectivity at 200°C with higher conversion level without deactivation compared with tungstated zirconia (Pietruszka et al., 2005). The CNTs/WO$_3$ nanocomposite has been applied for the degradation of dye. It could be concluded that doping WO$_3$ into CNTs enhances the photocatalytic activity as well (Wang et al., 2008). For example, Silva et al. (2015) studied the use of CNTs/TiO$_2$ composite for the photocatalytic production of H$_2$ from biomass-containing aqueous solutions, namely from methanol and saccharides. Another example, Liu et al. (2014) studied the photocatalytic activities of CNT/TiO$_2$ composites fabricated by the hydrothermal method using titanic acid nanotubes as the TiO$_2$ precursor for the degradation of methylene blue (MB). The results showed that the hydrothermal reaction time is equal to or greater than 6 hours; titanate nanotubes in the composites entirely transform into anatase TiO$_2$ nanoparticles. This is due to CNT facilitating visible light absorption, so it plays an important role in injecting electrons into TiO$_2$ conduction bands and triggering the formation of very reactive radicals—superoxide radical ion O$_2$ and hydroxyl radical OH—which are the essential species for the degradation of MB (Chen et al., 2011).

Electrocatalyst is a catalyst depending on electrochemical reaction. It is a specific form of catayst that accelerates the reactions on the interface between electrode and electrolyte. As an electrocatalyst, two properties are required: (1) electrical conductivity and electron transfer freely; (2) efficient catalytic activity towards target substrate. Metal oxide such as TiO$_2$ and SnO$_2$ are widely used as electrocatalysts due to some advantages such as low cost and facile preparation. Carbon nanotubes endow CNT/metal oxide composite enhancements on electrocatalysis due to electric conductivity and fast electron transfer (Hu and Guo, 2011). For example, Yang et al. (2017) examined the electrocatalytic activity of MWCNTs filters for remediation of aqueous phenol in a sodium sulfate electrolyte. MWCNTs were loaded with antimony-doped tin oxide and bismuth- and antimony-codoped tin oxide via electrosorption. This study showed that CNT filters can be used as anodes for electrocatalytic water treatment, and their performance is significantly improved simply by loading them with metal-doped SnO$_2$ particles. Another example, Mo et al. (2009) prepared ZnO/MWCNTs nanocomposite via a hydrothermal process and found remarkable electrocatalytic activity towards H$_2$O$_2$ by comparing it with bare MWCNTs. Subsequently, Ma and Tian (2010) discovered that ZnO-MWCNTs/Nafion film showed fast and excellent electrocatalytic activity towards H$_2$O$_2$ and trichloroacetic acid.

10.7.2 Chemical Sensors

Semiconductor metal oxide has been a prominent example of a sensing material used in gas sensors since 1962. It has been widely used for detection of different gases due to their electrical properties that are highly affected by the surrounding gas environment. Gas sensors were made from metal oxides such as bismuth (III) oxide (Bi$_2$O$_3$), SnO$_2$, ZnO, TiO$_2$, indium (III) oxide In$_2$O$_3$, gallium (III) oxide Ga$_2$O$_3$, WO$_3$, and Fe$_2$O (Aroutiounian, 2007). For example, WO$_3$ shows sensitivity to pollutants such as sulfur dioxide (SO_2), hydrogen sulfide (H_2S), nitric oxide (NO), and NH_3, while SnO_2 is sensitive to NOx, carbon monoxide (CO), ethanol, and acetylene (C_2H_4). The advantages of semiconductor metal oxide gas sensors are their rather high sensitivity, simple design, and low cost. However, these sensors continue to suffer from high temperatures from the pre-heating of the sensor body; further degradation is due to growth and aggregation, as well as lack of selectivity, which limits their applications. For example, conventional SnO_2 sensors do not perform well when operated at room temperature. SnO_2 sensor operates at temperatures between 200°C and 500°C (Berry and Brunet, 2008). Many commercial SnO_2-based sensor devices have been realized to detect organic compounds and hazardous gases such as CO and NO. These gas sensors often operate at high temperatures up to 400°C in order to realize high gas sensitivity (Aroutiounian, 2015).

In contrast with metal oxide sensors, CNTs show potential for use in gas sensor because of their high adsorption properties due to their high specific surface area and high aspect ratio, which provides a large number of active surface sites. This enables CNTs gas sensors to detect smaller concentrations of gas molecules than metal oxide sensors. A rich π-electron conjugation is formed outside of the CNTs, making them electrochemically active, sensitive to charge transfer, and chemical doping effects by various molecules (Meyyappan, 2004). When electron withdrawing molecules such as NO_2 or O_2, or electron donating molecules such as NH_3 interact with p-type semiconducting CNTs, they change the density of holes in the nanotube and its conductivity. The CNT gas sensor can be operated at temperatures close to room temperature. The use of CNTs as gas sensors are mainly restricted to SWCNTs because the MWCNTs are not very sensitive to ambient gas (Sun et al., 2012). CNTs/metal oxides hybrid composite sensors have been fabricated and investigated widely. Leghrib et al. (2011) investigated the gas sensing performance made from doped CNTs/SnO_2 composites for NO_2 detection at room temperature. The CNTs/SnO_2 hybrid sensor showed at least a 10 times higher response towards NO_2 at room temperature in comparison with the pristine SnO_2. The main reason for such an enhanced response is a co-existence of two different depletion layers and associated potential barriers: one at the surface of the metal oxide grains and the other one at the CNTs/metal oxide interface. Nitrogen or boron-doped CNTs were added into the SnO_2 matrix and enhanced the conductivity of the nanotube (Chen et al., 2006).

Some of CNTs/metal oxide nanocomposites sensors were investigated to apply in liquid sensing, especially biosensors including hydrogen peroxide, glucose, hydrazine sensor and dopamine. The biosensors have significant use in biological applications. For instance, Thandavan et al. (2015) prepared a biosensor based on catalase using a CNT/Fe_3O_4 interface for detection of H_2O_2 in milk samples and the determination of quality of milk. In another example, Jiang et al. (2010) fabricated cupric oxide (CuO) nanoparticle-modified MWCNTs array electrode for sensitive nonenzymatic glucose detection by the sputtering deposition method. The CuO/MWCNTs electrode exhibits an enhanced electrocatalytic property, low working potential, high sensitivity, excellent selectivity, good stability, and fast amperometric sensing towards oxidation of glucose, thus is promising for the future development of nonenzymatic glucose sensors.

10.7.3 Supercapacitors

Electrical energy storage is required in many applications demanding local storage or local generation of electric energy. A storage device must meet all the requirements in terms of higher energy and power density, light weight, small size, cost, and long life to be suitable for a particular application. Electrochemical capacitors are energy storage devices filling the gap between batteries and conventional dielectric capacitors, covering several orders of magnitude both in energy and in power densities. They are an attractive choice for energy storage applications in portable or remote apparatuses where batteries and conventional dielectric capacitors have to be over-dimensioned due to unfavorable power-to-energy ratio. In electric, hybrid electric, and fuel cell vehicles, electrochemical capacitors will serve as a short-time energy storage device with high power capability and allow storing the energy from regenerative braking. Increasing applications also appear in telecommunications such as cellular phones and personal entertainment instruments (Pan et al., 2010).

Electrochemical capacitors are generally classified into two types: (1) pseudocapacitors or supercapacitors, based on the pseudocapacitance of faradaic processes in active electrode materials such as transition metal oxides and conducting polymers, and (2) electrochemical double layer capacitors (EDLC), based on double-layer capacitance due to charge separation at the electrode/electrolyte interface, which thereby need materials with high specific surface area (Hu and Guo, 2011). A hybrid electrode consisting of CNTs and metal oxide incorporates a nanotubular backbone coated by an active phase with pseudocapacitive properties, which fully utilize the advantages of the pseudocapacitance and EDLC. The open mesoporous network formed by the entanglement of nanotubes may allow the ions to diffuse easily to the active surface of the composite components and to lower the equivalent series resistance and consequently increase the power density. A wide range of metal oxides have been investigated for use in CNT hybrids, including NiO, Bi_2O_3, Co_3O_4, MnO_2, Fe_3O_4, vanadium (V) oxide (V_2O_5), and RuO_2 (Hu and Guo, 2011, Abdalla et al., 2017). Among these metal oxides, RuO_2 has been proven to be one of the important materials in oxide supercapacitors. The electrostatic charge storage and the pseudofaradaic reactions of RuO_2 nanoparticles can be affected by the CNTs surface functionality due to the increased hydrophilicity. The specific capacitance of pristine CNTs/RuO_2 nanocomposites based on the combined mass was about 70 F/g (RuO_2: 13 wt% loading). However, the specific capacitance of hydrophilic CNT (nitric acid treated) RuO_2 nanocomposites based on the combined mass was about 120 F/g (RuO_2: 13 wt% loading) (Pan et al., 2010).

Due to high costs and environmental issues concerning RuO_2, and despite the lower specific capacitances, there has been a clear trend toward the use of hydrous manganese oxides in the past few years. Manganese oxide is one of the most promising pseudocapacitor electrode materials with respect to both its specific capacitance and cost effectiveness. Xie and Gao (2007) used in situ coating techniques to prepare the MWCNTs/MnO_2 composite, where the nanosized MnO_2 uniform layer (6.2 nm in thickness) covered the surface of the MWCNT and the original structure of the pristine MWCNT was retained during the coating process. The specific capacitance of the composite electrode reached 250.5 F/g, which was significantly higher than that of a pure MWCNT electrode.

To maximize the electrochemical utilization of the supercapacitor, it is desirable to (a) use a low loading of metal oxide and (b) increase the interfacial area between CNTs and metal oxide. Consequently, the preparation of a thin, uniform, connected coating on CNTs is expected to improve the specific capacitance significantly. This was confirmed by Nam et al. (2008), who observed increasing capacitances with decreasing NiO content and thus coating thickness. There have been several studies on ternary electrodes involving CNTs and two types of pseudo materials. Li et al. (2014) report a three-component, hierarchical, bulk electrode with tailored microstructure and electrochemical properties. Supercapacitor electrode consists of a three-dimensional CNT network (also called sponge) as a flexible and conductive skeleton, an intermediate polymer layer polypyrrole (PPy) with good interface, and a metal oxide MnO_2 layer outside providing more surface area. These three components form a well-defined core-double-shell configuration that is distinct from simple core-shell or hybrid structures, and the synergistic effect leads to enhanced supercapacitor performance including high specific capacitance (even under severe compression) and excellent cycling stability. The mechanism study reveals that the shell sequence is a key factor; in system, the $CNT-PPy-MnO_2$ structure shows higher capacitance than the $CNT-MnO_2-PPy$ sequence. Our porous core-double-shell sponges can serve as freestanding, compressible electrodes for various energy devices.

10.8 CONCLUSIONS

Owing to their excellent intrinsic properties, CNTs have been widely used in many applications. A key challenge is how to break the cohesion of aggregated CNTs in order to obtain a fine dispersion in the selected solutions or matrices. Several methods have been developed including acid treatment, amine functionalization, and amino functionalization. Depending on the functionalization methods used, functional groups can be introduced onto the surface of nanotubes. A functionalized nanotube might have mechanical, optical, or electrical properties that are different from those of the original nanotube. In this chapter we discuss and summarized the main methods (in situ and ex situ) for synthesis of the CNTs/metal oxide hybrid composites that offer great potential applications in different fields. Combinations of CNTs with metal oxides such as Al_2O_3, TiO_2, ZnO, Fe_2O_3 and etc. and their application in catalysis, sensor, and supercapacitors were also summarized in this chapter.

REFERENCES

Abdalla, A. M., Sahu, R. P., Wallar, C. J., Chen, R., Zhitomirsky, I. and Puri, I. K. 2017. Nickel oxide nanotube synthesis using multiwalled carbon nanotubes as sacrificial templates for supercapacitor application. *Nanotechnology*, 28, 075603.

Abjameh, R., Moradi, O. and Amani, J. 2014. The study of synthesis and functionalized single-walled carbon nanotubes with amide group. *International Nano Letters*, 4, 97.

Ahmad, M., Ahmed, E., Hong, Z. L., Ahmed, W., Elhissi, A. and Khalid, N. R. 2014. Photocatalytic, sonocatalytic and sonophotocatalytic degradation of Rhodamine B using ZnO/CNTs composites photocatalysts. *Ultrasonics Sonochemistry*, 21, 761–773.

Amr, I. T., Al-Amer, A., Al-Harthi, M., Girei, S. A., Sougrat, R. and Atieh, M. A. 2011. Effect of acid treated carbon nanotubes on mechanical, rheological and thermal properties of polystyrene nanocomposites. *Composites Part B: Engineering*, 42, 1554–1561.

Antonietti, M., Bandosz, T., Centi, G., Costa, R., Cruz-Silva, R., DI, J., Feng, X., Frank, B., Gebhardt, P. and Guldi, D. M. 2014. *Nanocarbon-Inorganic Hybrids: Next Generation Composites for Sustainable Energy Applications*. Berlin, Germany: Walter de Gruyter GmbH & Co KG.

Aroutiounian, V. 2007. Metal oxide hydrogen, oxygen, and carbon monoxide sensors for hydrogen setups and cells. *International Journal of Hydrogen Energy*, 32, 1145–1158.

Aroutiounian, V. 2015. Metal oxide gas sensors decorated with carbon nanotubes. *Lithuanian Journal of Physics*, 55.

Berry, L. and Brunet, J. 2008. Oxygen influence on the interaction mechanisms of ozone on SnO_2 sensors. *Sensors and Actuators B: Chemical*, 129, 450–458.

Bethune, D., Klang, C., de Vries, M., Gorman, G., Savoy, R., Vazquez, J. and Beyers, R. 1993. Cobalt-catalysed growth of carbon nanotubes with single-atomic-layer walls. *Nature*, 363, 605–607.

Bikiaris, D., Vassiliou, A., Chrissafis, K., Paraskevopoulos, K. M., Jannakoudakis, A. and Docoslis, A. 2008. Effect of acid treated multi-walled carbon nanotubes on the mechanical, permeability, thermal properties and thermo-oxidative stability of isotactic polypropylene. *Polymer Degradation and Stability*, 93, 952–967.

Boukhalfa, S., Evanoff, K. and Yushin, G. 2012. Atomic layer deposition of vanadium oxide on carbon nanotubes for high-power supercapacitor electrodes. *Energy & Environmental Science*, 5, 6872–6879.

Bozbag, S. E., Sanli, D. and Erkey, C. 2012. Synthesis of nanostructured materials using supercritical CO2: Part II. Chemical transformations. *Journal of Materials Science*, 47, 3469–3492.

Brinker, C. J. and Scherer, G. W. 2013. *Sol-gel Science: The Physics and Chemistry of Sol-gel Processing*. Cambridge, MA: Academic Press.

Byl, O., Liu, J. and Yates, J. T. 2005. Etching of carbon nanotubes by ozone a surface area study. *Langmuir*, 21, 4200–4204.

Bystrzejewski, M., Rümmeli, M., Lange, H., Huczko, A., Baranowski, P., Gemming, T. and Pichler, T. 2008. Single-walled carbon nanotubes synthesis: A direct comparison of laser ablation and carbon arc routes. *Journal of Nanoscience and Nanotechnology*, 8, 6178–6186.

Caglar, B. 2010. Production of carbon nanotubes by PECVD and their applications to supercapacitors. Master Thesis, Universidad de Barcelona.

Cao, L., Scheiba, F., Roth, C., Schweiger, F., Cremers, C., Stimming, U., Fuess, H., Chen, L., Zhu, W. and Qiu, X. 2006. Novel nanocomposite $Pt/RuO_2 \cdot$ x H_2O/Carbon nanotube catalysts for direct methanol fuel cells. *Angewandte Chemie International Edition*, 45, 5315–5319.

Carlsson, J.-O. and Martin, P. M. 2010. Chemical vapor deposition. In *Handbook of Deposition Technologies for Films and Coatings: Science, Applications and Technology* (3rd ed.). P. M. Martin (Ed.), pp. 314–363. Elsevier, Oxford.

Chen, H., Yang, S., Yu, K., Ju, Y. and Sun, C. 2011. Effective photocatalytic degradation of atrazine over titania-coated carbon nanotubes (CNTs) coupled with microwave energy. *The Journal of Physical Chemistry A*, 115, 3034–3041.

Chen, M.-L., Zhang, F.-J. and OH, W.-C. 2009. Synthesis, characterization, and photocatalytic analysis of CNT/TiO_2 composites derived from MWCNTs and titanium sources. *New Carbon Materials*, 24, 159–166.

Chen, R. J., Zhang, Y., Wang, D. and Dai, H. 2001. Noncovalent sidewall functionalization of single-walled carbon nanotubes for protein immobilization. *Journal of the American Chemical Society*, 123, 3838–3839.

Chen, X., Wang, J., Lin, M., Zhong, W., Feng, T., Chen, X., Chen, J. and Xue, F. 2008. Mechanical and thermal properties of epoxy nanocomposites reinforced with amino-functionalized multi-walled carbon nanotubes. *Materials Science and Engineering: A,* 492, 236–242.

Chen, Y., Zhu, C. and Wang, T. 2006. The enhanced ethanol sensing properties of multi-walled carbon nanotubes/SnO_2 core/shell nanostructures. *Nanotechnology,* 17, 3012.

Chiang, W.-H. and Sankaran, R. M. 2008. In-flight dimensional tuning of metal nanoparticles by microplasma synthesis for selective production of diameter-controlled carbon nanotubes. *The Journal of Physical Chemistry C,* 112, 17920–17925.

Chinta, J. P., Waiskopf, N., Lubin, G., Rand, D., Hanein, Y., Banin, U. and Yitzchaik, S. 2017. Carbon nanotube and semiconductor nanorods hybrids: Preparation, characterization and evaluation of photocurrent generation. *Langmuir,* 33, 5519–5526.

Dettlaff, A., Sawczak, M., Klugmann-Radziemska, E., Czylkowski, D., Miotk, R. and Wilamowska-Zawłocka, M. 2017. High-performance method of carbon nanotubes modification by microwave plasma for thin composite films preparation. *RSC Advances,* 7, 31940–31949.

Dhand, C., Dwivedi, N., Loh, X. J., Ying, A. N. J., Verma, N. K., Beuerman, R. W., Lakshminarayanan, R. and Ramakrishna, S. 2015. Methods and strategies for the synthesis of diverse nanoparticles and their applications: A comprehensive overview. *RSC Advances,* 5, 105003–105037.

Díaz Silva, N., Valdez Salas, B., Nedev, N., Curiel Alvarez, M., Bastidas Rull, J. M., Zlatev, R. and Stoytcheva, M. 2017. Synthesis of carbon nanofibers with maghemite via a modified sol-gel technique. *Journal of Nanomaterials,* 2017.

Dresselhaus, M. S., Dresselhaus, G. and Saito, R. 1995. Physics of carbon nanotubes. *Carbon,* 33, 883–891.

Eatemadi, A., Daraee, H., Karimkhanloo, H., Kouhi, M., Zarghami, N., Akbarzadeh, A., Abasi, M., Hanifehpour, Y. and Joo, S. W. 2014. Carbon nanotubes: Properties, synthesis, purification, and medical applications. *Nanoscale Research Letters,* 9, 393–393.

Eder, D. and Windle, A. H. 2008. Morphology control of CNT-TiO_2 hybrid materials and rutile nanotubes. *Journal of Materials Chemistry,* 18, 2036–2043.

Ershadifar, H., Akhond, M. and Absalan, G. 2017. Gold nanoparticle decorated multiwall carbon nanotubes/ionic liquid composite film on glassy carbon electrode for sensitive and simultaneous electrochemical determination of dihydroxybenzene isomers. *IEEE Sensors Journal,* 17, 5030–5037.

Fejes, D., Pápa, Z., Kecsenovity, E., Réti, B., Toth, Z. and Hernadi, K. 2015. Super growth of vertically aligned carbon nanotubes on pulsed laser deposited catalytic thin films. *Applied Physics A,* 118, 855–861.

Frank, O., Kalbac, M., Kavan, L., Zukalova, M., Prochazka, J., Klementova, M. and Dunsch, L. 2007. Structural properties and electrochemical behavior of CNT-TiO_2 nanocrystal heterostructures. *Physica Status Solidi (b),* 244, 4040–4045.

Gabriel, G., Sauthier, G., Fraxedas, J., Moreno-Manas, M., Martinez, M., Miravitlles, C. and Casabo, J. 2006. Preparation and characterisation of single-walled carbon nanotubes functionalised with amines. *Carbon,* 44, 1891–1897.

Geckeler, K. E. and Premkumar, T. 2011. Carbon nanotubes: Are they dispersed or dissolved in liquids? *Nanoscale Research Letters,* 6, 136.

Georgakilas, V., Gournis, D., Tzitzios, V., Pasquato, L., Guldi, D. M. and Prato, M. 2007. Decorating carbon nanotubes with metal or semiconductor nanoparticles. *Journal of Materials Chemistry,* 17, 2679–2694.

Georgakilas, V., Tzitzios, V., Gournis, D. and Petridis, D. 2005. Attachment of magnetic nanoparticles on carbon nanotubes and their soluble derivatives. *Chemistry of Materials,* 17, 1613–1617.

Govindaraj, A. and Rao, C. 2006. Synthesis, growth mechanism and processing of carbon nanotubes. *Carbon Nanotechnology,* 15–51.

Gupta, V. and Saleh, T. A. 2011. Syntheses of carbon nanotube-metal oxides composites: adsorption and photo-degradation. *Carbon Nanotubes-From Research to Applications.* InTech.

Hernadi, K., Ljubović, E., Seo, J. W. and Forro, L. 2003. Synthesis of MWNT-based composite materials with inorganic coating. *Acta Materialia,* 51, 1447–1452.

Hu, Y. and Guo, C. 2011. Carbon nanotubes and carbon nanotubes/metal oxide heterostructures: Synthesis, characterization and electrochemical property. *Carbon Nanotubes-Growth and Applications.* InTech.

Huang, C.-H., Yang, Y.-T. and Doong, R.-A. 2011. Microwave-assisted hydrothermal synthesis of mesoporous anatase TiO_2 via sol–gel process for dye-sensitized solar cells. *Microporous and Mesoporous Materials,* 142, 473–480.

Iijima, S. 1991. Helical microtubules of graphitic carbon. *Nature,* 354, 56–58.

Iijima, S. and Ichihashi, T. 1993. Single-shell carbon nanotubes of 1-nm diameter. *Nature,* 363, 603–605.

Ismail, A., Goh, P. S., Tee, J. C., Sanip, S. M. and Aziz, M. 2008. A review of purification techniques for carbon nanotubes. *Nano,* 3, 127–143.

Jeong, N., Jwa, E., Kim, C., Choi, J. Y., Nam, J.-Y., Hwang, K. S., Han, J.-H., Kim, H.-K., Park, S.-C. and Seo, Y. S. 2017. One-pot large-area synthesis of graphitic filamentous nanocarbon-aligned carbon thin layer/carbon nanotube forest hybrid thin films and their corrosion behaviors in simulated seawater condition. *Chemical Engineering Journal,* 314, 69–79.

Jiang, L.-C. and Zhang, W.-D. 2010. A highly sensitive nonenzymatic glucose sensor based on CuO nanoparticles-modified carbon nanotube electrode. *Biosensors and Bioelectronics,* 25, 1402–1407.

Jina, F., Liu, Y. and Day, C. M. 2007. Barium strontium oxide coated carbon nanotubes as field emitters. *Applied Physics Letters,* 90, 143114.

Jitianu, A., Cacciaguerra, T., Benoit, R., Delpeux, S., Beguin, F. and Bonnamy, S. 2004. Synthesis and characterization of carbon nanotubes–TiO_2 nanocomposites. *Carbon,* 42, 1147–1151.

José-Yacamán, M., Miki-Yoshida, M., Rendon, L. and Santiesteban, J. 1993. Catalytic growth of carbon microtubules with fullerene structure. *Applied Physics Letters,* 62, 657–659.

Karimi, M., Solati, N., Amiri, M., Mirshekari, H., Mohamed, E., Taheri, M., Hashemkhani, M. et al. 2015. Carbon nanotubes part I: Preparation of a novel and versatile drug-delivery vehicle. *Expert Opinion on Drug Delivery,* 12, 1071–1087.

Keshri, A. K., Huang, J., Singh, V., Choi, W., Seal, S. and Agarwal, A. 2010. Synthesis of aluminum oxide coating with carbon nanotube reinforcement produced by chemical vapor deposition for improved fracture and wear resistance. *Carbon,* 48, 431–442.

Khani, H. and Moradi, O. 2013. Influence of surface oxidation on the morphological and crystallographic structure of multi-walled carbon nanotubes via different oxidants. *Journal of Nanostructure in Chemistry,* 3, 73.

Kim, H. and Sigmund, W. 2002. Zinc oxide nanowires on carbon nanotubes. *Applied Physics Letters,* 81, 2085–2087.

Koo, J. H. 2006. *Polymer Nanocomposites.* New York: McGraw-Hill Professional.

Kroto, H. W., Heath, J. R., O'brien, S. C., Curl, R. F. and Smalley, R. E. 1985. C60: Buckminsterfullerene. *Nature,* 318, 162–163.

Leghrib, R., Felten, A., Pireaux, J. and Llobet, E. 2011. Gas sensors based on doped-CNT/ SnO_2 composites for NO_2 detection at room temperature. *Thin Solid Films,* 520, 966–970.

Li, H. 2011. Fabrication and applications of carbon nanotube-based hybrid nanomaterials by means of non-covalently functionalized carbon nanotubes. *Carbon Nanotubes-From Research to Applications*. InTech.

Li, P., Yang, Y., Shi, E., Shen, Q., Shang, Y., Wu, S., Wei, J., Wang, K., Zhu, H., Yuan, Q., Cao, A. and Wu, D. 2014. Core-double-shell, carbon nanotube@polypyrrole@MnO$_2$ sponge as freestanding, compressible supercapacitor electrode. *ACS Applied Materials & Interfaces*, 6, 5228–5234.

Li, Y., Moon, K.-S. J. and Wong, C. 2010. Nano-conductive adhesives for nano-electronics interconnection. *Nano-Bio-Electronic, Photonic and MEMS Packaging*, 19–45.

Lin, Y., Cui, X., Yen, C. H. and Wai, C. M. 2005. PtRu/Carbon nanotube nanocomposite synthesized in supercritical fluid: A novel electrocatalyst for direct methanol fuel cells. *Langmuir*, 21, 11474–11479.

Liu, Y., Hughes, T. C., Muir, B. W., Waddington, L. J., Gengenbach, T. R., Easton, C. D., Hinton, T. M., Moffat, B. A., Hao, X. and Qiu, J. 2014. Water-dispersible magnetic carbon nanotubes as T2-weighted MRI contrast agents. *Biomaterials*, 35, 378–386.

Ma, P.-C., Siddiqui, N. A., Marom, G. and Kim, J.-K. 2010. Dispersion and functionalization of carbon nanotubes for polymer-based nanocomposites: A review. *Composites Part A: Applied Science and Manufacturing*, 41, 1345–1367.

Ma, W. and Tian, D. 2010. Direct electron transfer and electrocatalysis of hemoglobin in ZnO coated multiwalled carbon nanotubes and Nafion composite matrix. *Bioelectrochemistry*, 78, 106–112.

Mallakpour, S. and Khadem, E. 2016. Carbon nanotube–metal oxide nanocomposites: Fabrication, properties and applications. *Chemical Engineering Journal*, 302, 344–367.

Mamat, M., Malek, M., Hafizah, N., Asiah, M., Suriani, A., Mohamed, A., Nafarizal, N., Ahmad, M. and Rusop, M. 2016. Effect of oxygen flow rate on the ultraviolet sensing properties of zinc oxide nanocolumn arrays grown by radio frequency magnetron sputtering. *Ceramics International*, 42, 4107–4119.

Martin-Gallego, M., Verdejo, R., Khayet, M., de Zarate, J. M. O., Essalhi, M. and Lopez-Manchado, M. A. 2011. Thermal conductivity of carbon nanotubes and graphene in epoxy nanofluids and nanocomposites. *Nanoscale Research Letters*, 6, 610.

Meunier, V., Souza Filho, A., Barros, E. and Dresselhaus, M. 2016. Physical properties of low-dimensional *sp*^{2}-based carbon nanostructures. *Reviews of Modern Physics*, 88, 025005.

Meyyappan, M. 2004. *Carbon Nanotubes: Science and Applications*. Boca Raton, FL: CRC Press.

Min, Y. S., Bae, E. J., Jeong, K. S., Cho, Y. J., Lee, J. H., Choi, W. B. and Park, G. S. 2003. Ruthenium oxide nanotube arrays fabricated by atomic layer deposition using a carbon nanotube template. *Advanced Materials*, 15, 1019–1022.

Mo, G.-Q., Ye, J.-S. and Zhang, W.-D. 2009. Unusual electrochemical response of ZnO nanowires-decorated multiwalled carbon nanotubes. *Electrochimica Acta*, 55, 511–515.

Murty, B., Shankar, P., Raj, B., Rath, B. and Murday, J. 2013. Nanostructured materials with high application potential. *Textbook of Nanoscience and Nanotechnology*, 176–213.

Nam, K.-W., Kim, K.-H., Lee, E.-S., Yoon, W.-S., Yang, X.-Q. and Kim, K.-B. 2008. Pseudocapacitive properties of electrochemically prepared nickel oxides on 3-dimensional carbon nanotube film substrates. *Journal of Power Sources*, 182, 642–652.

Oweis, R. J., Albiss, B., Al-Widyan, M. and Al-Akhras, M.-A. 2014. Hybrid zinc oxide nanorods/carbon nanotubes composite for nitrogen dioxide gas sensing. *Journal of Electronic Materials*, 43, 3222–3228.

Owens, G. J., Singh, R. K., Foroutan, F., Alqaysi, M., Han, C.-M., Mahapatra, C., Kim, H.-W. and Knowles, J. C. 2016. Sol–gel based materials for biomedical applications. *Progress in Materials Science*, 77, 1–79.

Pan, H., LI, J. and Feng, Y. 2010. Carbon nanotubes for supercapacitor. *Nanoscale Research Letters*, 5, 654.

Pandurangappa, M. and Raghu, G. K. 2011. Chemically modified carbon nanotubes: Derivatization and their applications. *Carbon Nanotubes Applications on Electron Devices*. InTech.

Paul, R. and Mitra, A. K. 2016. Nanocomposites of Carbon Nanotubes and Semiconductor Nanocrystals as Advanced Functional Material with Novel Optoelectronic Properties. *Functionalized Nanomaterials*. InTech.

Paulusse, J. M., Van Beek, D. and Sijbesma, R. P. 2007. Reversible switching of the sol–gel transition with ultrasound in rhodium (I) and iridium (I) coordination networks. *Journal of the American Chemical Society*, 129, 2392–2397.

Peng, B., Locascio, M., Zapol, P., Li, S., Mielke, S. L., Schatz, G. C. and Espinosa, H. D. 2008. Measurements of near-ultimate strength for multiwalled carbon nanotubes and irradiation-induced crosslinking improvements. *Nature Nanotechnology*, 3, 626–631.

Pietruszka, B., di Gregorio, F., Keller, N. and Keller, V. 2005. High-efficiency WO₃/carbon nanotubes for olefin skeletal isomerization. *Catalysis Today*, 102, 94–100.

Rani, S., Suri, P., Shishodia, P. and Mehra, R. 2008. Synthesis of nanocrystalline ZnO powder via sol–gel route for dye-sensitized solar cells. *Solar Energy Materials and Solar Cells*, 92, 1639–1645.

Ritala, M. and Leskelä, M. 2002. Chapter 2–Atomic layer deposition A2–Nalwa, H. S. *Handbook of Thin Films*. Burlington, NJ: Academic Press.

Rossi, M. and Meo, M. 2009. On the estimation of mechanical properties of single-walled carbon nanotubes by using a molecular-mechanics based FE approach. *Composites Science and Technology*, 69, 1394–1398.

Saifuddin, N., Raziah, A. and Junizah, A. 2012. Carbon nanotubes: A review on structure and their interaction with proteins. *Journal of Chemistry*, 2013.

Salam Hamdy, A. 2010. Corrosion protection performance via nano-coatings technologies. *Recent Patents on Materials Science*, 3, 258–267.

Shen, J., Huang, W., Wu, L., Hu, Y. and Ye, M. 2007. The reinforcement role of different amino-functionalized multi-walled carbon nanotubes in epoxy nanocomposites. *Composites Science and Technology*, 67, 3041–3050.

Shirai, T., Kato, T. and Fuji, M. 2013. Effects of the alumina matrix on the carbonization process of polymer in the gel-casted green body. *Journal of the European Ceramic Society*, 33, 201–206.

Silva, C. G., Sampaio, M. J., Marques, R. R., Ferreira, L. A., Tavares, P. B., Silva, A. M. and Faria, J. L. 2015. Photocatalytic production of hydrogen from methanol and saccharides using carbon nanotube-TiO₂ catalysts. *Applied Catalysis B: Environmental*, 178, 82–90.

Šljukić, B., Banks, C. E. and Compton, R. G. 2006. Iron oxide particles are the active sites for hydrogen peroxide sensing at multiwalled carbon nanotube modified electrodes. *Nano Letters*, 6, 1556–1558.

Su, J. 2011. Optimization of TiO₂ thin film growth at different temperatures by atomic layer deposition.

Sulong, A. B., Muhamad, N., Sahari, J., Ramli, R., Deros, B. M. and Park, J. 2009. Electrical conductivity behaviour of chemical functionalized MWCNTs epoxy nanocomposites. *European Journal of Scientific Research*, 29(1), 13–21.

Sun, J. and Gao, L. 2006. Attachment of inorganic nanoparticles onto carbon nanotubes. *Journal of Electroceramics*, 17, 91–94.

Sun, J., Gao, L. and Iwasa, M. 2004. Noncovalent attachment of oxide nanoparticles onto carbon nanotubes using water-in-oil microemulsions. *Chemical Communications*, 832–833.

Sun, Y.-F., Liu, S.-B., Meng, F.-L., Liu, J.-Y., Jin, Z., Kong, L.-T. and Liu, J.-H. 2012. Metal oxide nanostructures and their gas sensing properties: A review. *Sensors,* 12, 2610–2631.

Sun, Y.-P., Fu, K., Lin, Y. and Huang, W. 2002. Functionalized carbon nanotubes: Properties and applications. *Accounts of Chemical Research,* 35, 1096–1104.

Sun, Z., Liu, Z., Han, B., Miao, S., Du, J. and Miao, Z. 2006. Microstructural and electrochemical characterization of RuO_2/CNT composites synthesized in supercritical diethyl amine. *Carbon,* 44, 888–893.

Sun, Z., Zhang, X., Han, B., Wu, Y., An, G., Liu, Z., Miao, S. and Miao, Z. 2007. Coating carbon nanotubes with metal oxides in a supercritical carbon dioxide–ethanol solution. *Carbon,* 45, 2589–2596.

Susantyoko, R. A., Wang, X., Xiao, Q., Fitzgerald, E. and Zhang, Q. 2014. Sputtered nickel oxide on vertically-aligned multiwall carbon nanotube arrays for lithium-ion batteries. *Carbon,* 68, 619–627.

Szabó, A., Perri, C., Csató, A., Giordano, G., Vuono, D. and Nagy, J. B. 2010. Synthesis methods of carbon nanotubes and related materials. *Materials,* 3, 3092–3140.

Tempel, H., Joshi, R. and Schneider, J. J. 2010. Ink jet printing of ferritin as method for selective catalyst patterning and growth of multiwalled carbon nanotubes. *Materials Chemistry and Physics,* 121, 178–183.

Thandavan, K., Gandhi, S., Nesakumar, N., Sethuraman, S., Rayappan, J. B. B. and Krishnan, U. M. 2015. Hydrogen peroxide biosensor utilizing a hybrid nano-interface of iron oxide nanoparticles and carbon nanotubes to assess the quality of milk. *Sensors and Actuators B: Chemical,* 215, 166–173.

Thostenson, E. T., Ren, Z. and Chou, T.-W. 2001. Advances in the science and technology of carbon nanotubes and their composites: A review. *Composites Science and Technology,* 61, 1899–1912.

Vairavapandian, D., Vichchulada, P. and Lay, M. D. 2008. Preparation and modification of carbon nanotubes: Review of recent advances and applications in catalysis and sensing. *Analytica Chimica Acta,* 626, 119–129.

Vanalakar, S. A., Agawane, G. L., Shin, S. W., Suryawanshi, M. P., Gurav, K. V., Jeon, K. S., Patil, P. S., Jeong, C. W., Kim, J. Y. and Kim, J. H. 2015. A review on pulsed laser deposited CZTS thin films for solar cell applications. *Journal of Alloys and Compounds,* 619, 109–121.

Viet Quang, D. and Hoai Chau, N. 2013. The effect of hydrothermal treatment on silver nanoparticles stabilized by chitosan and its possible application to produce mesoporous silver powder. *Journal of Powder Technology,* 2013.

Wang, S., Shi, X., Shao, G., Duan, X., Yang, H. and Wang, T. 2008. Preparation, characterization and photocatalytic activity of multi-walled carbon nanotube-supported tungsten trioxide composites. *Journal of Physics and Chemistry of Solids,* 69, 2396–2400.

Wang, Y., Huang, R., Liang, G., Zhang, Z., Zhang, P., Yu, S. and Kong, J. 2014. MRI-visualized, dual-targeting, combined tumor therapy using magnetic graphene-based mesoporous silica. *Small,* 10, 109–116.

Wu, X., Liu, Z., Zeng, J., Hou, Z., Zhou, W. and Liao, S. 2017. Platinum nanoparticles on interconnected Ni_3P/Carbon nanotube–carbon nanofiber hybrid supports with enhanced catalytic activity for fuel cells. *ChemElectroChem,* 4, 109–114.

Xiao, J. R., Gama, B. A. and Gillespie Jr, J. W. 2005. An analytical molecular structural mechanics model for the mechanical properties of carbon nanotubes. *International Journal of Solids and Structures,* 42, 3075–3092.

Xie, X. and Gao, L. 2007. Characterization of a manganese dioxide/carbon nanotube composite fabricated using an in situ coating method. *Carbon,* 45, 2365–2373.

Yan, X.-B., Tay, B. K. and Yang, Y. 2006. Dispersing and functionalizing multiwalled carbon nanotubes in TiO_2 sol. *The Journal of Physical Chemistry B,* 110, 25844–25849.

Yang, K., Gu, M., Guo, Y., Pan, X. and Mu, G. 2009. Effects of carbon nanotube function-alization on the mechanical and thermal properties of epoxy composites. *Carbon*, 47, 1723–1737.

Yang, S. Y., Vecitis, C. D. and Park, H. 2017. Electrocatalytic water treatment using carbon nanotube filters modified with metal oxides. *Environmental Science and Pollution Research*, 1–8.

Yi, J., Xue, W., Xie, Z., Chen, J. and Zhu, L. 2015. A novel processing route to develop alumina matrix nanocompositesreinforced with multi-walled carbon nanotubes. *Materials Research Bulletin*, 64, 323–326.

Yoshimura, M. and Byrappa, K. 2008. Hydrothermal processing of materials: Past, present and future. *Journal of Materials Science*, 43, 2085–2103.

Yu, J., Fan, J. and Cheng, B. 2011. Dye-sensitized solar cells based on anatase TiO_2 hollow spheres/carbon nanotube composite films. *Journal of Power Sources*, 196, 7891–7898.

Yu, Y., Jimmy, C. Y., Yu, J.-G., Kwok, Y.-C., Che, Y.-K., Zhao, J.-C., Ding, L., Ge, W.-K. and Wong, P.-K. 2005. Enhancement of photocatalytic activity of mesoporous TiO_2 by using carbon nanotubes. *Applied Catalysis A: General*, 289, 186–196.

Yuen, S.-M., Ma, C.-C. M., Lin, Y.-Y. and Kuan, H.-C. 2007. Preparation, morphology and properties of acid and amine modified multiwalled carbon nanotube/polyimide composite. *Composites Science and Technology*, 67, 2564–2573.

Zhang, T., Kumari, L., Du, G. H., Li, W. Z., Wang, Q. W., Balani, K. and Agarwal, A. 2009. Mechanical properties of carbon nanotube–alumina nanocomposites synthesized by chemical vapor deposition and spark plasma sintering. *Composites Part A: Applied Science and Manufacturing*, 40, 86–93.

Zhang, W.-D. 2006. Growth of ZnO nanowires on modified well-aligned carbon nanotube arrays. *Nanotechnology*, 17, 1036.

Zhu, J., Kim, J., Peng, H., Margrave, J. L., Khabashesku, V. N. and Barrera, E. V. 2003. Improving the dispersion and integration of single-walled carbon nanotubes in epoxy composites through functionalization. *Nano Letters*, 3, 1107–1113.

11 Formation of SiO$_2$ Nanowires by Local Anodic Oxidation Process via AFM Lithography for the Fabrication of Silicon Nanowires

Khatijah Aisha Yaacob, Siti Noorhaniah Yusoh, Nurain Najihah Alias, and Ahmad Makarimi Abdullah

CONTENTS

11.1 INTRODUCTION

Silicone nanowires (SiNWs) have been used in the development of electronic devices because they are easy to prepare in small dimensions with high aspect ratios. Silicone nanowires can be prepared using "bottom-up" and "top-down" approaches. Atomic force microscopy (AFM) lithography is an example of a top-down approach. Local anodic oxidation (LAO) is an electrochemical process that can be done using AFM equipment. The AFM tip can locally oxidize a nearby substrate surface such as

313

metal or semiconductor forming an oxide layer (Tseng 2011). When the bias voltage is applied on the AFM tip, an electric field is created. If the formation of the electric field is greater than 10^9 V/m, adsorbed water molecules on the surface will be decomposed into H^+ and OH^- as expressed in equation (11.1) (Červenka et al. 2006). OH^- ions will penetrate into substrate surface at 2 to 3 nm depth and upon its reaction with the substrate, oxide will form (Chang et al. 2004, Červenka et al. 2006). When Si is used as a substrate, LAO can result in the formation of SiO_2. This happens when the OH^- reacts with Si as shown in equation (11.2) (Dagata et al. 1998). Figure 11.1 is a sketch to show how SiO_2 is formed on Si from SOI substrate using a LAO mechanism utilising a cantilever tip in AFM. The formation of the thin oxide layer was first developed by Cabrera and Mott and then adapted for LAO mechanism by Stiévenard (Cabrera and Mott 1949, Stiévenard, Fontaine, and Dubois 1997). AFM-LAO can be used to fabricate fine structures suitable for the creation of nanoelectronic devices. Local anodic oxidation is a simple and cost-effective technique for fabrication of 1-D nanostructures; nevertheless, there are parameters that need to be controlled to ensure the success in the formation of NWs by this oxidation technique. In here, SiO_2 NWs fabricated by AFM-LAO are reviewed. The SiO_2 NWs can be removed to then produce SiNWs structure. The process can be considered as a mask-making process to fabricate SiNWs (Alias 2015).

$$H_2O \rightarrow H^+ + OH^- \tag{11.1}$$

$$Si + 4h^+ + 2OH^- \rightarrow SiO_2 + 2H^+ \tag{11.2}$$

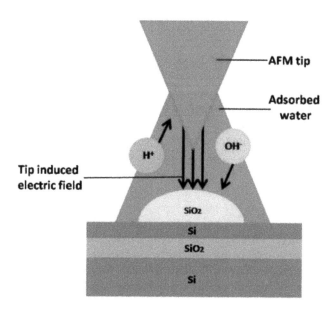

FIGURE 11.1 Concept of LAO process on SOI substrate.

AFM-LAO can also be used to fabricate a variety of oxide compounds with different geometries on metal or semiconductor substrates. For example sputtered titanium substrate can be subjected to AFM-LAO to form titanium dioxide (TiO$_2$) nanostructures (Ávila Bernal and Bonilla 2012). The LAO process can be used to fabricate SiNWs field effect transistors (Abdullah et al. 2010), double-lateral-gate junctionless (Larki et al. 2012), lateral silicon diodes (Rouhi et al. 2012) and gas sensors (Asmah et al. 2015). However, to produce high quality SiNWs for the above-mentioned applications, the AFM-LAO technique must be optimized by considering all parameters involved in the process. Parameters that would influence the properties of silicon oxide nanowires during the AFM-LAO fabrication are applied voltages, writing speed, relative humidity and temperature (Rouhi et al. 2012, Dehzangi et al. 2013, Hutagalung et al. 2014). These parameters are to be reviewed and discussed here.

11.2 PARAMETERS FOR AFM-LAO

11.2.1 TYPE OF CANTILEVER TIPS COATING MATERIALS

The AFM probe consists of a cantilever with a sharp tip that scans a sample surface. When the tip of the AFM probe travels near to the sample surface, the force between the tip and sample deflect the cantilever. The changes of the deflection are measured with a laser beam placed at the top of the cantilever. The changes of laser reflection from the cantilever are measured with a photodiode light detector and termed as laser reflectivity voltage.

The AFM probe is made of monolithic silicon which has been micromachined to micro size. The tip on the cantilever is coated with a highly refractive index material to increase the reflected laser signal and prevent interference of light reflected from the surface of the cantilever. There are three different types of cantilever tip coating materials studied and reported here: gold (Au) coated tip, chromium/platinum (Cr/Pt) coated tip and aluminum (Al) coated tip. Each of them shows different results during the patterning process of silicon substrate. Figure 11.2 shows AFM images for oxide patterns using Au, Cr/Pt, Al tips on SOI surface. Five different voltages were studied: 5, 6, 7, 8 and 9 V with

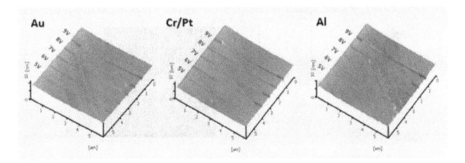

FIGURE 11.2 SiO$_2$ nanowires using different types of tip (Au, Cr/Pt and Al) with different applied voltages (5, 6, 7, 8, and 9 V) at 0.3 μm/s writing speed. (From Yusoh and Yaacob 2015.)

writing speed fixed at 0.3 μm/s. As seen, oxide patterns using Au and Cr/Pt coated tips are more uniform compared to Al coated tip. Besides tips coating materials, the formation of SiO_2 nanowires is also very much dependent on environmental factor such as room temperature and humidity. Because of this reason the repeatability of the same size silicon oxide nanowires is very difficult (Yusoh and Yaacob 2015).

Figure 11.3 indicates the relation of SiO_2 width and applied voltage for different types of cantilever tip coating materials. An Au-coated tip gives thicker oxide compared to Al- and Cr/Pt-coated tips. From this result, it is shown that different types of cantilever tip coating material affected the oxide width due to different laser reflectivity voltage of each coating material. Figure 11.4 shows the schematic of laser

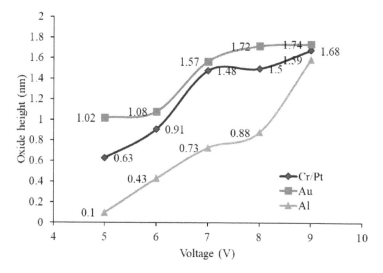

FIGURE 11.3 Relation of SiO_2 height at different applied voltages for different types of AFM cantilever tips. (From Yusoh 2018.)

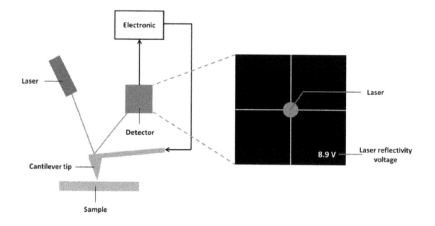

FIGURE 11.4 Schematic of laser reflectivity voltage for local anodic oxidation fabrication.

reflectivity voltage for the AFM-LAO process. The Au-coated tip reflects a higher laser reflectivity voltage than Cr/Pt and Al tips. For Au tip laser reflectivity voltage is between 7 and 9 V, whereas the laser reflectivity voltage for Cr/Pt and Al are between 3 and 5 V. The higher value and stable laser reflectivity are needed to have a better formation of SiO$_2$ nanowires (Yusoh and Yaacob 2015). This indicates that laser reflectivity voltage of cantilever tips influences the formation of SiO$_2$ nanowires.

11.2.2 APPLIED VOLTAGES

It is obvious from Figure 11.3 that applied voltage influences the thickness of the SiO$_2$ nanowires formed. It is known that in LAO, the bias voltage is applied to supply the electric field at the AFM tip for oxidation process. Oxidation occurs when the electric field is greater than 10^9 V/m to dissociate water molecules from the environment that are adsorbed on the substrate surface (Červenka et al. 2006). The electric field is calculated using equation (11.3) where E is electric field, V represents applied voltage and x is the height of oxide (Dagata et al. 1998). Dehzangi and co-workers reported on that the threshold voltage for oxidation to occur is 5 V since there is no formation of SiO$_2$ nanowires on Si substrate when the applied voltage is below 5 V at 6 μm/s of writing speed (Dehzangi et al. 2013).

$$E = \frac{V}{x} \tag{11.3}$$

Table 11.1 summarizes the electric field generated as a function of applied voltage relative to the height of the grown SiO$_2$. From the results obtained, SiO$_2$ can be grown if the electric field is in range of 10^{10} V/m, which is greater than 10^9 V/m for applied voltage of 5 to 9 V. The electric field generated at 5 V is known as a critical electric field to decompose the water into H$^+$ and OH$^-$ ions for SiO$_2$ to grow.

TABLE 11.1

The Electric Field Generated at Varies Applied Voltage Relative to SiO$_2$ Height

Applied Voltage (V)	Height of SiO$_2$ (nm)	Electric Field (V/m)
1	—	—
2	—	—
3	—	—
4	—	—
5	0.17	2.94 × 10^{10}
6	0.34	1.76 × 10^{10}
7	0.45	1.56 × 10^{10}
8	0.52	1.54 × 10^{10}
9	0.57	1.58 × 10^{10}

Source: Yusoh 2018.

Previous reported work claimed that larger applied voltage gives a thicker SiO_2 pattern (Dehzangi et al. 2013, Hutagalung et al. 2014). Hutagalung et al. (2014) reported that the oxide thickness grows by using Au-coated AFM tips are 1.5, 3.5 and 5.5 nm which is increased when the voltages of 7, 8, and 9 V were applied (Hutagalung et al. 2014). Figure 11.3 indicates the relation between oxide height and applied voltage for Au, Cr/Pt and Al-coated tips. It shows the oxide height is increased with increase applied voltage for all tips.

11.2.3 WRITING SPEED

Writing speed of the tip is also one of the major parameters in the optimization of the LAO process. It has been reported that the oxide height is decreased with an increased writing speed (Mendoza et al. 2016). According to Dehzangi et al. (2013), when the writing speed is increased, there are only few numbers of diffusible OH^- ions that could enter Si/SiO_2 interface that cause the decrease of oxide width and height.

Figure 11.5 presents the AFM image profile and width and height of SiO_2 nanowires at different writing speeds using Au tip at 8 V of applied voltage. As can be seen

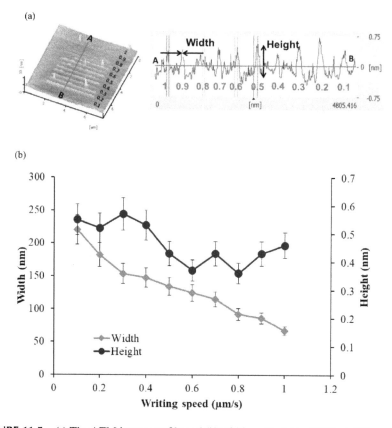

FIGURE 11.5   (a) The AFM image profile and (b) width and height of SiO_2 at different writing speed using Au coated tip at 8 V. (From Yusoh 2018.)

from Figure 11.5a, a lower writing speed from 0.1 to 0.7 µm/s shows continuous SiO$_2$ nanowires. However, at a higher writing speed, segmented or non-continuous SiO$_2$ nanowires are observed. Figure 11.5b shows that the width and height of oxide pattern decreased with an increase in writing speed from 0.1 to 0.6 µm/s. In addition, at a fast writing speed, the height of SiO$_2$ is decreased due to a shorter time available for the biased tip to reach the saturated oxide height (Mo et al. 2009). Furthermore, the accumulation of OH⁻ ions on the sample is reduced at fast writing speed decreasing the oxide height (Janoušek et al. 2010).

11.2.4 RELATIVE HUMIDITY AND TEMPERATURE

Relative humidity also plays an important role in LAO. The relative humidity is actually the source of water for the oxidation process. The thickness of water adsorbed on the surface of the substrate to be oxidized will depend on the relative humidity. At high relative humidity formation of SiO$_2$ NWs appears to be rather easy. A dry environment will affect the LAO performance whereby SiO$_2$ NWs do not grow uniformly. This is due to lack of water meniscus available to allow oxidation to occur. However, an excessive humidity will generate wider water meniscus and obviously will increase the amount of OH⁻ ions, therefore a larger area is oxidized. Figure 11.6 illustrates the condition of water meniscus for SiO$_2$ nanowires at (a) low and (b) high humidity. SiO$_2$ NWs formed in high humidity are larger compare to the SiO$_2$ NWs formed at low humidity (Rouhi et al. 2012). There are various ranges of relative humidity studied by

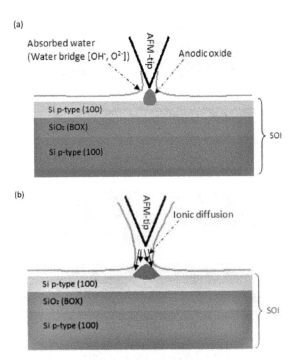

FIGURE 11.6   Schematic representation of LAO at (a) low and (b) high relative humidity.

researchers for LAO process such as 40% to 50% (Abdullah et al. 2010), 50% to 70% (Asmah et al. 2015) and 55.8% to 68.9% (Hutagalung and Lew 2010).

Dehzangi and co-workers studied the influence of relative humidity to the height of SiO_2 nanowires by Au- and Cr/Pt-coated tips at 8 V of applied voltage. The obtained result indicates that the oxide thickness is linearly increased with the increased relative humidity for both tips. In addition, Rouhi et al. (2012) studied the effect of relative humidity towards the width of SiO_2 NWs at 0.1 and 1 μm/s of writing speed. The study reveals the width of oxide is increased relative to the increased humidity for both writing speeds.

Table 11.2 shows the topographic images of the fabricated wire that give average thickness of oxide that was produced at 50%, 55%, 60%, 65%, and 70% of

TABLE 11.2

Summary of Results from AFM Images of SiO_2 Formed with LAO Process at Different RH%

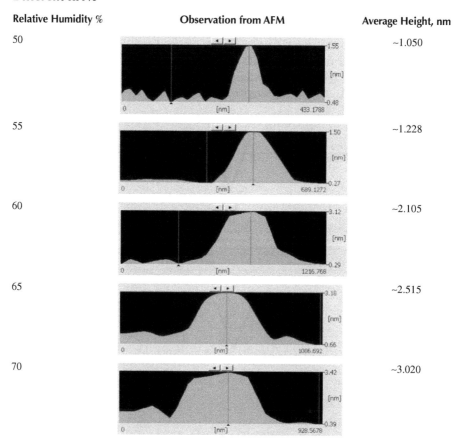

Relative Humidity %	Observation from AFM	Average Height, nm
50		~1.050
55		~1.228
60		~2.105
65		~2.515
70		~3.020

Note: AFM: Atomic force microscopy; LAO: local anodic oxidation.

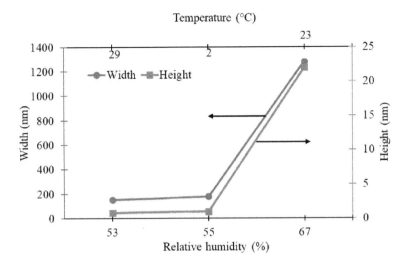

FIGURE 11.7 The width and height dimension of SiO₂ nanowire at different relative humidity and temperature.

relative humidity. In general, the images show that there are some changes in SiO₂ NWs height as the relative humidity change.

Environment temperature also affected the width and height of SiO₂ growth by LAO. The width and height measurements at different relative humidity and temperature conditions were recorded and presented in Figure 11.7. From here, the width and height are seen to increase when the relative humidity is increased. However, when temperature is increased, SiO₂ NWs grown have a smaller width and they are shorter. SiO₂ NWs grown at humidity greater than 65% and temperature less than 24°C have the largest width of 1278.21 nm and length of 21.95 nm. It is observed that with higher humidity more water molecules exist on substrate surface, which create wider water meniscus hence producing bigger pattern width. At humidity ranges 55% to 65% and temperatures 24°C to 26°C, SiO₂ NWs have 178.64 nm width and 0.99 nm height. Meanwhile, the humidity lower than 55 % and temperature greater than 27°C produced SiO₂ shorter NWs (0.79 nm) with smaller width (151.36 nm). Thus, the relative humidity below 55% and temperature more than 24°C are said to be the most suitable conditions to grow SiO₂ NWs with smaller width.

11.3 ETCHING

After the SiO₂ NWs patterns are drawn on the SOI wafer using the AFM-LAO process as discussed above, the SOI wafer needs to undergo two chemical etching processes to yield the desired SiNWs. The first etching process is used to remove unmasked silicon on SOI wafer and the second wet chemical etching is to remove the oxide pattern. Figure 11.8 shows the schematic diagram of the SiNWs fabrication steps.

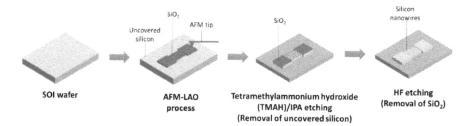

FIGURE 11.8   Schematic diagram of SiNWs fabricated using LAO by AFM lithography.

Wet etching can be either isotropic or anisotropic, depending on the silicon wafer orientation and the type of etchant that being used (Monteiro et al. 2015). In isotropic etching, the etchant removes the material uniformly at the same etching rate for all directions. On the other hand, anisotropic etching occurs when the material is removed uniformly in a vertical direction only. Furthermore, the anisotropic etching produced a trapezoidal cross-section due to the etching rate for (100) silicon plane is faster than (111) plane (Kong et al. 2012). In addition, anisotropic wet etching is mostly used to fabricate simple microstructures and nanostructures on a single crystal silicon on insulator (SOI) wafer (Abdullah et al. 2010, Larki et al. 2012, Hutagalung and Kam 2012, Asmah et al. 2015, Yusoh and Yaacob 2016).

11.3.1   SILICON ETCHING

Two-stepped etching processes are often employed to produce SiNWs from the AFM-LAO derived SiO_2 NWs. The first step is to remove Si and the second step is to remove the drawn SiO_2 pattern. Tetramethylammonium hydroxide (TMAH) (Shikida et al. 2001, Hutagalung et al. 2014, Tran et al. 2014), potassium hydroxide (KOH) (Abdullah et al. 2010), sodium hydroxide (NaOH) (Pakpum and Pussadee 2015), ethylenediamine-pyrocatechol (EDP) (Chung 2001, Dutta et al. 2011) and hydrazine water (Chung 2001) have been used to remove Si. However, certain etchants, such as EDP and hydrazine water, are not preferable because of their toxicity, instability and difficulty in handling. Meanwhile, NaOH is rarely used, unlike the KOH solution that is becoming more popular as an anisotropic etchant because of its good etching performance and it is rather benign. However, KOH is not CMOS compatible because of the mobile K^+ ion contamination (Merlos et al. 1993). By contrast, TMAH has attracted the interest of researcher because of its CMOS compatibility, good anisotropic etching and does not easily decompose at 130°C (Merlos et al. 1993, Jun et al. 2015). Although TMAH has a higher etching processing time compare to KOH, it can produce smooth-etched surfaces (Benjamin et al. 2014). There have been works reported on the success of TMAH as an etchant to remove silicon layers for fabrication of SiNWs devices (Hutagalung and Kam 2012, Tran et al. 2014, Asmah et al. 2015). Tetramethylammonium hydroxide has also been used to etch simple structure like cantilevers or pressure sensors (Zhang et al. 2010, Rouhi et al. 2012).

The etching rate of the TMAH etchant varied depending on the silicon wafer orientation according to the atomic organization of crystallographic planes (Shikida et al. 2001, Fashina et al. 2015). The etching rate is calculated by dividing the height of an etched structure over etching time as expressed in equation (11.4) (Shikida et al. 2000, Dung 2009).

$$\text{Etching rate} = \frac{\text{Etched structure height}}{\text{Etching time}} \qquad (11.4)$$

The etching rate can be increased when a higher temperature is applied and decreased as the etchant is more concentrated (Benjamin et al. 2014). Etching can result in an inhomogeneous surface. Improvement on the etched surface can however be obtained by adding additives such as isopropyl alcohol (IPA), triton x-100 polyethylene glycol (PEG), cationic ammonium of polyethylene glycol (ASPEG) and NC-200 (Mingqiu et al. 2015). It has been reported that IPA-added KOH or TMAH can result in a smooth etched surface (Zubel and Kramkowska 2001, Abdullah et al. 2010, Larki et al. 2012, Tran et al. 2014). Meanwhile, an additive such as Triton x-100 can also be used to enhance the smoothness of an etched surface (Mingqiu et al. 2015). Besides, additive such as IPA can also increase the etching rate. The etching rate is also a function of the volume of IPA added in TMAH. When an excess volume of IPA is added, the etching rate will be slow. This is because at high concentrations IPA molecules aggregated to form a monolayer on Si surfaces, as shown in Figure 11.9. In the etching solution, the Si surface is found to be hydrogen-terminated and the surface becomes

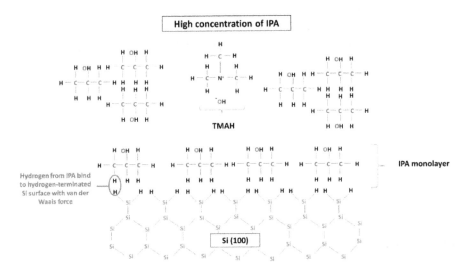

FIGURE 11.9 Formation of IPA monolayer due to high concentration of IPA in TMAH resulting slow etching rate.

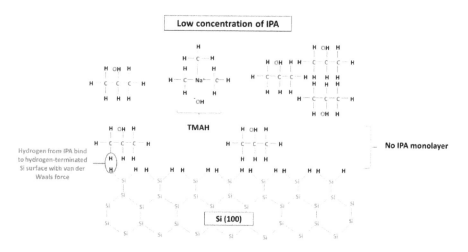

FIGURE 11.10 Formation of IPA monolayer due to low concentration of IPA in TMAH resulting fast etching rate.

hydrophobic (Wind and Hines 2000). This promotes the hydrocarbon from IPA molecules to bond with the hydrogen-terminated Si surface by van der Waals forces. The formation of an IPA monolayer on the Si surface can prevent OH⁻ ions from TMAH to access the Si surface thus reducing the etching rate. When less IPA is added to TMAH the etching rate is faster since there will be no aggregations of molecules on the surface. Then the OH⁻ ions from the TMAH solution can get access to silicon for the etching process to occur (Rola and Zubel 2013), as shown in Figure 11.10. After this etching process is completed, the SiO_2 NWs pattern will then be etched with hydrofluoric (HF) acid diluted with deionized water (DIW) with the ratio 1:20 of HF: DIW solution for 5 s. This process will yield Si NWs.

11.4   CONCLUSION

This chapter discussed the AFM-LAO process for SiO_2 nanowires formation as a template for SiNWs fabrication. The all-important parameters such as type of cantilever tips materials, applied voltage, writing speed and relative humidity used to grow the SiO_2 nanowires on SOI wafers are reviewed. A chemical etching process can be done to produce the SiNWs. Chemical wet etching used to develop the SiNWs is explained by discussing the role of TMAH as Si etchant.

REFERENCES

Abdullah, A. M., Lockman, Z., and Hutagalung, S.D. 2010. "Effect of KOH etchant concentration and initiator on the fabrication of silicon nanowire transistor patterned by AFM nanolithography." *Journal of Industry Technology* 19:197–217.
Alias, N. N. 2015. Fabrication of silicon nanowire arrays using atomic force microscopy (AFM) lithography, M. Sc Thesis, Universiti Sains Malaysia.

Asmah, M. T., Sidek, O., and Hutagalung, S. D. 2015. "Junctionless silicon-based device for CO_2 and N_2O gas detection at room temperature." *Procedia Manufacturing* 2:385–391.

Ávila Bernal, A. G., and Bonilla, R. S. 2012. "Local anodic oxidation on silicon (100) substrates using atomic force microscopy." *DYNA* 79:58–61.

Benjamin, J., Grace Jency, J., and Vijila, G. 2014. "A review of different etching methodologies and impact of various etchants in wet etching in micro fabrication." *International Journal of Innovative Research in Science, Engineering and Technology* 3 (1):558–564.

Cabrera, N., and Mott, N. F. 1949. "Theory of the oxidation of metals." *Reports on Progress in Physics* 12 (1):163.

Červenka, J., Kalousek, R., Bartošík, M., Škoda, D., Tomanec, O., and Šikola, T. 2006. "Fabrication of nanostructures on Si(100) and GaAs(100) by local anodic oxidation." *Applied Surface Science* 253 (5):2373–2378.

Chang, K. M., You, K. S., Lin, J. H., and Sheu, J. T. 2004. "An alternative process for silicon nanowire fabrication with SPL and wet etching system." *Journal of The Electrochemical Society* 151 (10):679–682.

Chung, G.-S. 2001. "Anisotropic etching and electrochemical etch-stop properties of silicon in TMAH:IPA:pyrazine solutions." *Metals and Materials International* 7 (6):643–649.

Dagata, J. A., Inoue, T., Itoh, J., and Yokoyama, H. 1998. "Understanding scanned probe oxidation of silicon." *Applied Physics Letters* 73 (2):271–273.

Dehzangi, A., Larki, F., Hutagalung, S. D., Goodarz Naseri, M., Majlis, B. Y., Navasery, M., Abdul Hamid, N., and Mohd Noor, M. 2013. "Impact of parameter variation in fabrication of nanostructure by atomic force microscopy nanolithography." *PLoS One* 8 (6):e65409.

Dung, D. V. 2009. "Research on optimal silicon etching condition in TMAH solution and application for MEMS structure fabrication." *Journal of Science: Mathematics-Physics* 25 (3): 161–167.

Dutta, S., Imran, M., Kumar, P., Pal, R., Datta, P., and Chatterjee, R. 2011. "Comparison of etch characteristics of KOH, TMAH and EDP for bulk micromachining of silicon (110)." *Microsystem Technologies* 17 (10):1621.

Fashina, A. A., Adama, K. K., Oyewole, O. K., Anye, V. C., Asare, J., Zebaze Kana, M. G., and Soboyejo, W. O. 2015. "Surface texture and optical properties of crystalline silicon substrates." *Journal of Renewable and Sustainable Energy* 7 (6):063119.

Hutagalung, S. D., and Lew, K. C. 2010. Electrical characteristics of silicon nanowire transistor fabricated by AFM lithography. *2010 IEEE International Conference on Semiconductor Electronics (ICSE2010)*, June 28–30, 2010.

Hutagalung, S. D., and Kam, C. L. 2012. "Effect of TMAH etching duration on the formation of silicon nanowire transistor patterned by AFM nanolithography." *Sains Malaysiana* 41 (8):1023–1028.

Hutagalung, S. D., Kam, C. L., and Darsono, T. 2014. "Nanoscale patterning by AFM lithography and its application on the fabrication of silicon nanowire devices." *Sains Malaysiana* 43 (2):267–272.

Janoušek, M., Halada, J., and J. Voves. 2010. "Lithography on GaMnAs layer by AFM local anodic oxidation in the AC mode." *Microelectronic Engineering* 87 (5):1066–1069.

Jun, K.-W., Kim, B.-M., and Kim, J.-S. 2015. "Silicon etching characteristics for tetramethylammonium hydroxide-based solution with additives." *Micro & Nano Letters* 10 (10):487–490.

Kong, T., Su, R., Zhang, B., Zhang, Q., and Cheng, G. 2012. "CMOS-compatible, label-free silicon-nanowire biosensors to detect cardiac troponin I for acute myocardial infarction diagnosis." *Biosensors and Bioelectronics* 34 (1):267–272.

Larki, F., Dehzangi, A., Abedini, A., Abdullah, A. M., Saion, E., Hutagalung, S. D., Hamidon, M. N., and Hassan, J. 2012. "Pinch-off mechanism in double-lateral-gate junctionless transistors fabricated by scanning probe microscope based lithography." *Beilstein Journal of Nanotechnology* 3:817–823.

Mendoza, C., Plata, A., Lizarazo, Z., and Chacón, C. A. 2016. "Analysis of the influence of the applied voltage and the scan speed in the atomic force microscopy local oxidation technique." *Journal of Physics: Conference Series* 687 (1):012039.

Merlos, A., Acero, M., Bao, M. H., Bausells, J., and Esteve, J. 1993. "TMAH/IPA anisotropic etching characteristics." *Sensors and Actuators A: Physical* 37–38:737–743.

Mingqiu, Y., Bin, T., Wei, S., and Gang, T. 2015. "Wet anisotropic etching characteristics of Si{100} in TMAH+Triton at near the boiling point." *Key Engineering Materials* 645–646:58–63.

Mo, Y., Zhao, W., Huang, W., Zhao, F., and Bai, M. 2009. "Nanotribological properties of precision-controlled regular nanotexture on H-passivated Si surface by current-induced local anodic oxidation." *Ultramicroscopy* 109 (3):247–252.

Monteiro, T., Kastytis, P., Gonçalves, L., Minas, G., and Cardoso, S. 2015. "Dynamic wet etching of silicon through isopropanol alcohol evaporation." *Micromachines* 6 (10):1437.

Pakpum, C., and Pussadee, N. 2015. "Design of experiments for (100) Si vertical wall wet etching using sonicated NaOH solution." *Applied Mechanics and Materials* 804:12–15.

Rola, K. P., and Zubel, I. 2013. Impact of alcohol additives concentration on etch rate and surface morphology of (100) and (110) Si substraytes etched in KOH solution. *Microsystem Technologies* 19 (4):635–643.

Rouhi, J., Shahrom, M., Hutagalung, S. D., Nima, N., Saeid, K., and Mat Johar, A. 2012. "Controlling the shape and gap width of silicon electrodes using local anodic oxidation and anisotropic TMAH wet etching." *Semiconductor Science and Technology* 27 (6):065001.

Shikida, M., Masuda, T., Uchikawa, D., and Sato, K. 2000. "Differences in anisotropic etching properties of KOH and TMAH solution." *Sensors and Actuators A: Physical* 80 (2):179–188.

Shikida, M., Masuda, T., Uchikawa, D., and Sato, K. 2001. "Surface roughness of single-crystal silicon etched by TMAH solution". *Sensors and Actuators A: Physical* 90 (3):223–231.

Stiévenard, D., Fontaine, P. A., and Dubois, E. 1997. "Nanooxidation using a scanning probe microscope: An analytical model based on field induced oxidation." *Applied Physics Letters* 70 (24):3272–3274.

Tran, D. P., Wolfrum, B., Stockmann, R., Offenhausser, R., and Thierry, B. 2014. "Fabrication of locally thinned down silicon nanowires." *Journal of Materials Chemistry C* 2 (26):5229–5234.

Tseng, A. A. 2011. "Advancements and challenges in development of atomic force microscopy for nanofabrication." *Nano Today* 6 (5):493–509.

Wind, R. A., and Hines, M. A. 2000. "Macroscopic etch anisotropies and microscopic reaction mechanism: A micromachined structure for the rapid assay of etchant anisotropy." *Surface Science* 460 (1):21–38.

Yusoh, S. N., and Yaacob, K. A. 2016. "Effect of tetramethylammonium hydroxide/isopropyl alcohol wet etching on geometry and surface roughness of silicon nanowires fabricated by AFM lithography." *Beilstein Journal of Nanotechnology* 7:1461–1470.

Yusoh, S. N., and Yaacob, K. A. 2015. "Contact mode atomic force microscopy cantilever tips for silicon nanowires fabrication." *International Journal of Electroactive Materials* 3:6–9.

Yusoh, S. N. 2018. "Development of dengue biosensor on silicon wire arrays using local anodic oxidation by atomic force microscope." PhD Thesis. Universiti Sains Malaysia.

Zhang, Y., Docherty, K. E., and Weaver, J. M. R. 2010. "Batch fabrication of cantilever array aperture probes for scanning near-field optical microscopy." *Microelectronic Engineering* 87 (5):1229–1232.

Zubel, I., and Kramkowska, M. 2001. "The effect of isopropyl alcohol on etching rate and roughness of (1 0 0) Si surface etched in KOH and TMAH solutions." *Sensors and Actuators A: Physical* 93 (2):138–147.

Index

Note: Page numbers in italic and bold refer to figures and tables respectively.

Index

Note: Page numbers in italic and bold refer to figures and tables respectively.